THE
Donkey
COMPANION

Selecting, Training, Breeding, Enjoying

& Caring for Donkeys

SUE WEAVER

Storey Publishing

*The mission of Storey Publishing is to serve our customers by
publishing practical information that encourages
personal independence in harmony with the environment.*

Edited by Deborah Burns and Nancy D. Wood
Art direction and book design by Cynthia N. McFarland
Text production by Liseann Karandisecky and Jennifer Jepson Smith

Front cover photographs by © Bob Langrish (top left), © Carien Schippers (top center, spine),
 © Lynn Stone (top right, bottom left) , and © Mark Barrett (bottom right)
Back cover photographs by © Sue Weaver (top left), © Ed Kosmicki (top right), and
 © Nigel Bean/naturepl.com (bottom)
Vintage postcards and photographs courtesy of the author
Interior photography credits appear on page 343

Illustrations by © Elayne Sears, except for donkey icons by © Leticia Plate
Maps and infographics by Ilona Sherratt
Indexed by Nancy D. Wood

The information in this book is true and complete to the best of our knowledge. All recommendations are made
without guarantee on the part of the author or Storey Publishing. The author and publisher disclaim any liability in
connection with the use of this information. For additional information, please contact Storey Publishing, 210 MASS
MoCA Way, North Adams, MA 01247.

Storey books are available for special premium and promotional uses and for customized editions. For further
information, please call 1-800-793-9396.

Printed in United States by Versa Press
10 9 8 7 6 5 4 3 2 1

Library of Congress Cataloging-in-Publication Data

Weaver, Sue.
 The donkey companion/Sue Weaver.
  p. cm.
 Includes index.
  ISBN 978-1-60342-038-9 (pbk. : alk. paper)
  1. Donkeys—Handbooks, manuals, etc. I. Title.
SF361.W43 2008
636.1'82—dc22
             2008026340

THE
Donkey
COMPANION

CONTENTS

501 A   Mauresque et son Bourriquot.

PREFACE

Why Keep Donkeys?

When I was five years old, my brother took me to the county 4-H fair. The year before there were pony rides; I could hardly wait! I'd spent all of my tickets riding my favorite red pony round and round the little track, and I intended to do it again. Then, as we rounded the corner by the Tilt-a-Whirl, I gasped. There, standing in line to accept wee riders, was a donkey. A *donkey* — like the donkeys in my storybooks and the one in the pictures our Sunday school teacher made us color at Easter and Christmastime. I dropped my brother's hand and ran to the donkey. I stopped, gingerly touched him, and gazed into his soft brown eyes. At that moment I was smitten; my heart was stolen away. I leaned close to the donkey and whispered, "I will have a donkey like you some day."

And so I did; many, in fact. As years passed, a steady stream of donkeys paraded through my life: pets, donkeys we bred, donkeys rescued, a saddle donkey, and back now to pets. And always, gazing into soft, patient donkey eyes takes me back to that first magical moment when a pony-ride donkey stole my heart.

Why donkeys? In my case because their serene, Zen-like demeanor brings me peace in this rushed, busy world. Others keep donkeys for adventure: the thrill of competing in a pack burro race, combined driving events, or at top-level donkey and mule shows. As a business: crafting donkey milk soaps, guiding wilderness travelers on donkey-trekking trips, or breeding fine Miniature Donkeys. To protect vulnerable creatures like sheep, alpacas, and goats. Or simply to share the love: through animal-assisted therapy or the living nativity at church.

For whatever reason, donkeys are for everyone, and that's what this book is about. If you don't already have donkeys, give them a chance. They're sure to steal your heart, too.

— Sue Weaver

This book is dedicated to
Paul and Betsy Hutchins
and Leah Patton
of the American Donkey and Mule Society,
and to
Dr. Elisabeth D. Svendsen M.B.C.
of The Donkey Sanctuary —
for all you've done for the donkeys.

ACKNOWLEDGMENTS

I'd like to thank Deb Burns for her infinite patience
and boundless encouragement; Deb, you're a treasure!
Thanks also to Nancy Wood, Cindy McFarland,
Liseann Karandisecky, Ilona Sherratt, Janet Jesso,
Elayne Sears, and all of the other wonderful people
at Storey who made this book what it is.
Finally, thanks to the donkeys in my life, beginning
with Chicatika in 1966 through my present donkey
girl, Ishtar. I couldn't have done it without you!

166 LOURDES. — *Types du Pays.* —

PART 1

Donkey Primer

Just a line from Cleethorpes.

CHAPTER 1

MEET THE DONKEY

To carry a load without resting, to be not bothered by heat or cold, and to always be content: These things we can learn from a donkey.

— Pakistani proverb

The charming creature that we call a donkey is rightfully called an ass. It wasn't until the late eighteenth century that English-speaking people substituted the word *donkey* for *ass* to differentiate it from the word *arse,* a rude word meaning "derriere." Because the words were pronounced the same, it made sense. Nowadays, however, "ass" is making a comeback as a respectable word for donkeys. We use it frequently throughout this book, so don't be surprised when you read it.

Asses in English-speaking countries are sometimes known by names like *burro* in the western United States or *moke, neddy,* and *cuddy* in Great Britain, and there are scores of other words for ass throughout

"Burro Transportation" is the caption on this postcard of a lady in a fancy hat, riding a patient burro at Castle Rock in Golden, Colorado. This sepia-tone image is from the 1900s.

the world. They all signify the same sweet animal we know and love as the donkey: *Equus asinus.*

A Basic Brayer Lexicon

There are a few terms that you should understand before you read any further, although there are many more defined in the glossary at the end of this book. These terms can be further qualified to describe an individual animal, such as a *weanling jack* (a young male weaned from his mother but not yet one year old), an *aged donkey gelding* (a castrated male donkey ten or more years of age), or a *yearling hinny* (a hinny that has celebrated his first birthday but isn't yet two years old).

You'll also want to recognize the parts of the donkey as we refer to them; there aren't a lot, so memorize them if you can.

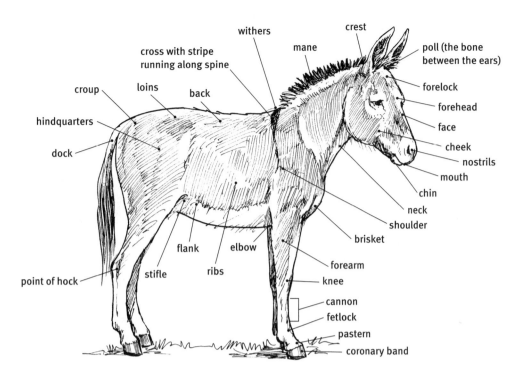

Terms to Know

ADMS. Stands for American Donkey and Mule Society, an organization devoted to registering and promoting donkeys and donkey hybrids of all sizes and types. Membership includes a subscription to *The Brayer,* the organization's thick, handsome bimonthly journal (*see* Resources). Members are eligible to compete in an array of award programs, including some that don't require showing. If you love donkeys, you'll love the American Donkey and Mule Society.

Ass. The proper label for a member of clan *Equus asinus.* Don't be shy about using it; most donkey owners do it with aplomb.

Bray. To make the loud *hee-haw* sound associated with donkeys and mules. In other countries donkeys are said to make other sounds (*see* box, page 4), but, in truth, braying sounds the same around the world.

Brayer. Many folks call donkeys and mules *brayers.* Another such term that you'll frequently encounter is *longears.*

Gelding. A castrated male equine of any age.

Hinny. Donkey hybrids produced by breeding a stallion (male horse) to a jenny (female donkey). Like mules, hinnies are sterile.

Jack. An intact (uncastrated) male donkey three years of age or older.

Jennet (pronounced JEN-et). The proper term for a female donkey three years of age or older; however, in common usage a female donkey is usually called a *jenny.* (I prefer "jenny," and in most cases that's the term I'll use in this book.)

John. Just as many people call a mare mule or mare hinny a *molly,* they tend to call a gelding mule or gelding hinny a *john.*

Mare. A mare is a female horse, three years of age or older. A female mule is called a *mare mule* and a female hinny is a *mare hinny,* although many folks call a female donkey-horse hybrid of either type a *molly.* In some parts of the world, female donkeys are referred to as mares instead of jennies.

Mule. Mules are donkey hybrids produced by breeding a jack (male donkey) to a mare (female horse). Mules are sterile, although rarely a molly produces a living foal (a *foal* is a baby equine).

Donkeys Aren't Horses

Although donkeys resemble horses, and indeed are closely related, there are notable differences between the two. All of the members of genus *Equus* descend from common prehistoric ancestors. However, during the Pleistocene era (stretching from approximately two million years ago to 10,000 years ago) four subgenuses evolved, perhaps to better suit the terrain and climates in which they lived. These four classifications are *Equus equus* (horses), *E. asinus* (asses), *E. doliichohippus* (Grevy's zebra), and *E. hippotigris* (the rest of the zebras and the now-extinct quagga); we'll discuss them fully in chapter 2.

I Am Donkey, Hear Me Bray

One undeniable difference between horses and donkeys is in their neighs and brays. Donkey vocalizations, say researchers David G. Browning and Peter M. Scheifele, are unique among members of the equine clan because sound is produced during both air intake (the *hee*) and air outflow (the *haw*). These vocalizations consist of a series of emanations that cease only when the donkey becomes short of breath. The acoustic character, duration, and sequence (some hee-haw, while others haw-hee) are unique to each animal.

How to Bray in 17 Languages

Albanian: i-a i-a	**Hebrew:** iya, iya
Arabic (Algeria): hiihan hiihan	**Hindi:** si:po:-si:po:
Bengali: chuuchuu	**Italian:** i-oo, i-oo
Catalan: i-haa	**Polish:** iha, iha
Croatian: i-ja, i-ja	**Russian:** ia-ia
Dutch: ie-ah	**Spanish (Costa Rica):** iii-aah, iii-aah
English: hee-haw	**Turkish:** a-iiii, a-iiii
French: hihan	**Ukrainian:** ii-aa, ii-aa
German: iaah, iaah	

It's a fact that people love or hate to hear donkeys bray; there is little middle ground. I love it. When my donkey Ishtar brays directly into my face (as she is wont to do because she knows it makes me smile) it gives me tingles — like standing 20 feet from the railroad tracks as the Amtrak thunders by.

If you don't appreciate the music of donkey braying (or your nearby neighbors don't), forgo owning a jack. Jacks sing for the sheer joy of hearing their lusty clarion calls echo across meadows and bounce off trees. They do it a lot, and their voices are *loud*.

Jennies and geldings tend to bray for a reason, ranging from, "Here comes the Boss — I bet she has a pocketful of peppermints!" to "Get out of bed and come feed us, we're starving!" To discourage braying, don't give up in desperation and give them what they want: for example, don't rush out the door in your jammies to feed them in hopes of shutting them up. If you do, their braying will escalate in frequency and volume and will happen earlier and earlier on successive mornings. Trust me on this.

Donkey song varies in pitch and volume not only between the sexes but between sizes as well. Miniature Donkeys, for all their adorable cuteness, have (to my ears) pathetically squeaky brays. For sheer tone and volume, Mammoths are the Pavarottis of the donkey world.

My daughter, Robyne, would agree. When she was a junior in high school, old Bluestone Maudeen took a special interest in the girl. As Robyne trudged aboard the school bus every school day morning, Maudeen would let loose a blast of mighty brays. Robyne, now nearing 40, still cringes when she recalls the other kids hooting and howling, "Your sister is calling you, Robyne!" Such is life with asses.

Maudeen was special in many ways. Her song began with a low-pitched *creeeeak* like a rusty iron cemetery gate being forced open. It gained volume with each quick inhalation until it reached maximum pitch, at which time she emitted a mighty *ii-aa, ii-aa* and passed gas at the very same time. The neighborhood kids in our small town adored Maudeen; they loved to bring her apples to make her bray.

Christian folklore claims that donkeys once neighed. After hiding in Egypt for several years, storytellers explain, Joseph and Mary decided to move back to Nazareth. They traveled by day, but at night they camped near the side of the road. One night while they slept, their donkey heard soldiers' horses approaching in the distance. Fearing the soldiers would kill young Jesus, the donkey neighed to alert the Holy Family. He whinnied and whinnied, again and again, but his voice was too soft to wake Joseph and Mary. Finally, as the soldiers drew near, the donkey prayed for a very loud voice to rouse his sleeping family. When he neighed again, he was rewarded with the loud, strident bray that donkeys have to this very day.

Key Differences

Though few of us will chance to work with wild asses or zebras, many horse owners branch out into donkeys or mules (and vice versa). Because they do, it's important to understand the physical and behavioral differences among donkeys, horses, and their hybrid offspring, the mule and the hinny, both of which inherit characteristics from each parent. (*See* Comparing Equine Characteristics *on page 6.*) The wide range of differences makes things tough for donkey newbies who think their existing horse tack and their horse-handling savvy are all they'll need to make the switch.

Consider saddles. Most saddles are constructed on saddle trees designed to accommodate a horse's comparatively prominent withers — a feature that asses and most mules are sorely lacking. Furthermore, today's horses tend toward roundness of back and wide spring of ribs, so that's the type most saddles are crafted to fit; try fitting a lean Mammoth riding gelding with a saddle built on Quarter Horse bars. It's a disaster!

A horse wearing a one-ear style bridle — not suitable for donkeys.

The same can be said for fitting donkeys with bridles designed to fit their short-eared kin. While a horse and an ass may have heads of the same length, the horse's browband will be way too narrow to fit Mr. Longears. Nor are "ear bridles," made for horses' short, slender ears, correct for donkey wear. Few donkeys tolerate having their delicate ears crammed through ear loops, and once in there, the base of the donkey's ear is likely to be painfully chafed.

Mule bridles made with snap-crown browbands are ideal for riding donkeys, too.

DONKEY

HORSE

MULE

COMPARING EQUINE CHARACTERISTICS

	DONKEY	HORSE	MULE/HINNY
Head	Broader forehead; much deeper jaw	Comparatively narrower forehead; smaller, shallower jaw	Intermediate in size and shape
Ears	Long (Mammoth donkeys' ears are longer in proportion to their heads than those of miniature and standard donkeys) and thick textured	Smaller, usually more shapely; thinner textured	Intermediate in size and thickness
Eyes	Larger; heavier, D-shaped, bony eye orbits (heavy brow ridges) set farther out on sides of head	Smaller in proportion to head; round eye orbits, not as bony	Intermediate with somewhat D-shaped orbits and brow ridges
Nasal passages	Smaller than a horse's; relatively nonflaring	Medium to large; flaring	Varies; generally smaller than a horse's
Withers	Doesn't have horselike withers (makes back appear longer)	Usually somewhat pronounced	Varies
Back	Lacks sixth lumbar vertebra in spinal column	Most have a sixth vertebra	Varies
Hindquarters	Longer, more steeply angled hip bones; hindquarters higher, steeper, narrower; much more smoothly muscled than horse	More muscle mass; generally much more muscle definition	Intermediate; generally smoother muscled than a horse but with more muscle mass than a donkey
Chestnuts	Front chestnuts only; flat, smooth, leathery	Front and rear; thick and horny	Varies
Hooves	Smaller in proportion to body size, boxy, more upright (hoof angles average 65 degrees), tougher, and more elastic; larger, well-developed frog with a thicker sole	Larger in proportion to body size; oval, less upright (hoof angles average 55 degrees); smaller frog with thinner sole	Usually boxy, steeper angle than a horse
Mane	Mammoths grow sparse, fall-over manes; smaller donkeys grow short, stand-up manes (if any at all)	Depends on breed; most have relatively long, finer-textured manes	Varies but most donkey hybrids have sparse, fall-over manes that are trimmed to stand up like a donkey's mane
Tail	Cowlike with long, coarse "swish" or "switch" on the lower one-third of the tailbone	Depends on breed; tends to be long and luxurious	Most donkey hybrids have a more horselike than donkey-like tail, usually with shorter hair near dock; coarser than most horses' tails

HINNY

	DONKEY	HORSE	MULE/HINNY
Male reproductive organs	Comparatively much larger than those of a stallion; rudimentary nipples on sheath; larger scrotal vessels and thicker scrotal skin; more prone to hemorrhaging during castration	Comparatively smaller than those of a jack's; no nipples	Intermediate; some have nipples, some don't
Chromosomes	62	64	63
Fertility	Yes; average conception rate of 78 percent	Yes; average conception rate of 65 percent	Nonfertile (although extremely rare exceptions have occurred)
Estrus (heat) cycle	23–30-day interval/average 6–9 days	21–25-day interval/average 3–7 days	Most mollies cycle, although erratically
Ovulation	5–6 days after onset of estrus	12–24 hours before the end of estrus	N/A
Gestation	360–375 days (or more)	335–345 days (or more)	Nonfertile (although mollies can be used as embryo recipients)

The above figures on estrus, ovulation, and gestation are taken from a paper by Debra J. Hagstrom titled *Donkeys Are Different: An Overview of Reproductive Variations from Horses.*

BRAY SAY: *When asses bray more than ordinary, particularly if they shake their ears as if uneasy, it is said they predict rain, and particularly showers. We have noticed that, in showery weather, a donkey, confined in a yard near the house, has brayed before every shower, and generally some minutes before the rain has fallen, as if some electrical influence, produced by the concentrating power of the approaching rain-cloud, caused a tickling in the windpipe of the animal, just before the shower came up.*

— Robert Merry's *Museum* (British children's magazine), November 1844

Edward, the author's riding donkey, expresses himself.

Still, these differences can be easily circumvented if you're willing to problem solve. Replace a bridle's standard browband with a longer one. Have one custom made or buy a longer browband from a tack shop or catalog dealing in dressage or hunt seat gear (English-style browbands designed for European warmbloods often fit bigger donkeys), or choose a bridle specially made for donkeys and mules. Browband-style "mule bridles" feature a snap-end browband that allows you to fasten it after the bridle is already on your ass or mule, so you needn't stuff his delicate ears through a standard opening. We'll talk more about riding and driving gear for donkeys in chapters 13 and 14.

Out of Africa

All domestic donkeys are descendants of the African Wild Ass, according to Albano Beja-Pereira and his associates from Université Joseph Fourier in France. They have gathered and extracted DNA from skin samples collected from donkeys in 52 European countries, Asia, and Africa, as well as fresh dung from wild asses in Sudan, China, and Mongolia. Altogether this represents a total of 427 asses. The skin samples, says Dr. Beja-Pereira, were "collected only from small and remote countryside taking great care to sample unrelated individuals," and the dung came from several representative herds.

Dr. Beja-Pereira and his associates extracted mitochondrial DNA from these samples, a process that allows scientists to track an animal's female descendants back through history. Based on their findings, Dr. Beja-Pereira's team concludes that today's domestic donkeys descend from two wild populations living more than 10,000 years ago and that two wild females, living far apart geographically, were the ancestors of virtually all of the donkeys living today.

Donkeys in Ancient Egypt

African Wild Asses were domesticated about six thousand years ago, most likely in Egypt. This event singularly altered the course of history. Horses didn't come to the Nile River Valley until about 1750 BC; water buffalo arrived around 600 BC; camels made the scene about 400 BC. The ass was the only riding, pack, and draft animal available in Egypt's Old Kingdom.

Asses in the ancient world excelled as beasts of burden for the same reasons that they shine in arid and semiarid developing countries today:

- ✘ Donkeys tolerate up to 30 percent dehydration and keep on working.
- ✘ They efficiently utilize low-quality, cellulose-rich food that would cause a horse or ox to starve.
- ✘ In most cases they can be turned out to scrounge for themselves on common land, where they not only survive but flourish.

Donkeys provided ancient families with transport (they could travel farther and faster than ever before), draft power to work their fields, tasty milk, and a ready supply of nourishing meat. Donkey caravans facilitated increased trade with distant tribes. Donkeys drew up water from deep desert wells, provided light tillage of crops, and trod farmers' valuable cache of seeds into the waiting furrows. Jennies provided nutritious milk

Groupe de baudets

No. 671   Vagnios & Zachos.   Cairo & Luxor

This vintage photo from Cairo and Luxor, titled "Groupe de baudets" *(group of donkeys) features the white riding donkeys that were so well regarded in Egypt.*

AN ASS BY ANY OTHER NAME

Afrikaans: groutjie, esel, donkie

Albanian: veshgjatë

American English: donkey, burro, jackass, mountain canary, desert contralto, Rocky Mountain nightingale

Arabic: himaar

Arapaho: bih'ihiihoox

Arikara (Sanish): xaawakáru'

Arvorec: rason

Aymara: urgu asnu

Basque: asto

Belarusian: asnowny

Blackfoot: omahksstooki

Catalan: ase

Chamorro: buliku

Cherokee: digalinvhidv

Chinese: lü

Cornish: asen

Czech: osel

Danish: æsel

Dutch: ezel

Egyptian: aa

Esperanto: azeno

Faeroese: asni, ásin

Finnish: aasi

French: âne, baudet

Frisian: ezel

Georgian: viri

German: esel

Great Britain: donkey, moke, neddy, cuddy

Guarani: mburika

Haitian Creole: bourik

Hausa: jaki

Hawaiian: kēkake

Hebrew: hamôr

Hittite: tarkasni

Hungarian: szamár

Indonesian: keledai

Iranian (Pashto): har

Irish: asal

Italian: asino, ciuco

Japanese: roba

Kannada: katte

Katcha: kisine

Khowar: gordogh

Kiowa: takai

Korean: nagwi

Kurdish: ker

Lakota (Sioux): susula

Latvian: ēzelis

Lithuanian: asilas

Malagasy: boriky

Maltese: hmar

Manx: assyl

Maori: haihe

Mongolian: eljigen

Navajo: tahilth-sapái

Norwegian: esel

Occitan: ase

Papiamen: buro, buriku, buriko

Persian: khur

Polish: osioł

Portuguese: burro, asno

Pueblo: T'ahhláahloon

Quechuan: b'uro, b'uur

Romansch: asen

Romany: hur, myla

Russian: osol

Sami: áse(n)

Sanskrit: khara

Sepedi: esela

Serbo-Croatian: magarac

Setswana: tonki

Sicilian: sceccu

Slovakian: somár; osol

Slovanian: osol

Somali: dameer

Spanish: burro, asno

Sranan: buriki

Sumerian: ansu

Swahili: punda

Swazi: i-mbóngolo

Swedish: åsna

Tagalog: buro; humento

Telug: gagaDida

Thai: laa

Tibetan: bon.bu.

Turkish: eşek

Ukrainian: osel

Venda: ndongi

Verdurian: osol

Volapük: cuk

Welsh: asyn

Yiddish: eyzl

to save the lives of orphaned infants and babies whose hungry mothers couldn't feed them. And in times of want, donkeys were eaten and their hides fashioned into sturdy leather.

Asses were once highly regarded in ancient Egypt. Sir E. A. Wallis-Budge recorded this passage in his 1909 translation of the Egyptian *Book of the Dead:* "Heart which is Righteous and Sinless say . . . come in peace, come in peace . . . for I have heard the word which was spoken by the Ass with the Cat."

However, from the New Kingdom period (1550–1070 BC) onward, ancient Egyptians looked upon asses with disfavor, because of the animals' association with the evil god Set, so much so that the war trumpet fell into disuse because its blast was said to resemble the braying of an ass.

Brayers on the Move

From Africa, domestic donkeys spread throughout the ancient world. The Israelites maintained huge herds of asses; hundreds of references in the Old Testament bear witness to their presence in Hebrew life. A tattoo on a mummified corpse dating to 430 BC depicts a stylized ibex, tiger, and ass, indicating that donkeys had reached Siberia by that early age.

Donkeys, and later mules, carried trade goods over the vast distance of the 2,500-mile Silk Road, stretching from Alexandria, Egypt, to the Pacific coast of China. Donkeys traveled north throughout the Roman Empire along with Caesar's legions as far away as parts of Britain, Gaul (France), and Lower Germany. During the Middle Ages, donkeys spread through the rest of Great Britain. Columbus carried asses aboard ship on his second voyage to the New World; burros (Spanish donkeys) swiftly spread throughout Spanish-held lands along with Spanish colonists and padres. Meanwhile, George Washington imported the forerunners of today's Mammoth Jackstock, the better to breed fine mules.

The history of our donkey friends is a rich dish worth savoring, so instead of serving it up as one fast-food meal, we've peppered the following chapters with trivia tidbits and snippets of long-ear historical lore.

Donkeys in Myth

The mythology and folklore of people the world over refer frequently to donkeys. Here are some typical references.

Epona. This Celtic goddess was protectress of horses, asses, and mules, as well as their human caretakers (she was also associated with dogs and birds, and she was a goddess of rebirth and fertility). Later, the Romans embraced Epona as protectress of the

---

Donkey Deities and Remedies

Set, the red-haired, ass-eared god of chaos, confusion, storms, wind, the desert, and foreign lands, was one of ancient Egypt's earliest deities. In later times, the worshippers of Horus overthrew the cult of Set, defaced his statues, and restructured the ancient myths. Set had the head of an animal; sometimes it was very asslike and sometimes that of a composite creature with a long snout, squared-off upright ears, slim body, and long legs — definitely something unknown to science!

The Egyptians also concocted several remedies using donkey parts. Don't try this at home!

- For hair loss, apply a mixture of five fats (horse, hippopotamus, crocodile, cat, and ibex) with the tooth of a donkey crushed in honey.
- To darken gray hair, apply the liquid from a putrid donkey liver steeped in oil. (The recipe as given in the *Hearst Papyrus:* "Donkey liver, leave in a pot until it is rotten. Cooked, put in lard. Rub in.")
- Worm blood was mixed with donkey dung and dabbed on a splinter.
- Donkey dung alone was rubbed into a wound to stop bleeding.

Roman cavalry. Her feast day fell on December 18. The worship of Epona is described in *The Golden Ass,* or *Metamorphoses* by Marcus Apuleius. He tells us that shrines to Epona were often erected in Roman stables, presumably to protect their equine inhabitants from harm.

Hephaistos (Roman: Vulcanus). This Olympian god of fire, metalworking, stonemasonry, and sculpture created the first woman, Pandora, at the command of Zeus, and also crafted Achilles' war armor. Because he was lame, Hephaistos is usually portrayed as a bearded man with a clubfoot, riding on a donkey.

DONKEY TAILS

Legend of the Donkey's Cross

A poor farmer near Jerusalem owned a donkey far too small to do much work at all. He felt that he couldn't afford to feed a worthless animal like this, one that could do him no good whatsoever, so at the supper table he told his family that he was going to kill the donkey.

His children, who loved the little donkey, begged him to sell it rather than harm it. But the farmer said, "It's wrong to sell an animal that can't do a good day's work."

Then his oldest daughter suggested, "Father, tie the donkey to a tree on the road to town, and say whoever wants it may take it for nothing." And the next morning, that's what the farmer did.

Soon, two men approached and asked if they could have the donkey. "It can carry almost nothing," the farmer warned them.

"Jesus of Nazareth has need of it," replied one of the men. The farmer couldn't imagine what a great teacher would want with such a useless donkey, but he handed it over.

The men took the animal to Jesus, who stroked the grateful donkey's face and then mounted it and rode away. So it was that on the day we call Palm Sunday, Jesus led his followers into the city of Jerusalem riding on the back of a small, common donkey.

The donkey so loved his gentle master that he later followed him to Calvary. Grief-stricken by the sight of Jesus on the cross, the donkey turned away but couldn't leave. It was then that the shadow of the cross fell upon the shoulders and back of the donkey, and there it stayed. All donkeys have borne the sign of the cross on their backs since that very day.

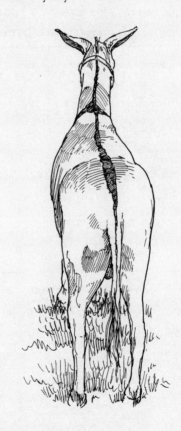

Most (but not all) donkeys have a dorsal stripe and shoulder cross.

During the 2nd century AD, Hyginus (*Astronomica* 2.23) wrote this passage about Hephaistos: "According to Eratosthenes, another story is told about the Asses. After Jupiter [Zeus] had declared war on the Gigantes, he summoned all the gods to combat them, and Father Liber [Dionysos], Vulcanus [Hephaistos], the Satyri, and the Sileni came riding on asses. Since they were not far from the enemy, the asses were terrified, and individually let out a braying such as the Gigantes had never heard. At the noise the enemy took hastily to flight, and thus were defeated."

Silenus (sometimes spelled Seilenos). The name of this Greek god of dancing and the wine press was derived from the words *seiô*, "to move to and fro," and *lênos*, "the wine-trough." Silenus invented the flute and was fond of playing one. He was also a god of drunkenness and was usually portrayed as a drunken

Did You 'Ear?

In 260 BC, there was a Roman consul named Gnaeus Cornelius Scipio Asina (*asina* in this case means "donkey"), who was so called because of his huge ears.

man mounted on his constant companion, a stalwart donkey.

Vesta (Greek: Hestia). The Roman goddess of fire (both sacred and domestic) and the hearth was one of the most widely worshipped of all the Roman deities. Daughter of Saturn and Ops, her sacred animal was the ass. Coronation of an ass was part of her festival, the Vestalia, held on June 7. Roman mythology holds that by braying a warning, an ass prevented Vesta from being raped.

CHAPTER 2

BORN TO BE WILD

An ass may bray a good while before he shakes the stars down.

— George Eliot, nineteenth-century author

There are two types of wild asses and they're worlds apart, both physically and geographically. One group is composed of true Wild Asses, and the others are feral donkeys, which are descended from domestic donkeys gone wild. *(See chapter 4 for more on feral donkeys, called "burros" by the United States Bureau of Land Management.)*

In ancient times, herds of graceful, slim wild asses once inhabited parts of northern Africa. Somewhat stockier wild ones ranged from the Middle East as far east and north as India and Tibet. All of these sub-species are imperiled today; some have already gone extinct, and others are sure to follow. Fortunately, most of the countries with surviving remnant herds have taken steps to preserve their remaining wild asses — nevertheless, they are critically endangered. This chapter contains some facts and figures on members of the ass family.

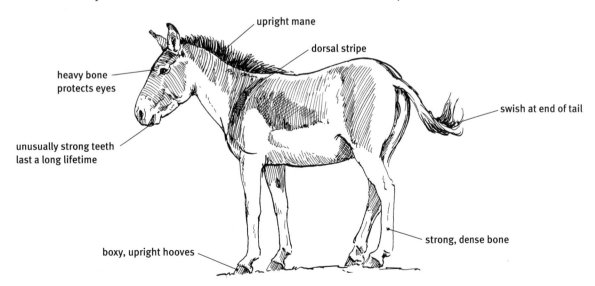

The wild ass is superbly adapted for his environment.

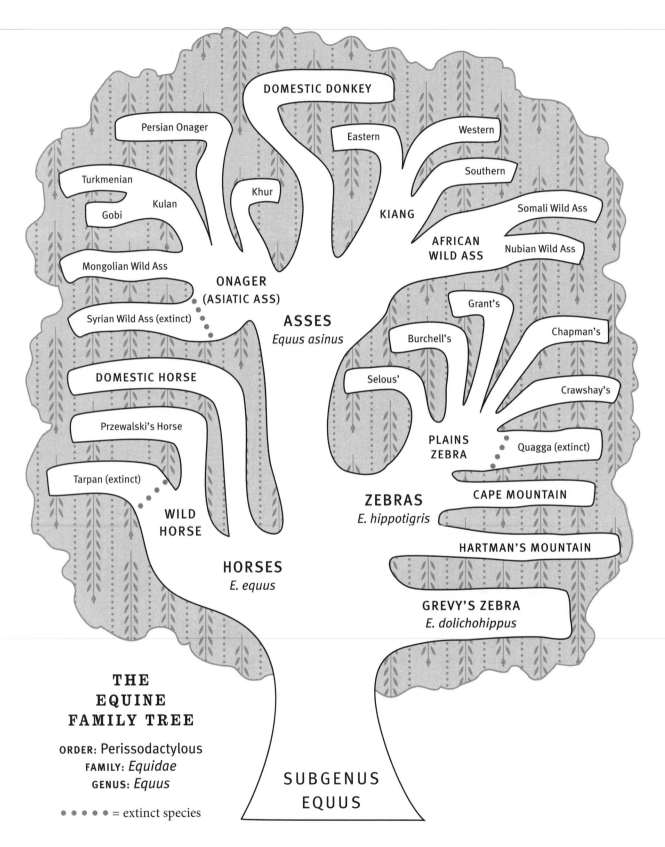

DOMESTIC DONKEY

Persian Onager

Eastern

Western

Turkmenian

Southern

Kulan

Khur

Gobi

KIANG

Somali Wild Ass

Mongolian Wild Ass

AFRICAN
WILD ASS

Nubian Wild Ass

ONAGER
(ASIATIC ASS)

Syrian Wild Ass (extinct)

Grant's

ASSES
Equus asinus

Chapman's

Burchell's

DOMESTIC HORSE

Selous'

Crawshay's

Przewalski's Horse

PLAINS
ZEBRA

Quagga (extinct)

Tarpan (extinct)

CAPE MOUNTAIN

WILD
HORSE

ZEBRAS
E. hippotigris

HARTMAN'S MOUNTAIN

HORSES
E. equus

GREVY'S ZEBRA
E. dolichohippus

**THE
EQUINE
FAMILY TREE**

SUBGENUS
EQUUS

ORDER: Perissodactylous
FAMILY: *Equidae*
GENUS: *Equus*

• • • • • = extinct species

THE AFRICAN WILD ASS

Somali Wild Asses (called *kiswahili* or *punda* in their native land) weigh about 440 to 510 pounds. They inhabit hilly, rock-strewn deserts and arid to semi-arid bush and grasslands in regions with a four- to eight-inch annual rainfall, where ground temperatures exceed 120 degrees. The Somali Wild Ass avoids sandy areas, so it has never populated the sandy dunes of the Sahara. Along with hunting, interbreeding with domestic donkeys is pushing the Somali Wild Ass to the brink of extinction. Some authorities believe that the remaining Somali "Wild Asses" in Africa are in fact feral animals descended in part from domestic donkeys. Nubian Wild Asses are now believed to be extinct in the wild.

SOMALI WILD ASS
(Equus africanus somalicus)

SPECIES: The African Wild Ass, *Equus africanus*
SUBSPECIES: Nubian Wild Ass *(Equus africanus africanus)* and Somali Wild Ass *(Equus africanus somalicus)*

COLORATION: Light gray to grayish fawn shading to white on the underbody, legs, muzzle, and eye areas; their ears are rimmed with black. Both species have a slender dark dorsal stripe along their spines. The Nubian Wild Ass (like the domestic donkey) has a shoulder stripe; the Somali Wild Ass has striped legs. Both species have stiff, upright mane hair tipped with black, and black tail swishes. African Wild Asses can be easily differentiated from their Asiatic counterparts by coat color (Asiatic Wild Asses' coats lean toward reddish brown) and their longer ears.

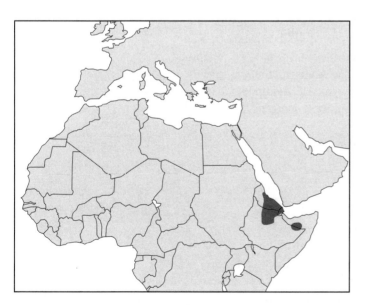

RANGE OF SOMALI WILD ASS

BRAY SAY: *Who is born a donkey won't die as a horse.*

— Italian proverb

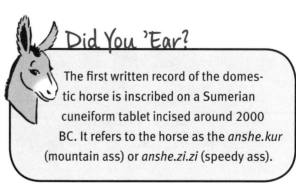

Did You 'Ear?

The first written record of the domestic horse is inscribed on a Sumerian cuneiform tablet incised around 2000 BC. It refers to the horse as the *anshe.kur* (mountain ass) or *anshe.zi.zi* (speedy ass).

501 A   Mauresque et son Bourriquot.

HISTORICAL RANGE: From the Moroccan Atlas across Saharan and possibly Sahelian Africa to the Sudanese and Somalian arid zones (and possibly the Arabian Peninsula).

CURRENT RANGE: Small numbers of Somali Wild Asses are still found in parts of Chad, Djibouti, Sudan, Ethiopia, Somalia, and Eritrea; numbers are sharply declining in Ethiopia and Somalia, while Eritrea has a small but stable population.

STATUS: Critically endangered. In 2002, the world population was fewer than 570 Somali Wild Asses. The World Association of Zoos and Aquariums reports that an additional 192 African Wild Asses are housed in zoos throughout the world.

The French caption on this card refers to the couple as Moors, so the photograph was likely taken someplace in North Africa.

CHROMOSOME COUNT

All members of the equine clan interbreed and produce viable offspring. However, unless both parents share the same diploid chromosome number, the resulting offspring are sterile. When domestic donkeys interbreed with African Wild Asses it contributes to the dilution of the wild species: their offspring are fertile and breed within the Wild Ass bands. That's not the case in Asia where the crossbred offspring of domestic donkeys and Asian Wild Asses are infertile.

SPECIES-SUBSPECIES	DIPLOID CHROMOSOME COUNT
Domestic donkey	62
African Wild Ass	62
Persian Onager	56
Kulan	54
Kiang	52
Hartmann's Mountain Zebra	33
Grant's Zebra	44
Grevy's Zebra	46
Przewalski's Horse	66
Domestic horse	64

THE ASIATIC WILD ASS

Onager means "half-ass." Onagers are true asses, but they're decidedly more horselike in appearance than their African cousins. Forty thousand years ago during the Late Pleistocene era, *Equus hemionus* ranged as far north as western Germany, where it lived alongside woolly mammoths, bison, aurochs, and wild horses.

SPECIES: Asiatic Wild Ass or Onager *(Equus hemionus)*

SUBSPECIES: Persian Onager *(Equus hemionus onager)*, Khulan or Mongolian Wild Ass *(Equus hemionus hemionus)*, Gobi Kulan or Dziggetai *(Equus hemionus luteus)*, Turkmenian Kulan *(Equus hemionus kulan)*, and Khur or Indian Wild Ass *(Equus hemionus khur)*. These subspecies are very closely related; however, having adapted to different climates and terrain, they differ slightly in size and outward appearance. All are frequently lumped under the catchall name of onager.

COLORATION: All Asiatic Wild Asses are a shade of reddish brown during the summer and yellowish brown

Did You 'Ear?

Onagers gallop at incredibly high speeds (they've been clocked running 43 miles per hour). This is alluded to in the Qur'an (Sura 74:50) when the narrator describes the retreat of unbelievers "as if they were asses fleeing before a lion."

in the winter, shading to a white or buff undercarriage and white shadings on the face and legs. All have dorsal stripes running the length of their backs. The dorsal stripes of Turkmenian Kulan, Persian Onagers, and Khur are bordered with white; the Gobi Kulan's dorsal stripe isn't.

HISTORICAL RANGE: One or another subspecies of Asiatic Wild Ass has at one time inhabited most of the steppe and desert areas between the Black Sea and Yellow River of China.

CURRENT RANGE: Over half of the total population of *Equus hemionus* dwells in southern Mongolia.

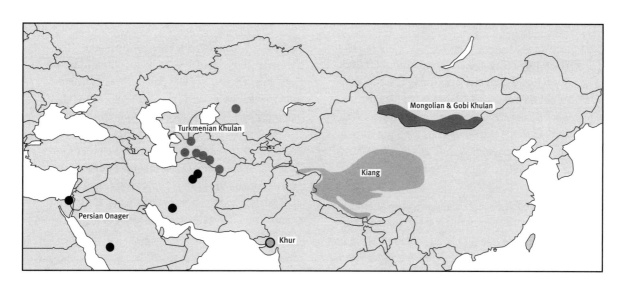

RANGE OF ASIAN WILD ASSES

- Historical range
- Kiang, Tibet
- Mongolian Khulan and Gobi Khulan, China and Mongolia
- Persian Onager, Northern Iran
- Turkmenian Khulan, Turkmenistan
- Khur, India

THE ANCIENT ONAGER

The onager was the wild ass of the Old Testament. It was found in Palestine and the surrounding countries for about two thousand years after the events of the Old Testament but was almost extinct in the area by the 1850s. A few lived on in Iraq and Jordan until well into the twentieth century, but those are now thought to be gone as well.

Faisal A. Dean's paper, "Lost Forever: The Onager of Arabia" (*see* Resources), offers readers a fascinating glimpse of this magnificent animal as it was written about in numerous Muslim texts; these are a few of the highlights:

- The word *onager* comes from the Greek *onagros*, meaning "wild ass." In Arabic the onager is called *al-himar, al-wahshi, al fara'*, and *al-'ir* (*al-'ir* is used for domestic asses as well).

- From earliest times onagers were considered game animals. Bas-reliefs excavated at ancient Ninevah depict members of one of Assyrian King Ashurnanipal's hunting parties (circa 650 BC) lassoing an onager. Peasants were more practical in their approach to onager hunting; they captured onager foals for the stewpot.

- The poet Al-Shammakh, born around AD 600, specialized in poetry describing the onager. About 43 percent (172 verses) of his life's work dealt with onagers and the hunters who stalked them. His poems describe the wild ones' anxiety, fear, jealousy, and anger, as well as their joy in the safety of their highland refuge.

- Several early authors described the onager's longevity. Al-Jahiz, writing in the ninth century, asserted that onagers lived considerably longer than domestic donkeys. A fourteenth-century author, Kamil al-Din al-Damari, postulated that in the mountainous Jabal al-Mudakhan wilderness area onagers lived for more than 800 years.

- The same writer, al-Damari, wrote of al-Akhdar, a male ass (perhaps an onager) that once belonged to the Persian shah Kisr Ardashir. This ass reverted to the wild and mated with the wild onagers, founding a strain considered to be the most beautiful and long-lived of wild asses.

- Writing during the same time period, Abu Yahya Zakariyya al-Qazwini wrote (in *The Wonders of Creation*) that male onagers are inclined to rip off a young foal's testicles to prevent him from competing for the old ones' jennies when they mature. To avoid this confrontation, an onager jenny going into labor will seek a secluded place to have her foal in case it's a male. Once the foal's hooves harden and he's able to run, she'll take him back to the herd.

"By 1850," writes Dean, "the Syrian onager (the Syrian Wild Ass) had nearly disappeared from the Syrian desert and in Palestine but was still common in Mesopotamia. It could be seen traveling in vast herds as far away as the Armenian mountains. . . . The last Syrian Wild Ass was shot in 1927. . . . Zoologist Otto Antonius wrote of the onager's passing, 'It could not resist the power of modern guns in the hands of nomads, and its speed, great as it might have been, was not always sufficient to escape from the velocity of the modern motor car.'"

PERSIAN ONAGER
(Equus hemionus onager)

Persian Onager (Gur, Goor)

Controversy surrounds ancient bas-reliefs and paintings (circa 2600–2000 BC) of onager-like equids hitched to chariots. Some authorities believe that these works of art prove that the ancient Sumerians domesticated the Onager; others think that the works depict domestic asses. The most plausible explanation might be that these are Persian Onager hybrids produced by exposing domestic jennies or mares to wild or captive Onager stallions, but no one knows for certain that this or either of the other theories is true.

CURRENT RANGE: Northern Iran

STATUS: Critically endangered. According to the International Union for Conservation of Nature and Natural Resources Red List, in 2002 there were an estimated 144 mature Persian Onagers in two protected groups in northern Iran, while in 2000 the World Association of Zoos and Aquariums estimated a wild population numbering 500 head (plus an additional 147 housed in zoos).

Khulan (Mongolian Wild Ass) and Gobi Kulan (Dziggetai)

Although these animals have different Latin names and some authorities consider them separate species, both the World Association of Zoos and Aquariums and the International Union for Conservation of Nature and Natural Resources suspect that the Khulan and the Gobi Kulan are essentially the same animal.

CURRENT RANGE: Khulan dwell in semidesert and grasslands in the border region between northern China and Mongolia; Khulan were previously found in Kazakhstan but have been extinct in that area since about 1930.

STATUS: Vulnerable. According to the International Union for Conservation of Nature and Natural Resources Red List, in 2002 there were between 5,766 and 11,866 mature individuals in the wild. According to a study conducted in 2005, due to massive livestock losses throughout the Khulan's range and a subsequent lack of affordable domestic meat, poachers kill as many as 4,500 Khulan every year.

MONGOLIAN WILD ASS
(Equus hemionus hemionus)

DZIGGETAI OR GOBI KULAN
(Equus hemionus luteus)

Turkmenian Kulan

A number of zoos throughout the world are working hard to establish captive breeding populations of Kulan. The World Association of Zoos and Aquariums reports that there are 1,022 Kulan registered in the International Studbook. A good place to see Kulan living under natural conditions is Canyon Colorado (*see* Resources), a United States–based wild equid conservation group that currently maintains 121 Kulan, 9 Persian Onagers, 3 Kiang, and 7 Somali Wild Asses along with Hartmann's Mountain and Grevy's zebras and Przewalski's horses.

CURRENT RANGE: According to various sources, between 600 and 6,000 Kulan range wild and protected within two wildlife preserves: Badkhys Nature Park in Turkmenistan, a semidesert preserve established in 1941, and Barsa Kelmes, a small island at the north end of the Aral Sea.

STATUS: Critically endangered. The International Union for Conservation of Nature and Natural Resources Red List estimates that in 2000 the world's wild Kulan population numbered roughly 650 head.

TURKMENIAN KULAN
(Equus hemionus kulan)

KHUR OR INDIAN WILD ASS
(Equus hemionus khur)

Khur
(Indian Wild Ass, Chor-khar)

The Rann of Kutch, 10,000 square miles of low-lying salt marsh in northwestern India and the Sind province of Pakistan, is the last remaining home of the Indian Wild Ass or Khur, most of which are found within the Indian Wild Ass Sanctuary in the Little Rann of Kutch (*rann* in Hindi means "salt marsh"). During monsoon, the entire Rann is flooded for about one month, during which time the asses congregate on grassy plateaus called "bets" until the waters recede. They forage enough salt grasses to survive.

CURRENT RANGE: Khur inhabit semidesert salt plains on the India-Pakistan border.

STATUS: Endangered. The World Association of Zoos and Aquariums reports that, according to a 1999 census, 2,840 Khur occupied Wild Ass Wildlife Sanctuary near Gujarat, India, in the saline desert of the Little Rann of Kutch; Indian figures set the current population at about 2,100 animals.

THE KIANG

There is little variation of color, size, and type between the Kiang's three subspecies. In fact, the Kiang is genetically and physically very similar to the Asiatic Wild Ass (the mitochondrial DNA divergence between the two species is only 1 percent, and the divergence probably occurred less than 500,000 years ago). Kiangs are the largest of the Wild Asses, often standing as tall as 14 hands and weighing up to 950 pounds. These are robust asses with thick, Roman-nosed heads and massive bones. Compared to other Wild Ass species, Kiangs have relatively short ears.

KIANG
(Equus kiang)

SPECIES: Kiang (Tibetan Wild Ass, Gorkhar), *Equus kiang*

SUBSPECIES: Western Kiang, *Equus kiang kiang;* Eastern Kiang, *Equus kiang holdereri;* Southern Kiang, *Equus kiang polyodon*

COLORATION: Kiangs are reddish colored in their slick summer coats, darkening to brown with much longer coats during the winter months, always with white underparts and a gray muzzle edged with white. A thin stripe of brown extends down the front of the

WILD ASSES IN THE OLD TESTAMENT

The first five chapters of the Old Testament are revered by Christians, Jews (as the Torah), and Muslims (as the Tawrat) alike. Later chapters are held in high esteem as well. These are a few of the Old Testament's many references to wild asses.

- Psalms 104:10–11 — *He sendeth the springs into the valleys, which run among the hills. They give drink to every beast of the field: the wild asses quench their thirst.*

- Job 6:5 — *Doth the wild ass bray when he hath grass?*

- Job 39:5–8 — *Who hath sent out the wild ass free? or who hath loosed the bands of the wild ass? Whose house I have made the wilderness, and the barren land his dwellings. He scorneth the multitude of the city, neither regardeth he the crying of the driver. The range of the mountains is his pasture, and he searcheth after every green thing.*

- Isaiah 32:14 — *Because the palaces shall be forsaken; the multitude of the city shall be left; the forts and towers shall be for dens for ever, a joy of wild asses, a pasture of flocks.*

- Jeremiah 14:6 — *And the wild asses did stand in the high places, they snuffed up the wind like dragons; their eyes did fail, because there was no grass.*

- Jeremiah 2:24 — *A wild ass used to the wilderness, that snuffeth up the wind at her pleasure; in her occasion who can turn her away?*

legs, which are otherwise white; along the spine runs a dark dorsal stripe. Kiangs have short, dark brown, erect manes, and tails with a dark brown tuft and long hairs growing up the side of the tailbone.

HISTORICAL AND CURRENT RANGE: Kiangs inhabit high plateaus and steppes in Tibet, the Tsinghai and Szechwan regions of China, Nepal, and India.

STATUS: Although Kiangs aren't currently endangered (they're classified "Lower Risk" on the International Union for Conservation of Nature and Natural Resources Red List), ongoing hunting for meat and skins is having a negative impact on Kiang numbers. There are currently an estimated 67,000 Kiangs in the wild, and, as of 2005, 116 individuals in zoos.

The Jawbone of an Ass

The most famous donkey jawbone in history was probably the wild ass jawbone that Biblical strongman Sampson used to slay a horde of enemy warriors. As the *Living Bible* tells us in Judges 15:15–17:

> Then he [Sampson] picked up a donkey's jawbone that was lying on the ground and he killed one thousand Philistines with it. Tossing away the jawbone he remarked,
>   Heaps upon heaps,
>   All with a donkey's jaw!
>   I've killed one thousand men,
>   All with a donkey's jaw!

This location in Palestine has been called "Jawbone Hill" ever since.

BRAY SAY: *"Khyang" is the name given by the Tibetans to the wild horse of their northern steppes. More accurately it is a species of ass, quite as large in size as a large Japanese horse. . . . To all appearance it is an ordinary horse, except for its tufted tail. It is a powerful animal, and it is extraordinarily fleet. . . . It has a curious habit of turning round and round, when it comes within seeing distance of a man. Even a mile and a quarter away, it will commence this turning round at every short stage of its approach, and after each turn it will stop for a while, to look at the man over its own back, like a fox. . . . When one thinks it has run far away, it will be found that it has circled back quite near, to take, as it were, a silent survey of the stranger from behind. Altogether it is an animal of very queer habits.*
— Ekai Kawaguchi, *Three Years in Tibet*, 1909

Donkeys and the Evil Eye

Boccaccio speaks of "the skull of an ass set up on a pole in a cornfield as a potent amulet against blight." As a modern parallel to this we are told that at Mourzak, in Central Africa, the people set up the head of an ass in their garden to avert the evil eye from their crops.

— *The Evil Eye,* Frederick Thomas Elworthy (1895)

In many cultures the evil eye is the name for misfortune transmitted, usually without intention, by someone who is envious or jealous. The old British and Scottish word for it is "overlooking," which implies merely that the gaze has remained too long upon the coveted object, person, or animal, and not that the look is evil in and of itself. Overtly praising someone (or something) can also inflict the dreaded evil eye. In a few parts of the world, particularly Southern Italy and Sicily, it is believed that some people can deliberately cast the evil eye. Here are some things done at times to avert or cure it:

- Nineteenth-century Bedouins would hang a blue bead and the tooth of a dead donkey around an animal's neck or on its bridle to avert the dreaded evil eye. Sometimes this amulet was hung between the animal's ears over its forehead, where it could be seen by anyone seeking to do injury to the animal.
- In Portugal, donkey skulls were once placed in prominent places to divert the evil eye.

- Ancient Greek peasants wore amulets made of pierced peridot strung on thongs made of plaited donkey hair.
- An amulet in the Arabic Folk Medicine and Magic archives at the Kelsey Museum of Archaeology in Ann Arbor, Michigan, is made of a crow's head and a bit of skin from a dead donkey's nose wrapped with seven different colored strands of silk. Seven pins are stuck inside the roll of skin. This amulet once was worn by an infant to protect the child against the evil eye, as well as to prolong his life.
- In old Hungary, a child who caught whooping cough (*szamár köhögés* in Hungarian literally means "donkey cough") was said to be cursed by the evil eye. As an antidote, the child was slipped around and under a donkey three times.
- In the Middle East, turquoise blue faïence beads (donkey beads) are still used to protect livestock from the evil eye.

CHAPTER 3

THE BREED YOU NEED

*Concerning the difference between man and
the jackass: some observers hold that there
isn't any. But this wrongs the jackass.*

— Mark Twain

There are scores of breeds of horses, cattle, sheep, and dogs to choose from, but that's not true of our long-eared friends. While there are hundreds of donkey breeds around the world (China has 18 all her own), we have only four in North America to select from, and only two of those are true American breeds.

What Is a Breed?

Wikipedia defines *breed* as "a domesticated subspecies or infrasubspecies of an animal." For a type to be recognized as a breed, there should be a viable true-breeding population. In the case of equine breeds, a registry or other governing body maintains a studbook containing pedigrees and other information about each individual of each given breed. In the United States, registries and studbooks are sometimes created for types of equines that aren't yet fully true-breeding; the American Council of Spotted Asses is a typical example.

The four breeds of ass commonly found in North America are the American Mammoth, the Spotted Ass, the Poitou Ass, and the Mediterranean Miniature Donkey. Of those four, the first two are true American breeds, while the latter two are fascinating and historic breeds from France and Italy.

As you look at the following photographs of various donkey breeds and read their histories, think carefully about your own individual needs and circumstances. Consider which breeds and types are most likely to fit your situation. If you long for a riding donkey and you're six feet tall, a Standard donkey won't do the trick. If girth, rather than height, is an issue, however, wide-set Large Standard donkeys are better weight-carriers than the narrower, lankier type of Mammoth. If driving is your forte, any breed or type will do, but if you want to haul multiple passengers along on your ramblings you'll need to drive a multiple hitch of Miniatures or choose a bigger type or breed. If you want to make donkey milk soap, bigger jennies tend to give more milk. If visiting shut-ins or doing living nativities is your venue, you can haul several Miniatures in the back of a van; they're easy to manage in close quarters and not as intimidating to the folks you plan to visit.

DONKEY BREEDS

SOMALI WILD ASS.
Today's domestic donkeys descend solely from African Wild Asses. While Nubian Wild Asses are extinct in the wild, small groups of critically endangered Somali Wild Asses still roam parts of Chad, Djibouti, Sudan, Ethiopia, Somalia, and Eritrea.

ONAGER. *(above and left) Now critically endangered, Persian Onagers were the wild asses of the Old Testament. In 2000, the World Association of Zoos and Aquariums estimated their population at around 500 head, plus an additional 147 housed in zoos.*

KULAN. *(below) In 2000, the International Union for Conservation of Nature and Natural Resources Red List estimated the world's wild Turkmenian Kulan population at 650 head. Kulan range is protected in two wildlife preserves: Badkhys Nature Park in Turkmenistan and on Barsa Kelmes, a small island at the north end of the Aral Sea.*

KIANG. *Also called Tibetan Wild Asses or Gorkhar, Kiang inhabit high plateaus and steppes in Tibet, the Tsinghai and Szechwan regions of China, Nepal, and India. There are currently an estimated 67,000 Kiang in the wild.*

MAMMOTH DONKEY. *(top left)* A rangy Mammoth ass shares grass and companionship with a smaller, Standard donkey.
(top right) Many Mammoth asses are tall and strong enough for the biggest men to ride.

POITOU ASS. *(center and bottom row)* Huge and hairy, Poitou asses hail from the French province of Poitou. They are critically endangered worldwide. If you look closely at the picture at lower left, you'll spy a cute Miniature Donkey peering out from under his massive, long-haired Poitou ass friend.

MINIATURE DONKEY. *Miniatures are arguably the world's cutest donkeys, especially when they're young and fluffy like the winsome foals on this page.*

SPOTTED ASS. *(above and left) Though spotted individuals occur in domestic donkey populations worldwide, the registered Spotted Ass is an all-American breed.*

WILD BURRO. *(below) Wild burros are feral donkeys descended from asses brought to our American Southwest by intrepid Spanish explorers and settlers.*

STANDARD DONKEY. *(top left and below) Standards are the midsize animals most people picture when they hear the word "donkey."*
(top right) Large Standard donkeys are bigger than Standards but smaller than Mammoths. They're a perfect size for riding and driving.

SARDINIAN ASS. *The Sardinian ass was an ancestor of the Miniature Mediterranean Donkey. This breed originated — and still lives — in Sardinia, a large island off the coast of Italy.*

Which Breed?

Before deciding which breed you need, take some time to answer the following questions, to help you address your own particular needs and preferences. Don't settle for less than you really want. Figure out your needs and buy the type of donkeys that meets those needs so you needn't switch breeds or upgrade later on.

Q *Should I buy registered animals or grade (unregistered) donkeys?*

A If you plan to breed, pay a little more and buy registered stock; registered foals fetch higher prices than unregistered youngsters, and with pedigreed stock you have records of your animals' ancestry to help you to formulate important breeding decisions. If you simply want donkeys to ride, drive, or love, but not to breed, then pedigrees aren't a major consideration.

Q *Am I willing to seek out (and spend more) for an unusual or rare breed or type of donkey?*

A No matter where you live, Mammoths are relatively scarce; Miniatures and Large Standards are more common than Mammoths, and Standards are the most common of all. Poitou Asses are very rare and expensive. Miniature prices vary tremendously, from $100 for an unregistered jack to many thousands of dollars for a top-flight show and breeding donkey. Large Standards, especially those toward the high end of the height limit, generally cost more than smaller Standard donkeys.

Q *If I'm looking for something unique, would I like to help to preserve and promote an endangered breed?*

A The American Rare Breeds Conservancy (ARBC) (*see* Resources) is a nonprofit organization that protects and promotes more than 150 breeds of rare and heritage breeds of livestock. The organization lists three breeds of asses in its Conservation Priority List:

- ✗ The Miniature Donkey is listed as **Recovering:** "Breeds that were once listed in another category and have exceeded Watch category numbers but are still in need of monitoring."
- ✗ The Poitou Ass is listed as **Critical:** "Fewer than 200 annual registrations in the United States and estimated global population of less than 2000."
- ✗ American Mammoth Jackstock is considered **Threatened:** "Fewer than 1000 annual registrations in the United States and estimated global population of less than 5000."

The Poitou Ass and the Mammoth definitely need additional dedicated breeders if they're to survive.

How Tall Is a Donkey?

The following information is provided by the American Donkey and Mule Society (ADMS).

Miniature Mediterranean: 36" and under
Standard: 36.1 to 48"
Large Standard: 48.1 to 54" (jennets); 36.1 to 56" (jacks and geldings)
Mammoth: 54.1" and above (jennets); 56.1" and above (jacks and geldings)

MAMMOTH

LARGE STANDARD

STANDARD

MINIATURE

Closing In

Let's assume you've narrowed your search to two or three breeds or types. What next? Seek out equine expos where breeders gather to showcase their animals. Talk with breed representatives, pick up handouts and price lists, and compare your favorite breeds and types.

Call or write the registries involved. Request their promotional literature and breeder directories, subscribe to their magazines and newsletters, and read everything cover to cover. Carefully examine each registry's materials and periodicals to determine their policies and procedures. Then ask yourself these questions:

- ✘ Are the claims made about this breed realistic? Everyone thinks his breed is the best on earth, but if things sound too good to be true, they probably are. Equines are individuals, no matter their genetic background; expect to encounter variations in conformation, temperament, trainability, and everything else. Don't accept claims verbatim.

- ✘ Does this group's materials, including any periodicals, reflect a degree of professionalism you're comfortable with? Some folks are perfectly happy with photocopied newsletters and speed-printed handouts; maybe you're not.

- ✘ Does this registry actively promote the breed it represents? How? When? Where? Are cooperative advertising groups promoting this breed? Can I join them? If you plan to show these animals, you'll want folks to know what they are. If you breed them, there should be a market for their offspring.

- ✘ Does this group sanction activities of the sort I want to participate in? Saddle log and driving projects? Showing, breed, or open? Incentive programs? Futurities? Conventions? Photo contests? If they're important to you, find out.

- ✘ Are there fees, policies, or procedures I couldn't live with? Every year folks abandon breeds over dissatisfaction with their registry or its ways and means. Be informed before you commit.

Satisfied? Then ask the registry for the names of owners and breeders, if any, in your locale. Visit these folks, admire their animals, and ask lots of questions. If you're offered a chance to ride or drive a typical donkey, do it and ask for breed-specific pointers. Ask why they chose this breed, and listen carefully to their answers. Ask if they're satisfied with the registry representing it. Ask about prices and availability: theirs and other breeders'.

Donkeys in the New World

- On April 9, 1495, King Ferdinand II of Spain (Isabella's husband) sent livestock to Christopher Columbus, then on his second voyage to the New World. The livestock included horses, cattle, hogs, sheep, and goats, as well as four jacks and two jennies.

- The first donkeys and mules to set hoof on what is now United States soil came with the party led by Spanish colonist Juan de Oñate as it crossed the Rio Grande from Mexico in 1598. Indications are that more than 30 donkeys and mules made the trip. Oñate's group eventually settled at San Gabriel, near present-day Española, New Mexico. By 1609, trains of donkeys regularly packed goods between missions and settlements in New Mexico's Rio Grande Valley and supply sources in Chihuahua, Mexico.

- Colonel W. H. Emory, writing about life in Santa Fe, New Mexico, in the mid-1800s, stated, "Generally, the burros when ridden were mounted from behind, after the fashion of leap-frog."

- The 1769–1770 expedition to Upper California, headed by Caspar de Portola, Spanish governor of Baja California, included numerous mules, many of which were eaten by the hungry explorers during the long journey.

Finally, weigh all these factors and decide: which breed? After selecting a favorite, expect to search widely within that body for a donkey whose conformation, trainability, disposition, and other attributes meet your needs. Animals are individuals, never cookie-cutter creations. Individuals should be evaluated accordingly.

Popular North American Breeds

There aren't a lot of recognized donkey breeds to choose from (at least in North America), but the ones we have cover the spectrum from wee, cobby Miniature Donkeys at 30 inches tall to brawny Mammoth Jackstock in the 17-hand range. No matter what you want, there's a breed (or type) of ass just right for you!

THE MAMMOTH

Before the tractor, farm power equated with oxen, horses, and mules. Once established as working stock in the United States — by George Washington, no less (*see* Donkey Tails *on page 49*) — mules quickly became America's draft animal of choice. Heavy mules were also needed during wartime to draw the army's heavy artillery. Tall, heavy-boned jacks were used to sire these big, strong mules, and to that end our forefathers developed Mammoth jacks.

Charles Sumner Plumb, professor of Animal Husbandry at Ohio State University, wrote the following in his 1906 book, *Types and Breeds of Farm Animals:*

The introduction of the ass to America is said to date back to colonial times, when, soon after the Revolutionary War, the King of Spain sent as a gift to General George Washington, a Spanish jack and jennet. In Washington's time some early interest was shown in breeding mules, as a result of this gift. Henry Clay is said to have imported some very fine Spanish jacks to Kentucky. From the Clay stock Kentucky obtained the start, which has resulted in such fame for that state in mule breeding. One of the first jacks taken to Tennessee was imported about 1840 . . . and many importations from Spain, France, and the Balearic and Malta islands have been brought to America since 1884.

High prices for jacks prevail to an extent not generally supposed. In France the Poitou ass brings a large figure, common ones selling close to $1000 and the better class from $1500 to $2000. At a recent Paris exposition one is reported selling at $3200. Paragon 63, imported from Catalonia in 1886, was sold for $2000. . . .

Between 1830 and 1890, American mule breeders imported several thousand Mammoth-type asses of a number of breeds — primarily Andalusian, Maltese, Catalonian, Majorcan, and Poitou Asses — from Spain and other parts of Europe. Some distinctions between these breeds are summarized below.

Andalusian. The Andalusian Ass is an ancient breed native to Andalusia in southern Spain. Most Andalusians were (and are) gray, often with attractive reverse dapples. Individuals imported to North America in the nineteenth century ranged in height from 14.2 to 15.2 hands. Read about today's endangered Andalusian Asses on page 39.

Maltese. Maltese Asses originated on the island of Malta in the Mediterranean Sea. They were spirited, free moving, and invariably black or brown in color. Because these asses were smaller (they rarely exceeded 14.2 hands) and more refined than other imported breeds, they were never as popular as their larger, draftier peers.

Catalonian. Catalonian jacks of the day were enormous, 14.2- to 16-hand, black or brown asses with lighter points; they hailed from Catalonia in the extreme northeastern corner of Spain. These lively, fast-maturing jacks became very popular with mule breeders in Missouri, Kansas, and the eastern United States. They're now endangered; read about today's Catalonian Asses on page 39.

This photo, enlarged from its tiny size, shows a man seated sideways on a donkey — probably a Mammoth or at least a Large Standard. The donkey is wearing the man's hat.

What Is a Hand?

Some breed associations compute heights in inches, and some in **hands**. A hand in equine-speak is equal to 4 inches. If a donkey is 12.2 hands high, he is 12 hands (48 inches) plus 2 inches, or 50 inches. Measurements are always taken from the ground to the point of the withers.

This undated postcard features a Majorcan ass with an elderly couple in native dress.

Majorcan. Majorcan jacks came to America from Majorca, one of the Balearic Islands in the Mediterranean off the coast of Spain. They were drafty, with huge bone, and averaged 15.2 hands in height. Though heavily used in Spain for producing artillery mules, with numerous fine specimens imported to both North and South America, this breed never became a favorite of American mule-men.

Poitou. During the nineteenth century, a handful of Poitou jacks were imported to America from the French province of Poitou, a rich agricultural region bordering on the Sea of Biscay. Though these large jacks with their huge heads and massive bone never became wildly popular in North America, those who imported Poitou jacks swore by the massive mules that they sired. For information about this ancient, long-haired breed in today's world, see The Poitou Ass on page 38.

American Mammoths Today

Most of the breeds described by Professor Plumb are now extinct, or nearly so. At last count 120 purebred Catalonian asses were all that remained; a group of Spanish conservators are working to save the endangered Andalusian ass, and the Poitou breed is staging a slow but steady comeback.

The American Mammoth Jackstock Registry dates back to 1888, when it was incorporated as the American Breeders Association of Jacks and Jennets. Later, a second registry, the Standard Jack and Jennet Registry of America, was formed, and in 1923 the two registries merged. In 1988 the name of the organization was changed from the Standard Jack and Jennet Registry of America (SJJR) to the American Mammoth Jackstock Registry (AMJR), and thus it remains today.

Mammoths can be registered with the American Mammoth Jackstock Registry in two ways: by pedigree (when both parents are duly registered in the studbooks) or by measurement. According to the

> **'Ear's a Tip**
>
> **Don't Call Them Donkeys!** Mammoths should never be referred to as donkeys; they are Jackstock, Mammoths, Mammoth Jackstock, or Mammoth Asses. That said, some riders (such as the nice people at the Mammoth Donkeys list at YahooGroups) do call their big steeds donkeys. If you do it, you won't be alone.

registry's Rule of Entry, to qualify for registration by measurement:

- ✗ Jacks must measure no less than 14.2 hands, standard measure, in height, at least 61 inches around heart girth, and at least 8 inches at cannon bone (measured around cannon bone midway between the knee and fetlock).
- ✗ Jennets and geldings must measure no less than 14 hands, standard measure, in height, at least 61 inches around heart girth, and at least 7½ inches at cannon bone (measured around cannon bone midway between the knee and fetlock).
- ✗ When applying for registry by measurements, submit three (3) photos with the application showing the entire animal clearly, front, rear, and side view.

In either case, the registry reserves the right to inspect the animal visually at any time prior to registration, to require an affidavit from a veterinarian certifying correct height and age, and to require DNA testing to verify parentage. Asses with dorsal stripes and shoulder crosses are not eligible for registration in the AMJR studbook.

The American Donkey and Mule Society accepts Mammoths in the American Donkey Registry studbook. Its height limits are slightly different from those of the AMJR, reflecting the lower height limits formerly endorsed by the Standard Jack and Jennet Registry of America, and asses with dorsal stripes and shoulder crosses are eligible for registration.

> **BRAY SAY:** *Of all the extraordinary animals in this exhibition [the Paris Agricultural Show] was an indigenous ass of Poitou, a colossal donkey, shaggy with long, black, fluffy hair, and a head like a hairy fiddlecase, forming a grotesque caricature of a huge wild ass, and so singular as to have been bought for a fabulously large sum.*
>
> — *The Dublin University Magazine*, 1860

American Mammoth Jackstock are found mainly in the United States, with small populations in Canada and Australia. According to the American Livestock Breeds Conservancy (ALBC) there are between three and four thousand of these asses alive today, although only a few hundred historic-type black Mammoth Jacks remain, making conservation of those strains a priority for the ALBC conservators.

THE POITOU ASS

Poitou Asses were developed for a single purpose: to sire outstanding mules when bred to mares of a massive breed of draft horse called the Mulassier (jennets were kept to produce more breeding jacks). For more than two centuries Poitou mules were considered the best in Europe, often commanding astoundingly high prices.

The Poitou Ass is an ancient breed that was well established in its modern form by 1717 when one of King Louis XV's advisors wrote, "There is found, in northern Poitou, donkeys which are as tall as large mules. They are almost completely covered in hair a half-foot long with legs and joints as large as those of a carriage horse."

Poitou Ass Fact Sheet

HEIGHT
13.2 to 15 hands (54–60") — see box, page 36

COLOR
Rusty black or dark brown

COUNTRY OF ORIGIN
Poitou-Charentes, France

REGISTRIES
Association de Sauvegarde du Baudet du Poitou (SABAUD)
The American Donkey and Mule Society (ADMS)

SPAIN'S ENDANGERED ASSES

CATALONIAN

In comparison with other specimens of the ass, the Catalonian is an aristocrat, with beauty, style and action. The head is trim and neat, and the ears well carried. The bone, while not especially large, is very hard and fine of texture and free of fleshiness. It is a very tough, hardy breed and has found more favor in America, in mule-producing districts, such as Missouri, Tennessee, and Kentucky, than any other breed.

— Liberty Hyde Bailey,
Cyclopedia of American Agriculture, 1908

There were once an estimated 50,000 Catalonian asses in Catalonia, but with the mechanization of farm work the breed declined. Today, there are only 120 purebreds left. Half of them live on a ranch in Olvan (Bergueda), where the Gassó family, who began with only 30 asses, has labored to conserve the breed for three decades. They have donated several purebred asses to the l'Empordà Wetlands Natural Park and established an association to protect and guarantee the survival of the Catalonian ass.

One of the foundation breeds used in the development of American Mammoth Jackstock, the Catalonian ass is known for his endurance and ability to travel long distances without losing condition. Conservators say that the Catalonian can go three days without drinking, and he is strong enough to easily pack 220 pounds. Adult animals reach heights of 65 inches or 16.1 hands and weigh 1,100 pounds.

Catalonian asses' ears are typically long and stand upright and their coat is dark brown except for their bellies, lower legs and eyes, noses, and the area around their eyes; these are white.

Catalonian asses are endangered in their Spanish homeland; perhaps American conservators could give this worthy breed a hand. For information, see Resources for the Web site of the Local Domestic Breeds of Catalonia.

ANDALUSIAN

Like the Catalonian ass, imported Andalusian asses were used to develop America's Mammoth Jackstock, the first of which was Washington's Royal Gift.

A conservation group, the Asociación Nacional de Criadores de la Raza Asnal Andaluza (ANCRAA), has formed in an effort to snatch back this elegant breed from the brink of extinction.

- They range in size from about 13.2 hands to 15.2 hands and larger for a jack. (*Note:* Equines are never measured in feet. The standard measure for large animals is hands; a hand is four inches with the decimal denoting extra inches, thus 13.2 is 54 inches.)
- Their color is silver gray; some are dappled.
- They must conform to certain measurements.
- Normally they do not have a cross, but it's permissible.
- Andalusian asses are intelligent, gentle, and biddable.

Just as the Poitou Ass is recovering, thanks in part to international breeders who have embraced the breed, endangered Andaluz asses need assistance to survive. For more information about them, contact ANCRAA (*see* Resources).

The September 1901 issue of *The English Illustrated Magazine* has this to say about the Poitou Ass:

Some of the finest donkeys in the world are those of the Poitou district in France, where mule breeding has been a thriving industry for many generations; and here again we find proof of the animal's extraordinary hardiness. The donkey mares . . . are kept in the lowest possible condition under the curious idea that this treatment results in the male offspring, which are most desired. The poor wretches are mere skin and bone, and are supplied with nothing but hay and straw in just sufficient quantity to keep them alive. Such treatment of the mother, one would imagine, would do anything but promote the stamina of the breed, but it appears to produce no bad effect.

The Poitou Ass is a big and powerful beast, standing from 13 to 15 hands high. One very curious practice is in vogue among the Poitou mule breeders: from the day the male ass is foaled his coat is left severely alone; he does no work, and is kept practically a close prisoner all his life, and being thus prevented from casting his hair, as he would do under normal conditions, the result is an accumulation of coats all tangled and matted together till they nearly trail on the ground. He then becomes an object of great pride to his owner, who apparently sees no drawback in the circumstance that this extraordinary mat of hair which envelopes his ass is a fruitful source of the worst skin diseases. Fortunately for the breed, this power of retaining an accumulation of coats is not universal among the Poitou asses; if it were, we could hardly expect them to attain the splendid development for which they are famed.

In 1867 there were 94 asineries (breeding establishments) housing 465 baudets (jacks), 294 jennets, and 50,000 Mulassier mares in the Department of Deux Sevres alone. A studbook was established in 1884.

However, after World War II, mules were replaced by tractors and automobiles, which quickly rendered the Poitou Ass obsolete. By 1950 only 50 asineries were operating in the entire Poitou-Charentes area, housing 300 baudets and 6,000 mares. In 1977 a breed

Did You 'Ear?

Many of the Acadians who settled Nova Scotia (and later, New Brunswick) came from Poitou province in France. In 1755, a large percentage migrated south to Louisiana, where they became known as Cajuns.

census was taken, showing that only 44 purebreds (20 jacks and 24 jennets) remained in all of France. Conservation efforts began and by 2001, 71 jacks and 152 purebred jennets were recorded in the registry studbook. Though still critically endangered, the breed is making a comeback in France and abroad, including small populations in Britain, Australia, and North America.

MEDITERRANEAN MINIATURES

Without a doubt, America's favorite breed of ass is the Miniature Mediterranean Donkey. America's wee asses, also referred to as Miniature Donkeys, Miniatures, and mini-donks, were originally bred as working donkeys on Sicily and Sardinia, islands located in the Mediterranean Sea off the southeastern and eastern coasts of Italy.

Miniature Donkeys first set foot on American soil in 1929 when New York stockbroker Robert Green received six jennies and a jack that he purchased sight unseen during a trip to Europe. A year later marauding dogs attacked the herd, killing three jennies; the jack and the three remaining jennies comprised the first breeding herd of Miniature Donkeys in the United States. Green soon imported more Lilliputian donkeys from the Mediterranean area, and by 1935 he'd amassed a herd of 52 donkeys. Wealthy buyers like Henry T. Morgan and August Busch Jr. also imported new blood from the Mediterranean after beginning with Green-bred stock.

A third group, the National Miniature Donkey Association, was established in 1989, not as a registry but to promote Miniature Donkeys through the establishment of show rules and a comprehensive breed standard, as well as to sponsor an array of programs for owners of Miniature Donkey Registry registered donkeys.

There is a lot to like about Miniature Donkeys. They're large enough for tiny tots to ride and to pull a cart carrying an adult and one or two children; they're easygoing and gentle to a fault; they're economical to keep and feed; and they're arguably the cutest creatures on God's green earth. These diminutive donkeys are readily available throughout North America, and there are also fair-sized populations in Britain, Australia, and parts of Europe. In addition, they're the perfect choice for knowledgeable breeders who want to show a profit by breeding asses.

Miniatures can be purchased for next to nothing (unregistered geldings and jacks bring $100–$200 in some locales) or for hefty five-figure prices. There is a strong market for high-end Miniature Donkeys, especially of the color du jour (at the moment, solid black with dark points). For example, at the 2007 North American Miniature Donkey Sale, 68 head averaged a tickle over $1,600 each, with a jack bringing $8,000 and 14 jennies selling for $3,000 or more each.

Then, in the early 1950s, Daniel and Bea Langfeld of Danby Farm (already breeders of world-class Shetland ponies) purchased a Miniature Mediterranean Donkey for their daughter, who had cerebral palsy. Soon they were major breeders with as many as 225 donkeys in their herd. The Langfelds widely promoted their donkeys, charming the readers of several national horse magazines with ads documenting the ongoing adventures of Parader's Seventy-Six Trombones (a Shetland colt) and Ricardo (his Miniature Donkey buddy). In 1958, Bea Langfeld established the Miniature Donkey Registry, which she turned over to the American Donkey and Mule Society in 1987. Today there are more than 50,000 donkeys registered in the Miniature Donkey Registry studbook, some of which have up to 500 recorded ancestors tracing all the way back to the first Miniature Donkeys in America.

A second Miniature Donkey registry, the International Miniature Donkey Registry, was incorporated in 1992. It differs in that it registers donkeys in two divisions (Class A is 36" and under; Class B is 36.1" to 38" tall), and it offers two-, three-, and four-star ratings based on each donkey's conformation.

This postcard, sent from Santa Barbara in 1905, is titled, "I am Stuart. Who are you?"

Though Miniatures range in height from 26 to 36 inches (and 38 inches for International Miniature Donkey Registry stock), judges and breeders tend to prefer donkeys in the 32- to 34-inch range. Very small Miniatures sometimes carry a dwarfing gene, so beware of large heads set on short necks and hefty bodies with very short legs. You don't want this gene in your herd.

THE SPOTTED ASS

Spotted Asses may be any height and of any breeding, but they must be spotted. According to registry literature, an "Ass or Half-Ass [the group registers mules and hinnies too] must have at least two spots visible on a photograph, behind the throatlatch and above the legs."

Montana native Dave Parker began raising donkeys in 1962, but it wasn't until 1967 that he bought his first spotted animals. Two years later, Parker and his friend John Conter (also of Billings, Montana) incorporated

Bluestone Mordecai, the author's Spotted Ass jack.

The American Council of Spotted Asses. The two maintained a registry that they actively promoted through a line of clever Spotted Ass products such as baseball caps, t-shirts, and mugs bearing the distinctive ACOSA logo. They also formed the SAPs Club for "people who loved asses — Spotted or spotless — and either owned one or patted one on the back" for which the partners issued members wallet-size, engraved bronze cards that quickly became collector's items.

In 1974, Parker sold his Spotted Ass herd due to health problems and passed the reins to his partner, John Conter, and John's wife, Claire, who have managed the organization since then. Though clearly a fun-loving organization, ACOSA is an established breed registry, and the Spotted Ass is a recognized breed.

Spotted Ass Fact Sheet

HEIGHT
Any height from Miniature to Mammoth

COUNTRY OF ORIGIN
United States

REGISTRY
The American Council of Spotted Asses (ACOSA)

OTHER BREEDS

Standard and Large Standard Donkeys. The American Donkey and Mule Association registers Miniature Donkeys in the Miniature Donkey Registry and Standard, Large Standard, and Mammoths of all types and colors in the American Donkey Registry (established in 1967 by ADMS founders Paul and Betsy Hutchins).

Wild Burros. The United States Bureau of Land Management (BLM) removes thousands of free-ranging feral horses and burros from BLM-managed public lands each year. Feral burros come in all sizes and colors, though average height is 44 inches (11 hands) — a good size for driving, packing, or for children to ride. Though technically "wild," these burros are easily tamed by compassionate adopters just like you. We'll talk about adopting one (or more) in chapter 4.

Evaluating Donkeys

No matter which breed you choose, it pays to learn to recognize the good and not-so-good physical characteristics of all members of the donkey tribe.

BRAY SAY: Known commonly as the jackass, this long-eared little creature is respected throughout the southwest — roundly cursed yet respected — and here he is usually referred to by his Spanish name, burro. Because of his extraordinary bray, he is sometimes ironically called the "Arizona Nightingale."

— Arizona state administration,
U.S. public relief program (1935–1943)

The following is roughly adapted from the National Miniature Donkey Association's breed standard for Miniature Donkeys so that it applies to asses of every type and size.

GENERAL APPEARANCE. Donkeys should be attractive, sound, strong, and sturdy. All parts should blend together in a nicely coupled, compact picture. Jennets should look feminine and slightly more refined than jacks and geldings. Bone should be strong and in proportion to the size of the donkey and his muscle development.

A WELL-PROPORTIONED ASS

DONKEY PROFILES

Roman-nosed
(Poitou)

Slightly dished
(Miniature)

Slightly Roman-nosed
(Mammoth)

Straight (all types)

HEAD. The head should be in proportion to the donkey. Donkeys should have broad foreheads with plenty of width between the eyes. The muzzle should taper with firm, even lips and large, open nostrils. Depending on breed, the profile can be Roman-nosed (Poitou), slightly Roman-nosed (Mammoth jacks), straight (all breeds and types), or slightly dished (Miniature Donkeys). Eyes should be large, prominent, dark, and clear, with symmetry and a kind expression. Ears should be long, parallel, upright (Poitou ears are allowed a little leeway), and carried alertly. The head should be well balanced and carried in an upright position.

TEETH AND BITE

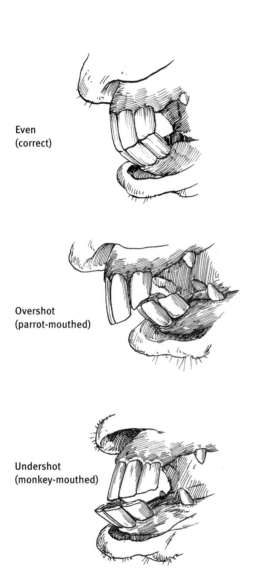

Even
(correct)

Overshot
(parrot-mouthed)

Undershot
(monkey-mouthed)

NECK PROPORTIONS

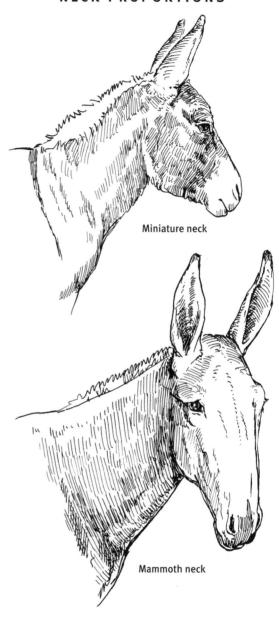

Miniature neck

Mammoth neck

TEETH. At maturity (after the third-year teeth have erupted) the teeth of the upper jaw should meet evenly with the teeth of the lower jaw — not overshot ("parrot-mouthed") or undershot ("monkey-mouthed" or "sow-mouthed"). Up to a ¼" variation from an even bite is acceptable for most registries.

NECK. The neck should be strong, straight, and in proportion to the head and body. It should not be too short and thick or too long and thin, or weak in appearance.

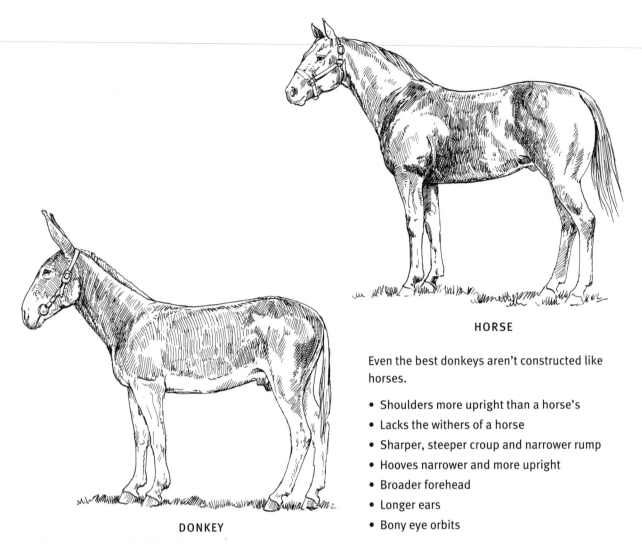

HORSE

DONKEY

Even the best donkeys aren't constructed like horses.

- Shoulders more upright than a horse's
- Lacks the withers of a horse
- Sharper, steeper croup and narrower rump
- Hooves narrower and more upright
- Broader forehead
- Longer ears
- Bony eye orbits

BODY. The body should have balance and symmetry. Each part should blend together in a well-proportioned, pleasing picture.

SHOULDERS. A donkey's shoulders are more upright than those of a horse but should not be excessively straight. An ideal shoulder angulation is approximately 45–50°.

HEART GIRTH AND RIBS. There should be a deep, generous girth. The ribs should be nicely rounded (well sprung). Watch for this especially when judging larger donkeys, as Mammoths, in particular, can be quite shallow and slab-sided.

BACK. The back should be of moderate length, strong, and level with a slight curve at the withers. The back should not dip or sway. The topline should be gently sloped from the withers to a level back and loin and then to a moderately long and sloping croup and well-placed tail. Avoid donkeys with long, weak backs.

LOINS. The loins, which are formed by the lumbar vertebrae, should be short-coupled, wide, and well covered. Loins should lead into a moderately long and gently sloping rump.

CROUP AND RUMP. As compared to a horse, the rump of a donkey has a sharper, more sloping croup. The pelvic bones are at a higher angle; therefore, the donkey croup is higher and the rump narrower. The rump should be strong and gently sloping. There should be good length from point of hip to point of buttock.

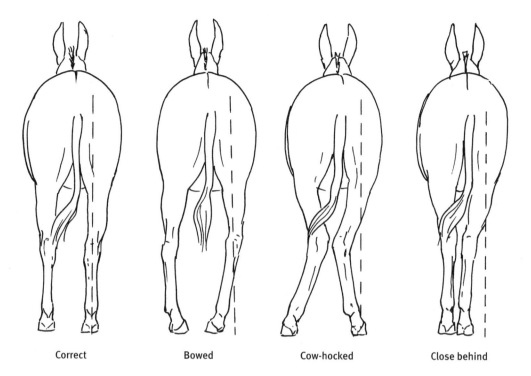

| Correct | Bowed | Cow-hocked | Close behind |

LEGS. Legs should be set squarely and straight when viewed from all angles, with adequate bone in the proportion to provide strength and balance to the animal. There should be ample space between the front legs, allowing good width of chest and room for lung expansion. The rear legs should have enough space between them so that the hocks don't touch.

FEET. The hooves should be even and of a uniform shape. A donkey's hoof is narrower, more oval, and more upright than a horse's hoof. (An exception: Poitou Asses' hooves are more horselike than those of other breeds and types.) Hooves should be well trimmed.

MOVEMENT. Movement should be active, smooth, straight, and free-flowing, covering the ground evenly with good length of stride. Each foot should be lifted completely off the ground and carried forward in a straight line.

WEAKNESSES. Any animal with severe or multiple weaknesses should not be selected as breeding stock. Keep an eye out for the following:

✗ **Head and neck.** Coarseness of the head and neck (except in Poitou Asses); short, thick neck and low head carriage.

✗ **Ewe-necked.** Thin, concave, weak neck.

✗ **Roach-backed.** Convex back; long, weak back; swayback (sagging).

✗ **Croup.** Short, high, or flat croup; hip too short (no depth of hip).

✗ **Goose-rumped.** Steep, sloping croup; narrow rump.

✗ **Proportion.** Insufficient bone in proportion to donkey.

✗ **Cow-hocked.** The hocks of the back legs turn inward toward each other when viewed from behind.

✗ **Bow-legged.** The hocks of the back legs turn outward away from each other when viewed from behind.

✗ **Close behind.** The back legs are too close together. This is common in a narrow-bodied, flat-ribbed animal.

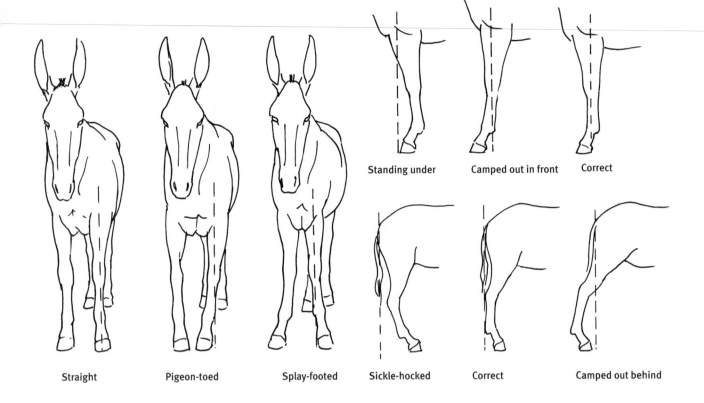

Standing under   Camped out in front   Correct

Straight   Pigeon-toed   Splay-footed   Sickle-hocked   Correct   Camped out behind

- ✗ **Sickle-hocked.** The back legs stand in and under the animal when viewed from the side.
- ✗ **Camped out behind.** The back legs are set too far back behind the body when viewed from the side.
- ✗ **Splay-footed.** Hooves of the front legs are turned outward and away from each other. This is common in narrow-chested animals.
- ✗ **Pigeon-toed.** Hooves of the front legs are turned inward and toward each other.
- ✗ **Under in front.** The front legs stand in and under the animal when viewed from the side.
- ✗ **Camped out in front.** The front legs are set too far in front of the body when viewed from the side.
- ✗ **Parrot-mouthed (overshot).** Upper teeth extend more than one-quarter inch beyond lower teeth.
- ✗ **Monkey-mouthed (undershot).** Also called sow-mouthed. Lower teeth extend more than one-quarter inch beyond upper teeth.

Faults Not Acceptable in Breeding Animals

Animals with any of the following faults should not be used for breeding.

- ✗ **Monorchid or cryptorchid.** A mature jack with one or both testes undescended into the scrotum.
- ✗ **Dwarfism in Miniature Donkeys.** A dwarfed animal has an overall stunted, thickset appearance. They can have multiple deformities in the legs and have disproportionate, heavy heads.
- ✗ **Extreme cow hocks.** Cow hocks that hinder the movement and the function.
- ✗ **Extreme malocclusion.** Teeth that miss an even bite by more than one-quarter inch.
- ✗ **Oversized or undersized height or bone measurement for the breed.** Keep in mind that few donkeys are mature until at least three years of age.

George Washington's Royal Gift

In 1783, George Washington retired to Mount Vernon, his 8,000-acre country estate 16 miles south of Washington, D.C., on the banks of the Potomac River. According to his adopted son and early biographer, George Washington Parke Custis, the former general "became convinced of the defective nature of the working animals employed in the agriculture of the southern States, and set about remedying the evil by the introduction of mules instead of horses, the mule being found to live longer, be less liable to disease, require less food, and in every respect to be more serviceable and economical than the horse in the agricultural labor of the southern States."

To this end, Washington vowed to purchase large jacks abroad to bring to Virginia to sire better mules. Instead, he was given two jacks — Royal Gift (an Andalusian jack) and Knight of Malta (a Maltese ass) — by foreign dignitaries King Charles III of Spain and Washington's friend the Marquis de Lafayette.

A later biographer, Paul Leland Haworth, expounded on these remarkable gifts:

Washington, according to his own account, was the first American to attempt the raising of mules. Soon after the Revolution he asked our representative in Spain to ascertain whether it would be possible 'to procure permission to extract a Jack ass of the best breed.' At that time the exportation of these animals from Spain was forbidden by law, but Florida Blanca, the Spanish minister of state, brought the matter to the attention of the king, who in a fit of generosity proceeded to send the American hero two jacks and two jennets. One of the jacks died on the way over, but the other animals . . . arrived on the fifth of December, 1785.

According to careful measurements . . . [Royal Gift] was fifteen hands high, and his body and limbs were very large in proportion to his height; his ears were fourteen inches long, and his vocal cords were good. He was, however, a sluggish beast, and the sea voyage had affected him so unfavorably that for some time he was of little use. Ultimately, however, 'Royal Gift' recovered his strength and ambition and proved a valuable piece of property.

[Royal Gift] was presently sent on a tour of the South. . . . No doubt the beast aroused great curiosity along the way among people who had never before set eyes upon such a creature. We can well believe that the cry, 'General Washington's jackass is coming!' was always sufficient to attract a gaping crowd. And many would be the sage comments upon the animal's voice and appearance.

In 1786 Lafayette sent Washington from the island of Malta another jack and two jennets . . . The new jack, the 'Knight of Malta,' as he was called, was a smaller beast than 'Royal Gift,' and his ears measured only twelve inches, but he was well formed and had the ferocity of a tiger. By crossing the two strains Washington ultimately obtained a jack called 'Compound,' who united in his person the size and strength of the 'Gift' with the courage and activity of the 'Knight.' The General also raised many mules, which he found to be good workers and more cheaply kept in condition than horses.

Henceforward the peaceful quiet of Mount Vernon was broken many times a day by sounds which, if not musical or mellifluous, were at least jubilant and joyous.

WHAT COLOR IS MY DONKEY?

The American Donkey and Mule Society recognizes these colors:

- Gray-dun (mouse-colored with a cross; called *grullo* or *grulla* in horses and mules; the quintessential donkey color)
- Brown (light, medium, dark)
- Black (with a cross/no cross)
- Red (includes sorrel, russet, red roan, "pink")
- White (includes ivory and spotted white)
- Spotted

There are, however, many names for the way these colors show up on a donkey or mule. Below is a basic color glossary. (For more information see Resources for books by D. Phillip Sponenberg.)

Apron face. A wide facial marking that completely encloses the eyes and chin; also called *bonnet-faced*.

Bald Face. A wide facial marking extending to or past the eyes but (usually) not the chin.

Bay. Black legs, mane, and tail on a red equine.

BEW. Blue-Eyed White (also called *ivory*).

Black. Nonfading (doesn't turn reddish brown when sun-bleached) or fading (turns reddish brown when sun-bleached).

Blaze. A wide white stripe down the middle of the face.

Boot. A white marking that extends into the fetlock, also called a sock.

Brown. A donkey ranging in color from very light tan to medium or chocolate brown to almost black.

Collar buttons. Black dots on the lower part of the neck below the jowls.

Coronet. White just above the hoof, around the coronary band, usually no more than one inch above the hoof.

Cross. A dorsal stripe paired with a stripe across the shoulders.

Dappled roan. A roan donkey with a lighter-colored face and reverse dapples (dark dapples on a light background).

Dapples. Rings of different-colored hair blending into an animal's coat.

Dark nosed. Having a black nose instead of the usual off-white nose in addition to an absence of white eye rings.

Dorsal stripe. A dark stripe with sharp, distinct edges running along an equine's back from withers to the hair of the tail; also called an *eel stripe* or a *race*.

Ermine spots. Marks the same color as an animal's body coat occurring within a white leg marking, usually just above or meeting the coronary band.

Flame. White hairs in the center of the forehead.

FSW. Frosted Spotted White: a white donkey with dark eyes and spotted pink and black skin.

Garters. Dark horizontal lines on the legs.

Glass eye. A whitish or blue eye; also called a *walleye, moon eye, china eye,* or *bird eye.*

High white. White stockings above the knee or hock, sometimes extending onto the flank or belly.

Irregular or crooked. An adjective qualifying a marking that does not take a more or less straight path.

Ivory. A blue-eyed white.

Masked spotted. A donkey with a blaze face and leg markings but no other spots on the body is genetically spotted, though it doesn't have body spots.

NLP. No Light Points; the donkey doesn't have a typical white nose, eye rings, and belly markings.

Partial. An adjective qualifying a leg marking that extends only up one side of a leg to the height indicated by the noun it qualifies; for instance: a partial coronet, a partial pastern, a partial sock.

Pastern. A white marking that extends above the top of the hoof but stops below the fetlock.

Points. The coloring of an equine's legs, mane, and tail.

Race. A narrow white stripe down the middle of the face.

Roan. An equine with white hairs mixed with black *(blue roan),* bay *(red roan),* or sorrel *(strawberry roan).*

Russet. A donkey with light sorrel shadings, a black/brown mane and tail, and light legs.

Shoulder stripe. A stripe running from shoulder to shoulder bisecting a dorsal stripe; together they constitute a cross.

Smokey black. A donkey who appears black, but whose cross and lower legs are visibly darker black.

Snip. A white marking on the muzzle between the nostrils.

Sock. A white marking that extends into the fetlock, also called a *boot.*

Sorrel. Red with legs the same color or slightly darker than the body coat.

Star. A white marking between or above the eyes.

Stocking. A white marking that extends at least to the bottom of the knee or hock, sometimes higher.

Strip. A narrow white stripe down the middle of the face.

This 1909 postcard addressed to Adams Mills, Ohio, is labeled "3 Jacks." The jack in the middle is a Mammoth breed donkey.

CHAPTER 4

A DONKEY
OF YOUR OWN

*ASS, n. A public singer with a good voice
but no ear.*

— Ambrose Gwinett Bierce,
nineteenth-century satirist

*This undated vintage postcard has no inscription,
but don't these boys look proud of their donkey!*

You've done your homework, and you're ready to
buy a donkey (or two or three or more). Remember,
the most important things are to find a sound, healthy
donkey and a seller who will stand behind the sale. But
where do you buy donkeys? And from whom? What
preparations are needed ahead of time?

Finding a Veterinarian

It's best to line up the services of a vet before bringing home your donkeys, and the best way to find a
really *good* one is to ask other donkey owners in your
locale for recommendations. At the same time, ask for
names of vets they wouldn't use and why they wouldn't.
Narrow it down to two to four vets most donkey people
like, then call their offices and ask the following pertinent questions:

✗ Does the doctor make routine farm visits or
  must you take your donkey to his clinic for treatment? What sorts of facilities are available if
  patients must be boarded overnight (or longer)?

✗ Will he come to your farm in an emergency? What about after hours, weekend, or holiday emergencies? To whom does he refer his clients when he's unavailable?

✗ How many veterinarians are associated with his practice? If more than one, can you stipulate which vet you want to see your animals?

✗ If you phone the clinic with a problem, will desk help connect you with the vet (within reason), or will they relay your concerns and return your call?

✗ How is payment handled? Is it cash up front for every call, or is the vet willing to carry a tab? Are credit cards acceptable? Does the clinic offer payment plans?

If you like what you hear, arrange for a time to visit the clinic. When you do, does the staff seem friendly and knowledgeable? Will they let you speak with the vet? Desk personnel and vet techs are your link to the vet, so if they make you feel uneasy, keep shopping.

Check out the practice's boarding facilities.

✗ Are they safe, clean, and arranged so that patients can't physically interact with one another?

✗ What sort of feed are equines eating?

✗ Can you provide your own feed if you like?

✗ Is water readily available, and is it clean?

If bedside manner matters to you, how do you like the vet? What is his take on clients performing routine health care procedures such as treating minor ailments and giving vaccinations? Once he knows that you're competent to administer them, will he dispense often-used prescription drugs like Banamine (flunixin meglumine) and epinephrine?

If he's not particularly donkey savvy, will he mind if you research a problem in books or online and take him your results to peruse?

Whatever you ask, you should never be made to feel inadequate or stupid. It's your donkey, and you'll be footing the bill, so if you're made to feel inferior, take your business elsewhere.

Once you've selected a vet, don't wait for an emergency to use his services. Schedule a routine farm visit and see how that goes. Does the vet arrive promptly, or does someone from his office phone to inform you of delays? How does he interact with your donkey? Are you comfortable with his attitude and his work? If so, congratulations — you've found a donkey vet! In turn, it's only fair to treat your new vet right:

✗ Have your donkey caught, cleaned up, and ready for treatment when the vet arrives. Chasing wayward equines across a 50-acre field is not in his job description, nor is wrestling with a mud-encrusted patient.

A New Owner's Checklist

Are you really ready to take possession of your first donkeys? You will need:

- Safe, roomy shelters to accommodate your donkeys, proper fencing, and a supply of bedding for any animals you plan to house. *(See chapter 6.)*
- Feed, feeders, water containers, and a clean, reliable water source. *(See chapter 7.)*
- Halters, leads, and grooming tools. *(See chapters 6 and 8.)*
- A first-aid kit; have it on hand before you need it. *(See chapter 10.)*
- If you plan to ride or drive your donkeys, you'll need appropriate tack. *(See chapter 6.)*
- If there are foals in your future, you'll need safe foaling quarters and a well-stocked foaling kit. *(See chapter 16.)*

And most important:

- Phone numbers of at least two or three veterinarians who treat donkeys (not all do). If the vet you want isn't available, you should know of at least two alternates whom you can call when you desperately need one.
- Phone numbers of knowledgeable donkey owners whom you can reach in an emergency. Sooner or later they'll come in handy.

- Learn to handle emergencies until your vet arrives. A country vet may be 50 miles away on another farm visit when you need him most. Check to make certain that your cell or cordless phone works in the barn in case your vet needs to talk you through a procedure. Stock well-equipped first-aid and birthing kits, and know how to use them.

- Never wait until a minor problem escalates into an after-hours or weekend emergency. Know what you can and cannot do by yourself, and involve the vet as soon as he's needed

- Be there. Your animals know you, and they'll behave better if you're on hand to help. If you don't understand a treatment, ask questions. If follow-up treatments you'll have to administer by yourself entail detailed instructions, write them down and follow them "to a T."

- Furnish a comfortable, weatherproof, well-lit place for your vet to work. Provide any restraints necessary to secure your donkey.

- A cold beverage on a sweltering summer afternoon or a steaming cup of coffee in the winter is always appreciated. And always settle your bill when payment is due.

Finding Breeders

If you plan to raise donkeys, you'll probably buy your foundation stock from other breeders. If you're lucky, you can buy fairly locally, and this is a good idea. Donkeys already acclimated to your region and spared the stress of long-distance travel stay healthier than animals trucked in from afar. However, if you plan to breed the highest-quality show and breeding stock, you might have to buy from distant breeders, perhaps through a production sale or the Internet.

Whatever type of donkey you're looking for — riding stock, Miniatures, Mammoths, or anything in between — it's important to locate reputable sellers of healthy

> ### Coggins Test
>
> In 1971 Dr. Leroy Coggins, a veterinary virologist at Cornell University, developed a simple, inexpensive blood test to check for the presence of equine infectious anemia antibodies (see chapter 10) in equines' blood. Nowadays most states require proof of a current Coggins test for every equine entering the state or attending shows, sales, or organized trail rides; some require that *all* equines be Coggins tested at least once in every 12-month period.

stock. Here are a few ways to find them.

Breed registries. Contact individual breed registries or the American Donkey and Mule Society. Visit their Web sites (*see* Resources) and peruse member-breeder directories. Phone or e-mail these organizations for additional information.

Online breeder directories. To locate online breeder directories, type *donkey directory* in the search box of your favorite search engine. Qualify it with a breed name (*miniature donkey directory*) if you like.

Online classified ads. Use your trusty search engine to find online classified ads (searches for *donkey ads* or *donkey classifieds* work well). Keep in mind that some major horse ad Web sites list donkeys, too.

Donkey magazines. Breed-specific journals such as *Asset* (published by the National Miniature Donkey Association) and all-donkey publications like *The Brayer* are packed with display ads, directories, and classifieds.

Breeder Web sites. Type the breed and the word *sale* into your search engine's search box; use, for instance: *Jackstock sale*. Qualify it, if you like, by state: *Jackstock sale California*.

E-mail groups. Join donkey-related e-mail groups. Breed and general interest donkey lists host "Friday sales" whereby subscribers post whatever they wish to sell. It's a great way to locate all sorts of donkeys and donkey-related goodies.

Donkey shows. Take in a donkey and mule show. Visit information booths and chat with exhibitors between classes. Breed associations and donkey and mule clubs sanction many shows each year; e-mail or call them for specifics.

Local resources. Check for "donkey for sale" notices on bulletin boards at feed stores and veterinarian practices, and pin up "donkey wanted" notices of your own. Monitor local classified ads. Talk to vets and County Extension agents in your buying area; they're sure to know who has donkeys for sale in your locale.

Evaluating a Seller

Whether dealing with somebody in the neighboring county or a showstock breeder three states distant, you'll want to assess a seller's reputation before you buy.

Simply because he markets a lot of donkeys or she runs flashy full-page ads in magazines doesn't mean a seller will sell you the kind of donkey you want and need, or stand behind him after the sale.

Fortunately in this day of Internet commerce the unscrupulous seller who fleeces buyers has little hope of sweeping his shady dealings under the rug. Fleeced buyers tend to go public, and they're likely to do it via industry-related e-mail lists. Satisfied customers sing sellers' praises via the same forums; listservs are an excellent tool, so use them.

Likewise, before you buy from local sellers, tap in to the local donkey and mule grapevine. Talk to other donkey and mule owners in your locale. Ask which sellers they would buy from, which they'd avoid, and their reasons why.

Once you've narrowed the field to a handful of individuals selling your type of donkeys, contact them and arrange to visit their farms.

INTERNET BUYING

Today, buying animals via the Internet is a way of life. Purchasers prefer Internet sales for numerous reasons. They can:

- Shop for donkeys anywhere in the world, at any time of day, seven days a week, from the comfort of their own homes.
- Quickly select from a vast pool of animals and breeders.
- Research animals and sellers before they deal, thus saving a great deal of time and money on farm visits.

When buying via the Internet, you should ask the same questions you would ask as part of an in-person farm visit. Of course, if a farm visit can be arranged, so much the better. But if not:

- Be especially careful to deal with reputable sellers. Request buyers' references and

check them out. Ask others in the industry for constructive feedback.

- Request video footage of the donkeys that interest you. If it's not available, ask to see additional photos taken from many different angles. Examine this material closely and address any issues before you buy.
- Always negotiate for a written guarantee.
- Be very clear about how you will transport the donkey to your home before you make a deposit. Who pays for interstate health papers and Coggins testing? How long will the seller hold the donkey after payment is made and you're working on lining up transportation? Who foots any vet bills incurred during that wait? What happens if the donkey should die? Get it in writing; don't leave anything to chance.

Shopping Etiquette

When visiting breeders and sellers, it's a good idea to mind your Ps and Qs. Be courteous and arrive at the designated time (if you're running late, call and let them know that you're on your way). When you arrive, look around. Farms needn't be showplaces, but they shouldn't be trash dumps, either. Some things to notice:

- ✗ Are all of the animals present (not just the donkeys) housed in safe, reasonably clean facilities?
- ✗ Are there droppings in the water tanks, or are animals eating hay off the ground?
- ✗ What are the donkeys fed?
- ✗ Are most of the donkeys in good flesh, neither snake thin nor roly-poly fat? In large groups, you're apt to spot a few that are a shade leaner or fatter than the norm, but the majority should be in just-right condition.

Don't be afraid to ask the seller plenty of questions, in particular: Why is the donkey you've come to see for sale? Is the seller changing bloodlines? Breeds? Downsizing? If they're culls, perhaps the trait he's culling for doesn't matter to you. For instance, maybe the larger Miniature Donkeys are being sold in favor of raising smaller stock. Other questions you could ask:

- ✗ What are the seller's management practices? For instance: Do they raise their animals in wide-open pasture or in pens or stalls? Do they work closely with a veterinarian or treat their animals themselves? Are they heavily into vaccinations and antibiotics or do they take a more natural approach to equine care? All of these things could matter to some buyers.
- ✗ Which vaccines and dewormers are used, how often, and why?
- ✗ Does the seller show you just the donkey you arranged to see or the entire herd? Insist on the latter; if there are problems out there, you want to know it before you buy.
- ✗ Specifically ask about a jenny's breeding and foaling habits. Is she easy to get in foal? Has she

had any foaling problems? Is she a good mom? How many foals has she produced?
- ✗ Is the seller willing to work with you after the purchase, should questions or problems arise? Ask up front to be certain.

If you like what you see, ask to examine the donkey's registration papers as well as his health, vaccination, deworming, and production records. Always ask about guarantees. Some producers give them and some don't; if there is one, get it in writing when you buy.

And trust your intuition. If at any time the seller seems evasive or otherwise makes you feel uneasy, thank him for his time and shop elsewhere. There are too many honest sellers out there to deal with someone whom you don't quite trust.

Production and Dispersal Sales

Production and dispersal sales are often ideal venues for buying quality donkeys at fair market prices. The best sales host an animal preview the day or evening before the event and a complimentary lunch during or just before the sale. They're publicized well before sale day, and printed catalogs highlight sale animals' pedigrees and production records. The best way to find out about production and dispersal sales is to peruse club newsletters and donkey magazines. These sales are donkey world social events and can

'Ear's a Tip

A Matter of Taste. A seller's bane might well be your joy. Perhaps his plain-Jane, standard-size gray-duns (which you adore) are being sold in favor of breeding registered Spotted Asses. And a donkey whose strident braying drives the seller nuts could be just the one you'd love to own.

Gender. If you're looking for a pet or a donkey friend to ride or drive, *don't buy a jack*. Jacks, like unaltered males of all livestock species, are undependable and best left to experienced breeders who need them and know how to handle them.

The major problem with jacks is that most are friendly, easygoing guys except when raging hormones suddenly cause them to behave out of character — and it happens more often than you think. Even a Miniature jack's powerful jaws can crush bone; think what a Mammoth's jaws can do. If donkeys are your pleasure, not your business, stick to jennies and geldings. You and anyone else who handles your donkeys will be safer and happier if you do.

Age. The world's oldest donkey was a British former racing donkey named Lively Laddie who was born in 1943 and lived over 60 years — *those* are donkey's years! However, barring unforeseen disease or accident, most any well-cared-for donkey will live to see his 25th birthday, so don't turn down a likely donkey because he's in his teens; donkeys live a long, long time.

be outstanding places to meet people and purchase quality donkeys.

Payment in full is expected on sale day. Donkeys sell with registration papers, health certificates, Coggins, and any other documentation needed for interstate shipment. Guarantees, if any, are stated in the sale catalog.

Always try to arrive well before the sale and do hands-on inspections of the donkeys you think you might bid on. Study the catalog to see which other lots were consigned by their owners and give those donkeys the once-over, too, to see how well they've been cared for.

Another sensible ploy: Mark your catalog, designating which animals you plan to bid on, and make a notation of your absolute top bid. Things get exciting when bidding runs hot and heavy, and you won't want to grossly overspend.

If you bring donkeys home from *any* sale, plan to quarantine them away from your existing herd. House newcomers in an easy-to-sanitize area at least 50 feet from any other equines. Deworm them, vaccinate them, trim their hooves, and keep them isolated for at least one month. Don't forget to sanitize the conveyance in which you hauled them home.

During that time, feed and care for your other animals first, so that you can scrub up after handling the new additions. Never go directly from the new group to the rest of your herd. If you can prevent it, don't allow dogs, cats, poultry, or other livestock to travel between one group and the other. When their time in quarantine is up, introduce the new donkeys to the rest of your herd and sanitize the isolation area and any equipment so that they are ready for next time.

Sale Barns

The first rule of donkey buying is to buy from individuals or at well-run dispersal and production sales, not from neighborhood sale barns. Avoid everyday livestock auctions; they're dumping grounds for other breeders' sick animals and culls.

When buying or adopting donkeys, keep in mind that these are very social creatures. No donkey should ever be kept by himself. Donkeys fare best in the company of at least one other donkey, but most will adapt to a companion of another sort, be it a horse, goat, sheep, cow, pet pig, or llama.

MEASURING DONKEYS

Donkeys are measured at their withers. To properly measure a donkey, stand him on a firm surface and lower his head, the better to ascertain the highest point of his withers (located near the end of his mane). Measure from there in a straight line to the ground; don't follow the curve of the donkey's sides!

It's always best to use an equine measuring stick if you have one; these can be purchased from horse supply retailers or on eBay. Barring that, recruit someone to hold a ruler straight across the donkey's withers at their highest point and measure down from the ruler's end; you'll get a more accurate measure that way.

If height truly matters, don't buy sight unseen or travel long distances unless a seller will guarantee height. Many sellers "guesstimate" rather than taking accurate measures; Miniatures often turn out to be considerably taller than sellers claim and Jackstock a good deal smaller.

If you buy at sales barn auctions, you won't know if the donkey you choose has been vaccinated, if she's pregnant and by what sort of jack, or what's going on in her herd of origin. She could be an escape artist extraordinaire or a jenny who gives birth and then wantonly abandons her newborn foal to his fate. A jack may be infertile — or so ornery that his owner is willing to see the last of him at any price.

The donkey you buy might have been exposed to or be infected with respiratory diseases like strangles or flu. And remember, donkeys who weren't exposed to disease *before* they're sold through an auction will likely be by day's end. The donkey you buy might have shared a pen with sheep or goats spilling pus from caseous lymphadenitis (CL) abscesses or that are limping cases of foot rot, which the donkey will drag home on her hooves to your own sheep and goats. Coughing horses, sheep with diarrhea, goats dropping soremouth scabs wherever they go: you'll find them at country sale barns. These serious maladies infect the animals you keep on your farm only if you bring in sick or carrier animals, so always err on the side of caution.

Always, without exception, before handing over your check, get all applicable guarantees and sales conditions in writing. Do it every time, even when dealing with friends. People misunderstand, people forget. A comprehensive written sales contract is an absolute must, so hammer out the details and *write them down!*

Adopting a Wild Burro

One of the best buys in the equine world is a burro (*burro* is the Spanish/Old West word for "donkey") from the United States Bureau of Land Management (BLM) Wild Horse and Burro Adoption Program. These are unbranded, unclaimed, free-ranging donkeys gathered from BLM-managed rangeland throughout the West. Although they're initially wild, almost any patient person can easily tame and train smart, willing creatures like BLM donkeys.

Wild burros range in height from Small Standard to Large Standard sizes; they average 44 inches in height and tip the scale at about 500 pounds. They come in all colors, though gray-dun is most common. The BLM adopts out only healthy animals; negative Coggins tests, health certificates, and individual medical histories are furnished. All burros are dewormed and freeze marked prior to adoption.

The best way to learn about wild burro adoption is to visit the Wild Horse and Burro Adoption Program Web site or call its adoption hotline (*see* Resources). In the meantime, here are some things to consider. To adopt a wild burro you must:

✖ Be at least 18 years old (parents or guardians may adopt a wild burro and allow younger family members to care for it)

✖ Have no prior conviction for inhumane treatment of animals or for violations of the Wild Free-Roaming Horses and Burros Act

✖ Demonstrate that you have adequate feed, water, and facilities to provide humane care for the number of animals requested

✖ Show that you can provide a home for the animals in the United States

Qualifications

You must provide a minimum of 400 square feet (20 feet by 20 feet) of corral space for each burro adopted. Fences must be constructed using sturdy pole, pipes, or planks and be at least 4.5 feet high for ungentled burros (barbed wire, large mesh woven, stranded, and electric fences are unacceptable). Posts should be at least 6 inches in diameter and spaced no farther than

---

DON'T BUY TROUBLE

Before you buy from any source, learn to recognize healthy donkeys.

A HEALTHY DONKEY	A SICK DONKEY
Is alert and curious	Is dull and disinterested in his surroundings; may isolate himself from his companions
Has bright, clear eyes	Has dull, depressed-looking eyes; may have fresh or crusty opaque discharges globbed in the corners of his eyes
Breathes in an easy and rhythmic fashion	May have thick, opaque, creamy white, yellow, or greenish nasal discharge (a trace of clear nasal discharge usually isn't cause for concern); may wheeze, cough, or breathe heavily or erratically
Has a clean, reasonably soft hair coat and pliable, vermin- and eruption-free skin	Has a dull, dry hair coat; his skin may show evidence of external parasites or skin disease
Moves freely and easily; stands with both front feet straight under him	Moves slowly, unevenly, or with a limp; may point (extend forward) one or the other front leg
Is of average weight for his size, breed, and age	May be thin or emaciated; an extremely obese donkey isn't necessarily sick, but his obesity may be difficult to control
Has a healthy appetite	Might refuse food; a donkey who won't eat is almost certainly ill (or will be, due to hyperlipidemia)
Has firm droppings; his tail and surrounding areas are clean	May have scours (a.k.a. diarrhea); his tail, tail area, and hair on his hind legs may be matted with fresh or dried diarrhea
Has a temperature in the normal range (97.2–100°F for an adult donkey; foals temperatures may run slightly higher)	Might be running a higher or lower than normal temperature; subnormal temperatures are generally more worrisome than fevers

Reading a BLM Freeze Mark

The BLM uses freeze marks, or freeze brands, to identify captured wild horses and burros because they're a permanent, unalterable, and relatively painless way to identify individual animals. BLM freeze marks incorporate a series of angles and alpha symbols known as the International Alpha Angle System. These marks are always applied to the left side of a burro's neck. Each BLM freeze mark begins with the BLM's unique symbol, followed by the individual burro's year of birth (with the two year numerals stacked), and then his registration number. A bar is placed under all or part of the numbers after the year to show where the actual bottom of each character ends. Close-clipping a burro's freeze brand can make blurred numbers easier to read.

KEY TO THE ALPHA ANGLE SYMBOL

Read each angle to determine the freeze-mark number

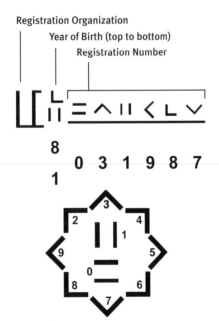

Wild burros are easily tamed. They make wonderful pet, performance, and livestock guardian donkeys.

8 feet apart. Using poles, there should be a minimum of five horizontal rails; using planks, there should be at least four rails. No space between rails should exceed 12 inches. Rails must be fastened to the inside of the posts using heavy nails or lag screws.

You must also provide shelter from inclement weather and temperature extremes. Shelters must have at least two sides and a roof. They must be well drained, adequately ventilated, and accessible to the burro(s). Tarps are not acceptable. Call the BLM hotline for further information about shelter requirements in your part of the United States.

If you meet the adoption requirements, call for an Application for Adoption of Wild Horses or Burros, or download one from the BLM Web site, fill it out, and mail it to the BLM office serving your area. The BLM will contact you during the application review process to verify that your facilities are adequate for the number of burros you wish to acquire.

You're generally allowed to adopt up to four burros in a 12-month period; however, you can apply to adopt more, in which case the agency requires additional verification of facilities and compliance checks. The minimum fee to adopt a wild burro is $125; most adoption events, however, use competitive bidding to establish adoption fees. Burros average about $135 each.

The federal government retains ownership of your adopted burro for one year, after which the BLM will send you a Title Eligibility Letter. Then, if you submit a signed statement from a veterinarian, County Extension agent, or humane official verifying that you've provided humane care and treatment, the BLM will send you a Certificate of Title for your burro.

Finding Your Burro

There are several ways to acquire a BLM wild burro. The BLM holds adoption events at hundreds of locations throughout the United States every year; there are sure to be events in or near your state (visit the program Web site or call for a schedule). (*See* Adoption Options, *page 62.*)

Whether adopting direct from a facility or at an adoption event, you're responsible for safely transporting your burro home. You should take a double-stitched nylon web halter and a 12- to 20-foot cotton or nylon lead rope for each animal you adopt. BLM employees will halter your burro and load him into your trailer.

This postcard was sent from Reno, Nevada, in 1946, with a message: "How would a horse like this be for you? Of course you would need a saddle and this water can. You have a bed roll, don't you. Be a good boy and get some wood for Mother. We are all O.K. Dad."

You're required to transport newly adopted burros in a stock-type trailer with rear swing gates (drop ramps, divided two-horse trailers, and trucks with stock racks are unacceptable). The trailer must have a covered top, sturdy walls and floor, a smooth interior free of any sharp protrusions, ample headroom, and adequate ventilation; the floor must be covered with nonskid material. Burros must be unloaded at least every 24 hours for food, water, and a minimum of five hours of rest. Only BLM-compliant layover corrals may be used.

It sounds like a lot of "musts," but by adopting a wild burro you're giving a home to a living, breathing piece of American history. In addition, wild burros are beautiful, healthy creatures. They tend to tame remarkably easily, and their hooves are arguably the toughest and soundest in the world!

Adopting a Rescued Donkey

Another way to obtain a fine donkey and do a good deed is to adopt a needy animal through a donkey rescue or an animal welfare organization. Keep in mind that you aren't buying these donkeys. Adoption fees are used to

Donkeys at Yahoo! Groups

When I want to learn a lot about any new topic in a short span of time, I subscribe to electronic mailing lists (also called listservs, e-mail groups, and e-mail lists) devoted to that topic. Today's major list server is Yahoo! Groups (*see* Resources). It hosts dozens of donkey- and mule-oriented listservs, and participation in Yahoo! Groups is always free.

ADOPTION OPTIONS

Most BLM adoption and holding facilities offer animals available for on-site adoptions year-round by appointment. Not all have burros on-site (a few specialize in wild horses only), but these do. *(See* Resources *for addresses and phone numbers.)*

- *Kingman Regional Wild Horse and Burro Facility in Arizona.* Wild burros gathered from public lands in Arizona, southern Nevada, and southwestern California make up the majority of the animals prepared for adoption at this site, which has a hold capacity of 240 animals. Adoptions are held on the third Saturday of every month.
- *Litchfield Wild Horse and Burro Corrals near Susanville, California.* This facility functions as a regional preparation center for wild horses and burros gathered from public lands in northern California and northwestern Nevada; it can hold more than 700 animals.
- *Ridgecrest Regional Wild Horse and Burro Corrals in California.* The wild horses and burros available at this center are gathered in Arizona, Nevada, and southern California.
- *Cañon City Facility in Colorado.* Serves as a regional preparation center for wild horses and burros gathered in Colorado and as a resting point for animals gathered throughout the West that are headed for adoption events in the central and eastern portions of the United States. Adoptions are held on two Fridays per month, by appointment.
- *Eastern States Wild Horse and Burro Facility near Mount Vernon, Illinois.* Serves as the easternmost holding and adoption center in the system. Adoption events are held at the facility five times per year and by appointment with an approved application.
- *Elm Creek Wild Horse and Burro Center in Nebraska.* Serves as a resting place for animals arriving from the West prior to shipping to adoptions in the central and eastern half of the United States. Approximately 3,000 to 5,000 wild horses and burros pass through the facility each year. Scheduled adoption events are held periodically at the facility.
- *The National Wild Horse and Burro Center near Palomino Valley, Nevada.* The largest preparation and adoption facility in the country, it serves as the primary preparation center for wild horses and burros gathered in Nevada (Nevada is home to more than 50 percent of the nation's wild horses and burros with approximately 102 herd-management areas spread throughout the state). Most of the animals at Palomino Valley are available for adoption by appointment, six days per week.
- *Pauls Valley Adoption Center, south of Oklahoma City.* Serves as a resting point for wild horses and burros arriving from the West who are en route to adoption event locations in the central and eastern United States. The facility holds up to 600 animals in 12 pastures stretching across 600 acres of land. Adoption events are held quarterly; call the BLM hotline for further information.
- *Salt Lake Regional Wild Horse and Burro Center near Herriman, Utah.* Serves as the primary preparation center for wild horses gathered in Utah but also holds animals gathered in Nevada and Idaho. Adoptions are by appointment only.
- *Delta Wild Horse and Burro Facility in Utah.* Serves as a preparation center for wild horses and burros gathered in Utah and as an overflow holding facility for several other BLM facilities. It holds approximately 300 animals. Adoptions are by appointment only.

maintain the group's other donkeys until they leave for new homes. Reputable rescue groups are concerned about the ongoing welfare of the animals they place; therefore, you must meet certain criteria to adopt, and in most cases, adoptive organizations retain ownership and the right to do farm checks on demand.

The beauty of these adoptions, apart from the satisfaction you'll derive from helping a donkey in need, is that rescue groups carefully prescreen both the animals they place and the homes in which they place them; thus, they're likely to find you a perfect match. If the adoption doesn't work out, the animal can be returned and re-homed. It's a win-win situation for everyone concerned.

To locate a donkey who needs your help, review the donkey rescues and sanctuaries listed at the back of this book. If there aren't any in your locale, chances are good that the ones listed can point you in the direction of animal shelters in your area that handle donkeys from time to time.

Other Ways to Help

Donkey rescue groups and donkey sanctuaries realize that not everyone can take in a donkey in need. If you can't adopt a donkey but you still want to help, in-house adoption programs exist whereby generous-hearted folks can sponsor or cosponsor an unadoptable animal who must remain on the organization's premises for life.

Another way to help is to donate money or to supply items on your favorite group's "wish list." A typical equine rescue wish list might include:

- ✖ Feed and bedding (good hay is always needed)
- ✖ Dewormers
- ✖ Vaccines and medical supplies
- ✖ Supplements, including milk replacer for orphan foals and weight builders for sick and elderly donkeys
- ✖ Blankets, turnout rugs, and waterproof sheets

- ✖ Halters in all sizes
- ✖ Lead ropes
- ✖ Treats
- ✖ Larger items like a used horse trailer, fencing, or materials to build another shelter
- ✖ Volunteer labor, specialized (veterinary assistance, farrier work, office help) or general (stall cleaning, training, grooming, hanging out with and loving the resident donkeys)

Transporting Donkeys

Once you've purchased or adopted your donkeys, you'll need to bring them home to your farm. Keep in mind that most equines don't handle changes very well; add the noise and confusion that are often part and parcel of hauling, and you're going to have stressed-out donkeys (think: colic and gastric ulcers). There are two types of stress factors:

Short-acting factors cause emotional stress:

- ✖ Unfamiliar surroundings
- ✖ Unfamiliar traveling companions
- ✖ Unstable footing

Long-acting factors have physical effects and tend to accumulate over the duration of a trip:

- ✖ Noise
- ✖ Vibration
- ✖ Being thrown against the vehicle or other animals
- ✖ Fatigue from standing
- ✖ Insufficient food and water
- ✖ Extremes of temperature and humidity

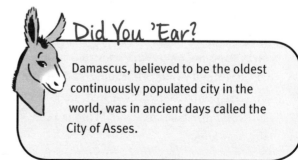

Did You 'Ear?

Damascus, believed to be the oldest continuously populated city in the world, was in ancient days called the City of Asses.

Make Travel Trouble-Free

To make hauling easier on your donkeys (and your-self), here are some things to consider before you hit the road:

1. Avoid hauling sick donkeys, injured donkeys, and late-gestation jennies; start out with sound, healthy animals and do your best to keep them that way.

2. Since equines are social creatures and a solitary donkey is a stressed one, take along a companion if you can.

3. Map the route in advance. Braking and cornering cause many in-transit accidents; crossing bumps and accelerating cause even more. Stop-and-start movement can drive the heart rate of an inexperienced traveler up to twice his norm. If the most direct route means dealing with rush-hour traffic or possibly hitting the red light at scores of stoplights, taking a longer but easier route is always a wiser choice.

4. Factor rest stops into longer journeys. Locate points along the route where you can safely offload your donkeys at least once every 24 hours. Offer familiar feed, water them, and carefully check each animal for signs of injury and excessive stress.

5. Pack along a well-appointed first-aid kit. To reduce the possibility of digestion-related catas-trophes, include enough tubes of probiotic and electrolyte paste to dose each donkey at least once a day.

6. Load reluctant or frightened donkeys with compassion and care. Allow enough time to do the job without rushing. If loading at night, pro-vide plenty of interior lighting, as equines move more easily from darkness into light than the other way around.

Hark, the Angels Sing

A story published in the June 2, 1871, edition of the *Denver Tribune* recounts a preacher's experience delivering a hell fire and damnation sermon to his rural Clear Creek County congre-gation. At the climax of the sermon, he paused for emphasis and raised his arms. Lifting his eyes toward heaven, in a clear, strong voice, he quoted from the Bible, "Hark, I hear an angel sing!" Just at that moment one of a herd of donkeys loafing outside the little church let loose with a prolonged and thunderous bray. The preacher had a difficult time regaining the attention of his greatly amused flock.

---

7. When hauling smaller donkeys in an unpar-titioned stock trailer, consider hauling them loose instead of tied. For short jaunts, each don-key should be allotted enough room to stand without constantly slamming into the rest of the load; on longer hauls, give them enough space to lie down comfortably.

8. Whenever possible, though, fit stock trailers with interior dividers so that you can partition animals into compatible groupings based on sex, size, age, and/or aggressiveness.

9. Whatever sort of conveyance you choose, cut down on vibration by using lots of bedding, and reduce noise levels by padding gates and partitions with pieces of rubber matting or old blankets.

10. Allow enough time to drive carefully. Accel-erate slowly and smoothly, and do your best to stop that way as well. Ease up on the gas well ahead of turns, and don't take corners too abruptly. Factor in load checks, too; stop 20 minutes after departure to check your load and at least once an hour after that.

Build a Traveling First-Aid Kit

Put together a traveling first-aid kit to augment the kit you keep at home in the barn. Pack it in a lidded five-gallon plastic pail (restaurants and fast-food joints will often give you one for free) and keep it in your truck or trailer at all times. If you use something, replace it as soon as you get home. Having a well-equipped first-aid kit and knowing how to use it can make the difference between life and death when you're on the road and far from the closest vet. At the bare minimum a kit should include:

- ✘ Betadine to flush fresh wounds
- ✘ Sterile gauze sponges
- ✘ Telfa nonstick absorbent pads to cover wounds
- ✘ Several individually wrapped sanitary napkins to use as pressure pads to stop heavy bleeding
- ✘ Several rolls of Vetwrap or comparable self-adhesive bandage
- ✘ A roll of 2½-inch-wide sterile gauze bandage
- ✘ 1-inch- and 2-inch-wide rolls of adhesive tape
- ✘ Antibiotic ointment
- ✘ Saline solution and topical eye ointment
- ✘ A rectal thermometer and a tube of lubricant
- ✘ Tweezers or a hemostat
- ✘ Probios or a comparable probiotic gel
- ✘ Electrolyte paste or gel
- ✘ Banamine (a prescription paste or injectable pain reliever and anti-inflammatory drug you must get from your vet) or children's aspirin for pain
- ✘ If you choose injectable Banamine, add disposable needles, syringes, and a vial of epinephrine to counteract anaphylactic shock
- ✘ Band-Aids and nonprescription pain relievers for yourself

Handling the Heat (or Cold)

Weather extremes head the list of long-term hauling stressors. When traveling in hot, steamy weather, pack along a cooler containing jugs of ice-cold water and bagged ice cubes. Wrap ice cubes in cloth and hold them in an overheated animal's armpits or groin.

To compensate for sizzling, steamy weather, reduce loading capacity by 15 to 20 percent; overcrowding rapidly leads to excessive heat buildup. Create additional ventilation by opening windows or replacing solid upper walls with sturdy, closely spaced pipe or heavy wire mesh. Travel only at night or during cooler morning hours and keep the number and length of stops to the barest minimum. Never park a loaded trailer in direct summer sunlight. Offer water to parched animals whenever and wherever you stop.

Extreme cold is equally dangerous. When the mercury plummets well below freezing, it's best to stay at home. If that's not an option, cover large openings to protect your donkeys from rain and wind chill, add more bedding, allow extra space so chilled individuals can move away from frigid drafts, and blanket vulnerable individuals such as foals and elderly or ailing donkeys.

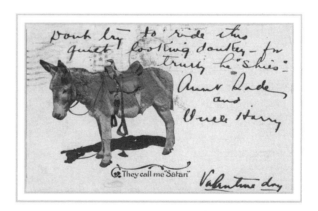

This postcard from the turn of the century, titled "They call me 'Satan'," was apparently a very popular card, as there are hundreds of copies for sale today on eBay.

Choosing a Hauling Conveyance

For best results, donkeys should be hauled in horse and stock trailers. They should never be hauled in enclosed cargo trailers or any sort of a conveyance with openings large enough for a donkey to bail out of. When choosing a donkey-carrying conveyance, keep these points in mind:

- ✘ Animals die quickly from carbon monoxide poisoning, so make absolutely certain that engine exhaust can't reach your donkeys.
- ✘ Adequate ventilation is an absolutely essential element of donkey-trailer design.
- ✘ Poor footing leads to scrambling, physical injuries, and rampant stress. Damp wood and some rubber flooring are as slick as glass once they're wet. Invest in nonslip trailer mats; they quickly pay for themselves many times over. If you're used to bedding your trailer to cushion road concussion, use dust-free disposable bedding on top of the mats; dusty bedding contributes to a host of respiratory ills.

Miniature Donkeys and some small Standards can be hauled in full-size vans, but don't try this unless you know that the donkey you'll be hauling is tame. If you do it, remove excess seating, place a plastic tarp on the floor, either fit the van with a backseat barrier or secure the donkey safely using a halter and a rope tied with a quick-release knot, and take along a helper whose

FAMOUS BRAYERS

Apollo, listed in the *Guinness Book of World Records* as world's tallest mule at a whopping 19.1 hands tall

Baba Looey, cartoon character Quickdraw McGraw's Mexican burro sidekick (1960s)

Benny Burro, costar of MGM cartoons (1940s)

Blossom, movie and TV Western sidekick Gabby Hayes's mule

Blue, the white mule belonging to the John Walton family in the long-running television series *The Waltons*

Brighty, title character of the popular children's book *Brighty of the Grand Canyon* by Marguerite Henry

Burrito, Tito's burro friend in several Screen Gems cartoons (1940s)

Charlie-O, former mule mascot of the Kansas City Athletics and Oakland A's (1963–1976)

Clopper, Mary and Joseph's donkey in *Clopper, the Christmas Donkey* by Emily King

Cupcake, food-loving, burping mule costar of the children's TV program *Thunder* (1977)

Dominick the Italian Christmas Donkey in Lou Monte's 1960s-era novelty record

Donkey, wisecracking, talking donkey star of the *Shrek* movie series

Duldul, the prophet Muhammad's white mule (called Fadda in some resources)

Edward (Edouard), title character in *The Story of Edward* by Philipe Dumas (English translation, 1977)

El Rucio, a.k.a. Dapple, Sancho Panza's donkey in *Don Quixote* by Miguel de Cervantes

Ernest, star of a series of charming early reader books written by Laura T. Barnes

Fanny, the jenny who modeled for Big Ass Fans' famous logo

Gus, the football-kicking star of a Walt Disney feature of the same name (1976)

Hoof, bucktoothed donkey with red hair and wearing a green fedora in the Disney cartoon series *Darkwing Duck*

Hoopoe, King Solomon's ass

sole responsibility is to keep an eye on the donkey at all times (a nice meal of hay in a hay net helps, too). We once hauled a 35-inch Miniature jack over 50 miles in a compact car with the back seat removed — not recommended, but certainly doable.

Working with a Livestock Transporter

In many cases, transporting your own donkeys isn't cost effective. If you're relocating to a new home a thousand miles away and taking your donkeys with you, or if you buy your dream jack from a breeder in another state, you may elect to pay a livestock transporter to haul them for you.

Most professionals and weekend transporters (such as owners headed to shows or transporting equines of their own and willing to carry an extra donkey or two along for pay) post their trips to equine-related listservs and e-mail lists — another great reason to subscribe. You can also locate pros by doing a Web search for livestock transport using your favorite search engine.

Transporters usually charge a hefty fee for the first animal picked up per stop and considerably less for others if you ship more than one. Obtain estimates from at least three or four companies; prices vary dramatically from hauler to hauler.

Hunky and **Spunky,** mama and son burros starring in several Fleischer Studio cartoons (late 1930s and early 1940s)

Jacko, the donkey who pulled aging Queen Victoria's carriage

Jenet, the donkey protagonist of *Tappan's Donkey* by Zane Grey

Missie, Kevin O'Hara's long-eared Irish companion in *Last of the Donkey Pilgrims*

Modestine, Robert Lewis Stevenson's long-eared companion in *Travels in the Cervennes with a Donkey*

Molly, the mule who played Francis in the *Francis the Talking Mule* movies (1950s)

Muffin the Mule, puppet star of BBC-TV children's programming (1946–1957) who returned to TV in animated form in 2005

Nestor, the donkey who carries Mary to Bethlehem in the perennial Christmastime TV special *Nestor, the Long-Eared Christmas Donkey* (1977)

Number 7, Mad Jack's burro in the television program *Grizzly Adams* (1970s)

Perico, beer-drinking donkey who wandered the streets of Santa Clara City, Cuba, during the 1930s and 1940s; a monument to Perico now stands on his gravesite in the heart of Santa Clara City

Platero, silver-colored donkey in *Platero y Yo* (Platero and I) by Nobel Prize-winning Spanish poet Juan Ramón Jiménez; also longtime ADMS administrator Betsy Hutchins's famous donkey pal

Puzzle, the donkey character in C. S. Lewis's book *The Chronicles of Narnia*

Ruth, handsome molly mule ridden by Festus on the long-running TV Western *Gunsmoke*

Spieltoe, Santa Claus's donkey, narrator of the perennial Christmastime TV special *Nestor, the Long-Eared Christmas Donkey* (1977)

Worthless, a mule ridden by Sally Fergus in the short-running *Gunsmoke* spin-off, *Dirty Sally*

Ya'fūr (sometimes spelled 'Ufayr), the prophet Muhammad's talking donkey

Unless you contract for the whole load, routes are rarely direct, and your donkey may be on board for days (or occasionally weeks). Because some donkeys in transit are easily stressed, it's important to choose a transporter who knows donkeys and goes the extra mile to keep them safe and well. When comparing rides, find out:

- ✖ What sort of trailer and hauling unit does the hauler use? What are his contingencies in case of breakdowns?
- ✖ Is he willing to partition your donkey away from other livestock? If so, will your donkey have any sort of up-close and physical contact with other animals en route or when off-loaded at rest stops?
- ✖ Will he take along a supply of your donkey's accustomed feed (and feed it to him), or does he insist that the animals he hauls eat what he provides?
- ✖ What is his policy for dealing with sick or injured animals?

- ✖ If something goes amiss, will he contact both shipper and receiver so that they know what's going on? Will he phone if he's running late? Will he make his cell phone number available to involved parties so that they can call him en route, if they wish?

When shopping for livestock transporters, ask for references and take time to check them out. Better still, post to listservs and ask other subscribers to contact you (off-list) with their recommendations or tales of woe.

Find out what sorts of tests and health papers are necessary for your donkey's interstate shipment, and have the paperwork in hand when the hauler arrives to pick them up (reputable haulers will usually refuse your pickup, otherwise).

And don't necessarily choose the cheapest transporter or the one who does the most advertising; find one you're comfortable with based on experience or other donkey owners' recommendations. Transporting strange equines is always fraught with risk, so hire the best livestock hauler you can afford.

BRAY SAY: *A workman who lives near Gobbet Mine, on the moor, once told me of an adventure that befell him . . . which, happening to a rustic prone to believe in the supernatural, might not improbably have been regarded as a genuine appearance of the Arch Enemy of mankind, but which our workman, quickly recovering from his fright, discovered to be caused by something of a very innocent character. He was — to use his own expression — returning from 'courting' one dark night, when he suddenly found himself sprawling at full length on the ground, and a dusky form with two immense ears bending over him. His first impression was that it was 'wishtness,' but this was quickly dispelled when on scrambling to his knees and boldly facing it, he discovered it to be a black donkey, which, having lain down to rest in his path, he had stumbled over.*

— William Crossings, *Tales of the Dartmoor Pixies*, 1890

The Pet Donkey

Once there was a chief's daughter who had a great many relations so that everybody knew she belonged to a great family.

When she grew up she married and there were born to her twin sons. This caused great rejoicing in her father's camp, and all the village women came to see the babes. She was very happy.

As the babes grew older, their grandmother made for them two saddlebags and brought out a donkey.

"My two grandchildren," said the old lady, "shall ride as is becoming to children having so many relations. Here is this donkey. He is patient and surefooted. He shall carry the babes in the saddle bags, one on either side of his back."

It happened one day that the chief's daughter and her husband were making ready to go on a camping journey. The father, who was quite proud of his children, brought out his finest pony, and put the saddlebags on the pony's back.

"There," he said, "my sons shall ride on the pony, not on a donkey; let the donkey carry the pots and kettles."

So his wife loaded the donkey with the household things. She tied the tepee poles into two great bundles, one on either side of the donkey's back; across them she put the travois net and threw into it the pots and kettles and laid the skin tent across the donkey's back.

But no sooner done than the donkey began to rear and bray and kick. He broke the tent poles and kicked the pots and kettles into bits and tore the skin tent. The more he was beaten the more he kicked.

At last they told the grandmother. She laughed. "Did I not tell you the donkey was for the children," she cried. "He knows the babies are the chief's children. Think you he will be dishonored with pots and kettles?" and she fetched the children and slung them over the donkey's back, when he became at once quiet again.

The camping party left the village and went on their journey. But the next day as they passed by a place overgrown with bushes, a band of enemies rushed out, lashing their ponies and sounding their war whoop. All was excitement. The men bent their bows and seized their lances. After a long battle the enemy fled. But when the camping party came together again — where were the donkey and the two babes? No one knew. For a long time they searched, but in vain. At last they turned to go back to the village, the father mournful, the mother wailing. When they came to the grandmother's tepee, there stood the good donkey with the two babes in the saddlebags.

— As told by Marie L. McLaughlin
in *Myths and Legends of the Sioux*, 1916

CHAPTER 5

DONKEY BEHAVIOR

The ass that thinks he is a stag discovers his mistake when he comes to leap over the ditch.

— Italian proverb

In order to truly enjoy your donkeys, it's important to know what makes the species tick. With that in mind, let's take a look at some donkey behaviors. Then we'll talk about putting this knowledge to work for you and your long-eared friends.

Donkey Behaviors, A to Z

One of the most rewarding things you can do with donkeys is simply to observe them. These are some typical behaviors to watch for:

Assurance

Donkeys are physical animals who reassure each other through body contact. A jenny comforts her foal by holding him gently between her chin and her chest. Older youngsters and adults drape their heads across another's back or rump to reassure or show affection; they rest their heads on a favorite human's shoulder for the same reason.

Breeding

Jennies in heat are anything but coy. To first-time owners' dismay, some even "show" to other species, including humans. Heat display can include increased braying, chomping (chomping is an exaggerated chewing movement with the sides of the mouth drawn back and teeth mostly covered), yawning, urinating, squatting, raising her tail, "winking" her vulva, mounting other jennies (or being mounted by other jennies or geldings), backing into other donkeys (or humans!) while bouncing her hindquarters, and physical signs such as a swollen vulva and vaginal mucous discharge.

Many jacks brutalize the jennies they breed, especially unfamiliar females brought to them to be bred. They're especially likely to savage the jenny's neck while breeding, and for that reason hand breeding *(see the glossary)* is preferable to pasture breeding.

Pastured jacks engage in a lot of chasing. Ears pinned back, chin lowered nearly to the ground, and upper lip extended, a breeding jack snakes his neck as he drives his jennies before him. Occasionally a jack homes in on an unwilling jenny and pursues her at a gallop until both are bathed in sweat and ready to drop.

According to a two-year study mentioned in "Pasture Mating Behavior of Donkeys at Natural and Induced Oestrus" (*see* Resources), the mean time from approach to ejaculation in domestic jacks engaged in pasture breeding varied from 12 to 26 minutes, during which partial drops of the penis, mounts without ejaculation, grazing, and prolonged flehmen episodes occurred.

The research focuses on two experiments conducted in 1989 and 1990 in which Brazilian jacks were observed at pasture, interacting with groups of 21 jennies; the jacks were housed at night and returned to the pasture at daybreak. Here are some things researchers concluded.

The jack's return to the herd invariably provoked braying by both sexes. The jack checked out each jenny in turn, and then began teasing his choice of the jennies. Teasing behavior included nose-to-nose contact; nibbling the jenny's head, neck, knees, and flanks; and sniffing parts of her body, especially the perineal area. Teasing generally culminated in one or two mounts without erection. After that, the jack rested.

A jenny in heat may be mounted by another jenny or a gelding.

A jack will sniff a jenny before mounting her.

A jack makes many false starts before actually breeding the object of his affection. This jenny signals her willingness by chomping, as does the one above.

BRAY SAY: *If the first thing you hear in the morning is a donkey's bray, make a wish and it will come true.*

— English proverb

WHY IS MY DONKEY LAUGHING?

When your donkey flips his lip back in a "horse laugh," he isn't amused, he's "flehmening." The flehmen response (from the German word *flehmen,* meaning "to curl the upper lip"), also called the flehmen position or flehmen reaction, is a grimace that most ungulates, cats, and a few other mammals make to explore scent.

When a donkey flehmens, he flips his upper lip back to expose and draw odorants into his Jacobsen's organ, a pheromone-detecting organ located in the roof of his mouth. Jacks do this to determine whether a jenny is in heat, but both sexes flehmen after sniffing unusual scents, especially the manure or urine of unfamiliar donkeys.

A flehmening jack

While the jack interacted with his chosen lady, a group of other jennies (particularly those coming into or already in heat) gathered to watch. The jack's braying appeared to attract them; 78 percent of the in-heat jennies in the second trial brayed at least once while watching the teasing in progress. The observing jennies engaged in mounting, herding/chasing, teasing, and flehmen behavior, all of which are rarely observed in female horses; during the second trial, two jennies even mounted the jack. From time to time throughout the experiment, one jenny prevented the jack from mating with another jenny by kicking, biting, striking, and pushing him off the jenny whom he was breeding.

Periods of teasing and rest continued until the jack achieved an erection and mated with a jenny (not necessarily the one he was teasing or the jenny closest at hand). Mating frequency and duration weren't influenced by the time of day.

The study indicates that domestic donkeys have a territorial, nonharem type of social organization; the jacks spent most of their time in the same area (surrounded by their admirers), while out-of-heat jennies grazed the rest of the field. When another jack approached the perimeter fence, the breeding jack rushed to confront him but, unlike stallions (horses), made no attempt to gather the jennies and herd them away.

What's Taking So Long?

It takes most hand-bred jacks considerably longer to achieve an erection and breed a mare or jennet (10 minutes to several hours; 30 minutes on average) than it does the average stallion (10 minutes or less).

Also, unless raised with horses — completely separate from other donkeys from weaning age on — most jacks refuse to breed mares, even under pasture-breeding conditions.

Elimination

Both male and female donkeys assume a specific pose to urinate, with legs braced wide, tail hoisted at a 45-degree angle, ears turned back, and a studious expression upon their faces. When one jenny in a group urinates, she's apt to be joined in quick succession by several more jennies. Jacks prefer to dung atop their own previous deposits, often creating veritable mountains of manure. Jennies tend to dung in approximately the same area, but don't pile their manure the way jacks and some stallions do.

THEY DON'T EAT DONKEYS — DO THEY?

Monsieur Isouard of Malta reports that, as a result of the blockade of the island of Malta by the English and the Neapolitans, the inhabitants were reduced to eating all the horses, dogs, cats, donkeys, and rats: "This circumstance," he says, "led to the discovery that donkey meat was very good; so much so, in fact, that gourmands in the city of Valetta preferred it to the best beef and even veal. Particularly boiled, roast, or braised, its flavor is exquisite."

— Alexandre Dumas, *Grande Dictionnaire de Cuisine,* 1873

While we definitely don't recommend eating our long-eared friends, in other times and in certain places even today, dining on donkey meat was (and is) de rigueur.

In Olden Days

- In ancient Athens, there was a separate market for the sale of donkey meat because it was food for the poor.
- Roman statesman Maecenas (1st century BC) arranged for donkey meat to be served at Roman banquets. He conducted experiments to determine the ideal feed for the asses he served; his conclusion: biscuits soaked in milk.
- Wild asses were once considered gourmet fare. At their sumptuous banquets, Persian royalty served whole roast onagers and rich red wine dished up in onager hoof cups.
- The prophet Muhammad dined on onager flesh, making the meat of wild asses acceptable for consumption by Muslims, while domestic donkey meat remained unclean.
- Donkey meat has long been used for stuffing European sausages. The famous saucisson of Arles was originally made from a mixture of donkey meat and bull meat, which gave it a special succulence.

History Repeats Itself

- There is a thriving market for donkey meat in Nigeria and Nambia. It's marketed locally as donkey meat but often passed off as beef outside the area.
- In Borgomanero, Italy, a local delicacy, *tapulon,* is made from donkey meat smothered in garlic.
- Author William Black travels around Italy munching anything and everything remotely resembling food. When in Mantova he rhapsodized over donkey stew and even supplied readers with the recipe.

The Asian Connection

- Lady Jade's Donkey Meat Restaurant at 50 Baochu Road in Shanghai serves a wide variety of donkey meat dishes that are "good for skin and for stomach."
- In the city of Jinan, a favorite local delicacy is donkey meat sandwiches.
- Chopped donkey meat is the basis of a recipe for Hua Tuo meatballs.

The Last Word

The authors of *Carcass Analysis and Meat Composition of the Donkey* tell us that donkey meat is higher in protein than beef and a rich source of essential vitamins and minerals. We don't care; we say donkeys are for loving, not eating!

Given the chance, donkeys eliminate away from their living and grazing areas, usually dropping manure in the same general areas. However, they eschew grass "spoiled" by urine or manure piles, even after considerable time passes.

Feeding (Adults)

Donkeys spend most of their waking hours alternating between feeding and periods of rest. Where temperatures soar during summer months, free-ranging donkeys relax during the heat of the day and are active from dusk until dawn.

In mixed herds, donkeys are usually subservient to bossy horses. When the two are fed together, donkeys may not get their share. Because they don't stick up for themselves in a mixed hierarchy, it's important to feed donkeys separately (or feed using properly fitted nosebags).

This jenny nudges her foal toward her udder, encouraging her new baby to take a sip of milk.

Feeding (Foals)

Newborn foals are wired to seek sustenance in dark, warm places; that's what they're doing when they snuffle and search along their dams' armpits and bellies before discovering the udder. Jennies help their newborns to find the teat by positioning their bodies just so, and sometimes flexing the hind leg on whichever side the foal is standing to make the udder more accessible. While her foal nurses, the jenny sniffs or nuzzles him, and she'll nip his rump if he gets too rough.

Domestic donkeys allow unrestricted suckling for three days to a week; after that, foals are expected to "ask." To do so, a donkey foal folds back his ears and wuffles as he approaches his dam, sometimes also tossing his head. If the jenny agrees, he pushes his head under her and grabs the teat; if she doesn't, she'll continue whatever she's doing or walk away. If he continues to make a pest of himself, she'll threaten to nip or kick.

On average, neonatal (newborn) foals suckle every 15 to 20 minutes for a minute or so at a time. Foals suckle at any time but especially after being frightened or separated from their dams, and after resting.

Foals start nibbling grass at about one week of age but don't begin drinking water and grazing in earnest until they're three to five weeks old. Neonates often sample their mothers' manure, a practice that rarely fails to amaze and disgust first-time donkey owners. However, manure contains digestive enzymes that populate a foal's gut and help pave the way for him to digest solid foods on his own.

Fighting

True fights (as opposed to play fighting) occur between jacks and always involve a lot of angry braying. Combatants clamp their teeth into the crests of one another's necks and attempt to bite each other's legs. In the wild, where such battles are apt to occur, badges of honor include scars, broken ears, and missing tails. Anyone

When two jacks battle, they fight to kill.

keeping two jacks should know enough to keep them separated at all times. A note of caution: Trying to separate a serious jack fight would be suicidal. Like fighting dogs, the jacks might turn their fury on the humans trying to stop the fight.

Flight or Freeze

Given a chance, horses run when faced with a scary situation, whereas donkeys tend to "freeze" in their tracks. Because of this trait, donkeys are unfairly pronounced stupid or stubborn, when the truth is they're confused or afraid. The tendency to freeze, not flee, is much safer for the handler.

Greeting

Equines of all species, including donkeys, greet newcomers muzzle to muzzle, cautiously sniffing and blowing into one another's nostrils.

Building a Donkey Dust Bath

A donkey loves nothing better than a satisfying roll in soft dirt. He does it to fluff his coat to restore its insulating qualities, to coat himself with dust to repel biting insects, to aid in shedding, and simply because it feels so good.

If you live where the soil base is sandy or loamy, your donkeys can and will make their own dust bowl. However, if you'd prefer they roll in a spot of your choosing, not theirs, or if rocky soil makes a dust bath painful, bring in a pickup load or two of soft sand (building contractors and gardening centers sell it by the truckload) and pile it in a 12- to 20-foot circle at least 8 inches deep. Be sure to check for rocks every few days, especially as the donkeys paw through packed sand to the earth below. In rocky locales, add sand as needed to maintain a nice, soft bath. Your donkeys will love you for it!

Mutual grooming is a favorite donkey pastime.

Grooming Habits

Skin care is important to donkeys. They "dust bathe," and, unlike horses, they don't shake to clear their coats of dust after rolling (unless his coat is drenched, patting a donkey raises a cloud of dust). Donkeys create dust-bathing areas of bare earth by pawing the ground with their forefeet. They tend to roll after rest periods, often in groups, with a number of animals rolling in quick succession.

Donkeys also groom themselves by rubbing body parts against trees and other solid objects, the better to soothe the itch of sweaty skin or insect bites and to shed molting winter hair.

They also groom one another. To initiate grooming, one donkey approaches a family member or friend with his neck slightly extended, ears forward, and mouth slightly agape, then he touches the intended partner's neck. If the partner agrees, grooming commences; if not, the second donkey walks away or threatens to bite or kick. Mutual grooming sessions last anywhere from a few minutes to half an hour or more. To groom one another, two donkeys stand head to tail and nibble or bite each other's necks, backs, and rumps.

Hearing

A donkey's hearing isn't quite as acute as a dog's but is much better than that of a human (or even a horse). When a donkey picks up a sound or sights something unusual, he pricks both ears in that direction to capture the minutest sounds and funnel them down to the ear canal and on to his brain. The donkey's huge ears can each swivel and capture sound independently; separate sounds from each ear are processed and analyzed at the same time.

Jennies (and Neonatal Foals)

Jennies tend to leave their herd prior to foaling, sometimes in the company of a female friend or family member. Foaling usually occurs in the dark of night and near water sources such as stock ponds, streams, or boggy ground. In pasture situations, jennies sometimes sequester themselves and their newborns away from the herd for up to a week or even longer.

A jenny licks her newborn foal to clean it and in doing so forms a strong bond with her young one. For the first few days postfoaling, jennies are especially protective of their foals, as owners may discover when their sweet, laid-back jenny attacks someone for interfering. She maintains close physical contact with her foal while she whispers to, nuzzles, and grooms him.

Within a week or two, friendly herd members and other foals are allowed to approach the foal, although she keeps him by her side for a few more weeks, protecting him while he rests. Often several jennies with foals band together, so that one jenny can stand guard while the others (and all of the foals) relax.

As time passes, foals are allowed to leave their dams to play with other foals; even so, in times of perceived danger each jenny calls to and searches for her foal until all are safely reunited.

Loyalty

Donkeys form strong bonds with companions and often suffer deep emotional pain, leading to depression and refusal to eat, when a much-loved companion is sold or dies.

Donkeys prefer the companionship of other donkeys, but when none are available, they often form lasting bonds with equines of other species, sheep, goats, llamas and alpacas, dogs and cats, and even humans.

Play

Equine behaviorists Sue M. McDonnell and Amy Poulin (*see* Resources) describe the ways equines, including asses and zebras, amuse themselves. Playful behavior is especially common among foals, but adults play, too.

Their study is based on observations of a 50- to 75-head herd of semiferal Shetland-type ponies kept at the University of Pennsylvania School of Veterinary Medicine. However, McDonnell and Poulin cross-referenced the works of more than a dozen equine behaviorists to note which of the 38 behaviors they catalogued also occurred in wild, feral, and semiferal

Did You 'Ear?

The word *onolatry* refers to the worship of asses.

Cephalonomancy (from Greek *kephalaion*, "head" + *onos*, "donkey" + *manteia*, "prophecy") was a form of divination using the head or skull of a donkey. It involved boiling the skull while reciting various phrases, often the names of criminal suspects. If the skull crackled or the jaw moved while a name was spoken, this was taken as evidence to identify the guilty party.

populations of similar equids. One of the experts consulted was P. D. Moehlman, who studied feral burros in Death Valley, California.

Catalogued play behaviors that were also observed among feral donkeys include:

Nibble. Donkey manipulates an object (a dropped glove or a bucket; the mane, tail, or body part of a companion; or an animal of another species) without using his teeth.

Mouth. Donkey then grabs the object between his teeth (a dam's tail is a common target), which often leads to . . .

Pick Up. Donkey raises his head and lifts the object from the ground a distance ranging from a few inches to several feet. This in turn sometimes leads to . . .

Shake. The object is grasped between Donkey's teeth and scraped along the ground, shaken, or swung in a circular movement, possibly leading to . . .

Drop or Toss. Away it goes!

Paw. Donkey paws or paws at an object, generally several times in succession.

Kick Up. Standing at right angles to a herd mate (usually baby Donkey's dam) and with his rump often touching the belly of his target, Donkey's weight is transferred to his front legs while his hind legs hop a few inches off the ground (but no actual kicking occurs).

To and From. Donkey trots or gallops away from an object (such as his dam), to a landmark and then returns.

Circle. Donkey circles an object (usually his dam) at any gait.

Resting Rear. Standing at right angles to his target, Donkey rears up against a companion (or his dam) so that one or both forelegs dangle across the other's body.

Mount. Donkey rears up against the back or hindquarters of a herd mate of either sex (or his dam) mimicking the mating of adult donkeys. An abbreviated form consists of merely resting the chin against the hindquarters of the target, as if to mount.

Frolic. Donkey exuberantly leaps straight into the air. This frequently leads to . . .

Run. Donkey runs from or toward nothing in particular, simply for the sheer joy of running.

Buck. Donkey lowers his lead and shifts his weight to his forelegs, then kicks up with both hind legs.

Jump. Donkey uses his hind legs to launch himself into the air.

Leap. In a combination of **Jump** and **Frolic**, Donkey leaps over, away from, or toward an object.

Male donkeys (both jacks and geldings) of all ages indulge in the following play-fighting maneuvers. Although jack foals may invite young jennies to play, jennies are usually disinclined to play-fight among themselves or with geldings or young male members of the herd.

Head/Neck/Chest Nip and Bite. Donkey's ears are up and his lips retracted as he playfully nips or bites a playmate.

Neck Grasp. Donkey clamps the crest of his playmate's neck.

Neck Wrestle. Donkey and his playmate spar using their necks and heads. One or both may drop to their knees; shoulder slamming is also included in this behavior.

Rear. Donkey rears, sometimes in a nearly vertical fashion.

Hindquarter Threat. With his ears back and his rump toward his playmate, Donkey raises one leg as if aiming to kick.

Kick. Donkey kicks, rather ineffectually, with one hind leg.

Evasive Balk. As the playmates approach one another, Donkey abruptly halts and spins away in a sweeping motion while his hindquarters pivot in place.

Although these are the only maneuvers the two studies found in common, donkey observers will recognize a few that P. D. Moehlman missed, such as **Carry** (an object), **Pull** (the dam's tail), and **Chase**. It's a fascinating study and it's available online for free; to find it, visit your favorite search engine and type the words *equid play ethogram* into the search box.

Rest

Donkeys rest standing on their feet or lying on the ground, either on their sides with legs extended or up on their sternums with their forelegs tucked under them and their hind legs extended to the side, sometimes with their necks extended. Young foals generally sprawl on their sides or rest by leaning against their mothers; older foals are more likely than young ones to rest while standing.

Self-preservation

Donkeys rarely put themselves in harm's way, much to the displeasure of handlers who think they're simply being stubborn or cantankerous when they refuse to do something that might endanger the donkey's life or limbs (at least to the donkey's way of thinking — like working beyond their capability or entering a dark, scary place where they've never been before).

Smell

The donkey's sense of smell is extremely acute. Donkeys refuse foods that smell different from their usual fare, making it difficult (short of trickery) to feed medication- or dewormcr-laced feed to an ass. They frequently sniff objects and each other (as well as each other's urine and dung): smell is one of their major senses. Jennies recognize their neonatal foals by smell. Old-timers say that asses can smell water from a mile away. Donkeys, especially jacks, often raise their heads to sniff the wind.

Social Order

A well-defined pecking order exists within every herd of equines, large or small, and every newcomer has to earn a place in the group. Where an individual stands in the hierarchy depends on his or her age, sex, personality, aggressiveness toward other herd members, and in the case of mixed herds, his species. Unweaned foals assume their dam's place in the order and often rank immediately below her after weaning.

Speed

Donkeys move between feeding areas at a walk. Adults trot when frightened or excited, while playing, and when approaching watering holes. They rarely gallop except as part of courtship chasing or play.

Stoicism

Suffering and sick donkeys generally soldier on instead of exhibiting symptoms of pain or distress. Because of this, early symptoms of diseases as serious as colic and laminitis frequently go undetected. Most donkeys are also remarkably tolerant of medical procedures when kindly treated. Donkeys are tough!

Vision

A donkey's vision is his primary detector of danger, so sight is very important to the ass. Both eyes can be used together to focus on an object using binocular vision or independently of one another using monocular vision. Donkeys have blind spots directly in front of and behind them. They can see quite well to the sides and reasonably well to the back, especially when the head is lowered. By raising or lowering his head or turning it to one side or the other, a donkey can see more clearly and focus on specific objects, near or far.

Killer Donkeys and Airline Crashes

Snopes.com (that great online debunker of scams, hoaxes, and urban legends) has investigated a misleading statement about donkeys that appeared in the London *Times* in 1987. The article's intent was to address concerns about the safety of flying:

> The statistics on the safety of flying are immensely comforting, despite recent reports of a near-miss between a 747 and an RAF Hercules over Carlisle, and the Boeing 747 captain who apparently had to be reminded to lower his craft's undercarriage before landing at Heathrow. One expert has estimated that more people in the world are kicked to death by donkeys than die in plane crashes.

This statement has often been repeated, but never with any facts to back it up. We knew it couldn't be true!

Donkeys can change their focus instantly from distant objects to those close by, and they can easily detect the smallest movement, even far away — a trait especially important to wild and feral asses.

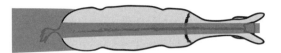

Donkeys can't see directly behind themselves nor can they see the ground when their heads are raised.

Donkeys also have superior night vision, though it's not as acute as that of cats. However, their eyes need time to adjust when moving from light into darkness, which explains a donkey's usual hesitation when loading into a gloomy horse trailer.

Donkeys discern some but not all colors (though researchers disagree about which colors donkeys actually see).

Vocalizations

Donkeys do not just bray. They communicate in a wide variety of snorts, grunts, wheezes, and wuffles that vary somewhat between individual donkeys but that humans with whom they spend time can quickly learn to understand.

Jacks bray most frequently and loudest. Wild, feral, and domestic jacks all tend to bray at dawn, and some bray at dusk; if there are other jacks in the area, they answer with raucous brays of their own. Jacks bray

HOW BLIND DONKEYS COPE

Smart, stoic, calm, and careful, donkeys adjust surprisingly well to full or partial blindness. In fact, owners of pet donkeys are sometimes startled to discover, due to some change in routine, that a donkey's encroaching blindness has gone virtually unnoticed.

Blind and near-blind donkeys often form strong bonds with another donkey who then serves as eyes for his sightless friend. These bonds should be encouraged, and if an animal is perceived to be going blind, obtaining a calm, sensible companion while the first donkey is still partially sighted is a very wise move.

Whether or not a blind donkey has a friend to help him about, his environment should be kept as much the same as possible, with water, feed,

and fences all in accustomed places. Movements around a sightless or partially blind donkey must be deliberate and unhurried; touch or talk to the donkey while you work, so he knows where you are.

One-eyed donkeys quickly adapt to their altered condition, although their depth perception suffers with the loss of an eye. Take care to talk to this donkey when working on his blind side. Donkeys who lose an eye as a foal and thus adapt in their formative years, and older donkeys who are already trained when blinded, often cope well enough to be ridden and driven, especially by riders and drivers they trust.

Owners of blind donkeys are advised to visit the BlindHorses Web site (*see* Resources). The information will educate and inspire you!

when rounding up jennies prior to a move and when courting jennies in heat. Domestic jacks bray simply because they like to. Don't keep jacks where noise creates discord.

Domestic jennies and geldings bray when calling for food, when separated from their herd, or in answer to another donkey's brays. Young foals bray rarely, if at all.

Grunts are antagonistic and usually accompanied by assertive body language like tail lashing, chin jerking, or stomping; snorting denotes excitement (or possibly a bug up one's nose); wuffling is used to call an intimate, such as a jenny to her wandering foal or when inviting another donkey to mutually groom. Jennies reserve an additional low-pitched, gentle wuffle for communicating with their newborn babes.

Water

Donkeys don't drink as much or as often as horses — they drink only when they need to rehydrate, and usually in groups rather than alone. In times of need, wild and feral donkeys can go as long as three days without water. In a presentation titled "A Donkey Is Not a Horse: The Differences from a Practical Veterinary Standpoint," Stephen R. Purdy, D.V.M., states that domestic donkeys can dehydrate and lose 30 percent of body weight without adverse affects and rehydrate within five minutes after drinking.

Donkeys are picky drinkers; if their water source is dirty, ice-packed in the winter months, or sizzling in the summer sun, many donkeys won't drink. Don't expect donkeys to break the ice covering frozen ponds or water troughs with their hooves the way horses do; breaking ice is foreign to most asses.

Whiskers

Whiskers near the eyes and on the muzzle act as touch receptors. Adult donkeys test electric fences by almost but not quite touching them with their whiskers, and foals with their still-crumpled whiskers tend to bump their noses into things. There is a large nerve supply from a donkey's whiskers to his brain; people who shave the whiskers for show purposes deprive their donkeys of an important secondary sensory organ.

Zoning Out

Donkey foals have been known to slip into a catatonic state when they're deeply frightened. The right thing to do is back off and allow the baby time to recover on his own. When he has, proceed with whatever you were doing but in a quieter, more donkey-friendly manner.

Donkey Whispering

"Horse whisperers" emphasize communicating with horses in a manner they understand: body language. Unfortunately, horse body language doesn't always translate into donkey talk. However, the concept is a good one. Here are some tips for speaking "donk-ese."

Greeting a Donkey

Donkeys greet newcomers through nose-to-nose snuffling, the best way to take in the strangers' scent while remaining far enough back to stay out of harm's way. To simulate this greeting, extend your hand, palm down and relaxed, so the donkey can smell you from a safe distance.

In most cases it's unwise to blow into a donkey's nostrils because some equines bite in perceived self-defense. However, when greeting a familiar donkey whose moods you can clearly assess, gently blowing air from your nostrils into his is acceptable.

Mutual Grooming

The best way to make points with a donkey is to scratch his itchy spots. The withers, shoulders, just above the tail, and "cheeks" of the rump are likely places to start, and some donkeys like to have the insides of their ears gently stroked.

Comfort

When a donkey foal is frightened or tired, his dam hangs her head across his back or gently nibbles his withers. Reassure a donkey of any age the way his mom did by leaning lightly into him and hanging your arm across

Studying Donkey Behavior

The best way to learn to read donkeys is to spend a lot of time among them. Lug a lawn chair out to the pasture, get comfortable, and observe. At first your donkeys will cluster around you, watching you watch them, but after a day or two the novelty wears thin and life returns to normal.

Another good way to get a feel for life among the donkeys is to camp in the barn in an out-of-the-way, unobtrusive spot. A good place is atop piled hay stacked where you're situated above the donkeys' normal field of vision. String colored Christmas lights around the barn beams to cast subdued light on your donkeys' lounging area. Add a sleeping bag and a flashlight and settle in.

Though little has been written about donkey behavior, horse books can suggest what to watch for. Especially good ones include *The Horse's Mind,* by Lucy Rees, and *How to Think Like a Horse,* by Cherry Hill (*see* Resources). Keep a notebook and jot down your observations; then join a donkey listserv and share your notes with interested, kindred souls (including me!).

The author's Bluestone Pooka reassures her newborn daughter, Bluestone Arielle.

his withers. Another ploy: dangle your hand across his withers and gently rub his shoulder or back.

Motivation

Motivating a donkey with force is counterproductive because a frightened or confused donkey simply freezes in his tracks. Instead, give a "stuck" donkey time to calm down, and then try again, praising him effusively for responding to your patience.

Donkeys love to eat, so consider motivating a frightened, balking donkey with food. It's not a bribe; consider it a means to help the donkey focus on something other than his fear or confusion. Once he does, he's likely to relax.

DONKEY TAILS

Honest Big-Ears

In his 1910 book, *Pueblo Indian Folk-Stories,* Charles Lummis recounts the following folk tale favorite of the Pueblo people of New Mexico. Until Europeans came, these natives had no horses, donkeys, sheep, goats, or cattle. When these animals appear in Pueblo legend, either the story is recent or older legends have been modified to include them.

Once Big-ears was coming alone from the farm of his master to Isleta, carrying a load of curd cheeses done up in buckskin bags. As he came through the hills he met a Coyote, who said: "Friend Big-ears, what do you carry on your back?"

"I carry many cheeses for my master, friend Too-wháy-deh," answered the Burro.

"Then give me one, friend, for I am hunger-dying."

"No," said the Burro, "I cannot give you one, for my master would blame me."

Many times the Coyote asked him, with soft words; but Big-ears would not, and went his way. Then Too-wháy-deh followed him behind, without noise, and slyly bit the bag and stole a cheese. But Big-ears did not know it, for he could not see behind.

When he came to the pueblo, the man who awaited him unloaded the cheeses and counted them. "There lacks one," he said, "Where is the other?"

"Truly, I know not," answered Big-ears, "but I think Too-wháy-deh stole it!"

So Big-ears went back to the hills and looked for the house of Too-wháy-deh. At last he found it, but the Coyote was nowhere, so he lay down near the hole, and stretched his legs out as if dead, and opened his mouth wide, and was very still.

Time passed, and the Old-Woman-Coyote came out of the house to bring a jar of water. But when she saw the Burro lying there, she cried:

"Hloo-hli! Come out and see! For a buffalo has died out here, and we must take in some meat."

So Too-wháy-deh came out, and was very glad, and began to sharpen his knife. But his wife said: "But before you cut him up, get me the liver, for I am very hungry."

Then Too-wháy-deh, thinking to please her, went into the Burro's mouth to get the liver; but Big-ears shut his teeth on Too-wháy-deh's head, and jumped up and brought the thief to his master.

When the master heard what had been, he killed the Coyote, and thanked Big-ears, and gave him much grass. And this is why, ever since, Big-ears strikes with his hind feet if anything comes behind him slyly, for he remembers how Too-wháy-deh stole the cheese.

By the same token, donkeys work hard for a tidbit of food, making them marvelous candidates for clicker training. Read all about it in chapter 12.

Crowding

When pushy donkeys crowd you at feeding time, warn them back with donkey body language.

THE TAIL SWISH. Donkeys warn their foals and herd mates not to tailgate by lashing their tails, stomping, and turning to glare. If you sense a donkey is getting too close, put one hand behind you at backside height and wag it back and forth like a grouchy donkey's tail. It looks funny, but it works!

THE WARNING KICK. If an individual still follows too close, warn him back by kicking behind you without actually making contact with the donkey or *lightly* thump the donkey's chest with your heel.

THE STOMP AND GLARE. If tail swishing and warning kicks don't convince a tailgater that you mean business, stop abruptly, turn quickly, extend your head toward the miscreant, and glare. For extra emphasis, stomp one foot as you turn.

This early tourist postcard reads: "We're on our way" and "Arkansas."

PART 2
Donkey Care

CHAPTER 6

TACK AND SHELTER

*When you go to the donkey's house, don't
talk about ears.*

— Jamaican proverb

*This card, the work of a traveling photographer, was sent
to Philadelphia in 1910. What a beautifully turned-out
donkey! He's groomed to the max and even wears a crupper
and a checkrein. This prevents him from lowering his head
to eat grass, thereby pulling the reins through the young
subject's hands.*

Before bringing your donkeys home, you should
have some basic tack on hand — and know how
to use it. We'll start with some basic information on
halters and leads, and address shelter options later in
this chapter.

Halters

Plan to buy a halter and lead for every donkey you
own, especially if you live where wildfires, hurricanes,
flooding, or other natural disasters sometimes neces-
sitate quick evacuation of area farms. Mark each with a
nameplate, lettering it with indelible ink (you'll have to
touch it up from time to time) or in another permanent
manner so that you can immediately tell which halter
fits each donkey. Store the halters where you can find
them quickly; hangers labeled with each donkey's name
are a perfect fix. Here are some pointers:

✗ Depending on size and type, a donkey might
  wear anything from a miniature-horse- to a draft-
  size halter. The best way to be certain that you get
  the proper fit with a new halter is to take his old
  halter with you when you shop for the new one.

This donkey is wearing a grooming halter.

- ✘ For leading and tying you need a strong, unbreakable halter, so that if your donkey yanks back while tied, his halter won't break.

- ✘ Unless there's an important reason to leave a halter on your donkey when he's turned out to graze or exercise, *don't do it.* Sooner or later he's going to catch his halter on something and panic, so it's a very dangerous thing to do.

Still, some equines simply can't be caught unless they're wearing halters. If that's the case and your donkey must wear a turnout halter, make certain it's a safe one. If your donkey wears a turnout halter and he snags it on a fence post or limb and panics, you want that halter to break. If not, he could break his neck in his struggle to free himself. The solution: Always turn him out wearing a flimsy, single-ply leather halter or a nylon breakaway halter designed for the task. Some ready-made breakaway halters feature lightweight leather crownpieces; others incorporate skinny leather "fuses" or special Velcro closures that give way when stressed.

A grooming halter has snaps sewn to both ends of its throat piece and another at the leading ring end of its *gullet:* the strap underneath your donkey's jaw that connects his halter's throat piece and chin straps. (To clarify: the bowed part of a saddle pommel is also called

Do-It-Yourself Safety Halters

Safe doesn't mean fancy (or expensive). Here's how you can make an effective homemade breakaway insert for your donkey's everyday halter.

1. Measure the width of the halter's crownpiece. Small halter straps are usually ¾ inch wide, while larger halters generally have 1-inch straps.

2. Now go to your closet (or maybe your favorite secondhand store or thrift shop) and find a lightweight leather belt of the same width; it has to be the type with a tongue buckle and a keeper loop.

3. Measure 5 inches from the buckle to make an insert for a Miniature Donkey's or Small Standard's halter or 6 inches for a larger donkey's halter, and snip the belt off at that length.

4. Round the cut end with scissors, then using a leather punch or a hammer and large nail, poke a single hole 1 inch below the keeper loop.

5. Now buckle the insert between the cheekstrap and the crownpiece of the halter. Be sure to remove the insert before tying your donkey; if he pulls back, this insert will break, as intended.

This easy-to-make breakaway insert could save your donkey's life.

a gullet. In this case the word refers to the strap nearest the animal's gullet, or esophagus.) With the lower parts removed, a grooming halter exposes more of a donkey's head for clipping. Not everyone uses them, but they're nice to have around.

Choosing a Halter

Any style or size halter should incorporate strong, rust-resistant fittings. Also called hardware, halter fittings are the metal rings and slotted squares that connect a halter's leather or nylon straps. If your donkey shies or purposely sets back while tied, what usually snaps first isn't part of the halter itself: it's one of those fittings.

The best, strongest, most rust-resistant hardware is crafted of solid brass or stainless steel. Second best is brass- or nickel-plated steel; however, when their plating wears off, these fittings rust. Cheap halters (and even some expensive ones) are made using brittle brass- or nickel-plated "pot metal" hardware that pops in a heartbeat when stressed. Plating is usually bright, shiny, gold- or silver-colored; solid brass fittings are duller, more satiny looking and often have the words "solid brass" stamped in teeny letters someplace on each ring or square.

If you're not sure what a particular halter's fittings are made of, ask a salesperson. If she can't tell, ask her to check the halter manufacturer's sales literature and find out.

FITTING A HALTER

Donkeys' heads aren't shaped like those of horses and ponies, so care must be taken when using a standard horse or pony halter on your long-eared friend. If his halter is too snug, it can rub his face raw, and if it's way too tight it can also make it hard for him to chew. Worn too loose, it's easy for him to snag his halter on things — even his back hoof when he scratches his chin. If it drops below his nasal bone, it can also restrict his breathing while he's being led. Here's how to fit most donkeys.

Hold up the first two fingers of one hand; lay them side-by-side against your donkey's face between his cheekbone (the big one that juts out about midway between his eye and nostril) and the halter's noseband.

Then stick them, thumb upward, between the ring at the bottom of the halter's noseband and your donkey's jaw. Do your fingers exactly fit into both spaces with maybe just a smidge of room to spare? That's perfect!

TOO LOOSE

TOO TIGHT

JUST RIGHT

Hand-knotted rope halters are nearly unbreakable and the best buy in donkey headgear.

Since Western-style knotted rope halters are made without any hardware at all, they're practically unbreakable; even donkeys who pop leather and nylon halters won't ruin one. A good rope halter is knotted from a single length of high-quality rope, and it's firm, not floppy. Be sure that it comes with a tag explaining how to adjust it. Knotted rope halters are inexpensive, sturdy, and long-lasting.

Leather halters should be sewn using neat, small stitches and have a heavy, substantial heft. Quality halter leather feels smooth and waxy; poorer grade leather seems grainy and dry. Leather halters are handsome and very traditional. However, leather requires constant upkeep, and donkeys who pull back can break them.

Premium nylon halters are soft and supple, never stiff and bulky. Sturdy ones are two or three nicely stitched layers thick. Buckles should buckle and unbuckle easily, without their tongues getting stuck in any holes. Nylon stable halters are most owners' pick because they're colorful, inexpensive, and easy to care for. Made with solid brass hardware, they're nearly as unbreakable as knotted rope.

Cleaning Halters

Leather halters require a lot of upkeep. To keep it from rubbing its wearer's face and to make it last a lot longer, halters must be kept pliable and clean. Every time you remove your donkey's leather halter, examine it carefully. If it's stiff or dry or it's caked with dirt or dried sweat, gloss on plenty of saddle soap or leather conditioner and scrub off the grime. Let it dry, then rub off any soapy residue before using the halter again.

Nylon and rope halters are easier to maintain. Quick-clean a rope or nylon halter by soaking it in a bucket of cool, soapy water, then scrub off surface sweat and dirt, dip it in a plain water rinse, and hang it up to dry. For a more thorough cleaning, stuff it into an old pillowcase, knot the pillowcase shut, and drop it into the washer using mild detergent and the machine's cold-water and delicate-wash-cycle settings.

Tossed down and forgotten, halters get tangled around animals' (and people's) legs, soiled and broken, and often just plain lost. Hang them up when they're not being used. Some people hang halters by

Did You 'Ear?

Because Jesus rode an ass into Jerusalem on Palm Sunday, early church dignitaries rode "humble" asses or mules instead of "warlike" horses. However, there were no donkeys in northern Russia until well into the Middle Ages. Their religious ceremonies required a donkey substitute. An English traveler touring Russia in 1557 writes, "On Palm Sunday they have a very solemn procession in the manner following: first, there is a horse covered with white linen cloth down to ye ground, his ears being made long with ye same cloth like an asses eares."

Donkey with a lead shank chain

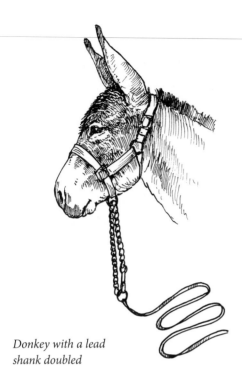

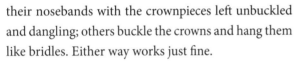

Donkey with a lead shank doubled

their nosebands with the crownpieces left unbuckled and dangling; others buckle the crowns and hang them like bridles. Either way works just fine.

Leads

Leads stored with halters should be strong, long, and fairly easy on hands because in an emergency things get harried, donkeys misbehave, and you may not have time to search for gloves. The best leads for this purpose are thick cotton rope leads with strong, brass snaps; flat-braided versions are especially nice.

For everyday use, I prefer polypropylene rope leads. They come in a wide array of single- and two-tone colors, and they feel really nice in the hand. They also have two shortcomings: If the donkey you're leading falls

> **BRAY SAY:** *Any jackass can kick down a barn, but it takes a good carpenter to build one.*
> — Attributed to both U.S. President Lyndon B. Johnson and former Speaker of the U.S. House of Representatives Sam Rayburn

back or rushes forward unexpectedly, polypropylene rope will burn your hands (wear gloves), and snaps on these ropes are clamped on, not woven on as with cotton lead ropes.

Expensive marine and mountain-climbing rope leads are the crème de la crème of rope leads. They usually come as part of a hand-knotted rope halter ensemble (in which case they attach with a loop-back instead of a snap), but snap-style versions are available, too (Linda Tellington-Jones's 8-foot TTEAM catch rope leads the pack!) (*see* Resources).

Some folks prefer flat nylon, cotton, or leather leads with a length of chain sewn into one end. These are properly called lead shanks or simply shanks. Lead shanks are primarily used to control a rowdy equine, in which case the chain is run in the near side halter square, over the nose, and snapped to the off side square as in the drawing at top right. If the animal misbehaves, the handler can get his attention by reefing on the chain (an action commonly referred to as "shanking"). Lots of people do this, but there are better and kinder means of control; we definitely don't recommend it. Leather lead shanks are also used for showing, in which case the chain is run through the lower halter ring and snapped back onto itself as in the drawing at top left.

Shelter Options

Donkey accommodations can be as extravagant as custom-made stables with cushy box stalls or as simple as three-sided field shelters with walls made of cattle panels. Donkeys need a means of staying dry in the winter and away from blazing sun and biting flies in the summer; if the shelter you already have or are planning to build provides the essentials (and it's roomy enough to prevent crowding), it's probably sufficient.

Donkeys in a Barn

Donkeys, like horses and ponies, can be housed in stalls in standard barns. Correct stall sizes range from 8-foot by 8-foot stalls for most Miniatures and up to 14 feet by 14 feet for Jackstock. If you breed donkeys, you'll need larger stalls at foaling time and for housing jacks. It's also nice to have a roomy sick bay and separate areas for feed storage and tack.

This said, donkeys, like their horsy kin, are happiest when they live outdoors, taking shelter only when they want or need it.

Field Shelters

Perhaps the best (and certainly the simplest) way to house donkeys is in inexpensive, three-sided field shelters. If you put the shelters on skids, so much the better; they're easier to clean that way. Exact sizes vary depending on the size of the donkeys that use them, but allow about 50 square feet of floor space per Miniature Donkey and up to 144 square feet of floor space for most Jackstock. A field shelter's open side should face away from prevailing winter winds, and a southwest exposure helps to utilize the winter sun as a source of heat.

In hot climates, simply build a roof and framework and enclose three sides with roof-high, welded-wire

A three-sided field shelter will protect your donkey from inclement weather while giving him plenty of healthy fresh air.

cattle panels. Open-air shelters provide essential hot-climate air flow, shade, and rain protection in the summertime, and then when winter comes, you can enclose the fenced sides with sturdy plastic tarps; it doesn't look elegant, but we do it and it works!

Whatever Stabling You Choose

Packed dirt or stone floors are easier on donkeys' legs and feet than concrete (if the building you refurbish for donkeys has concrete floors, top standing areas with rubber stall mats made for horses); wood floors are slippery when wet, and they tend to rot. Whatever floors are made of, cover them with 4 to 6 inches of absorbent bedding (sand, straw, poor-quality hay, shavings, rice or peanut hulls, even shredded paper) and clean it out as needed through the summer months. When winter comes, follow a deep-bedding system by removing surface wetness and badly soiled areas, and then adding more bedding atop the rest. Decomposing manure pack provides a source of winter heat and makes a fine addition to the compost heap come spring.

Building Shelters or Fences?

If you're starting from scratch, do your homework and get things right the first time; it's annoying (and expensive) to find out later that you messed things up and that you need to rebuild or fix them.

With that in mind, stop by your County Extension office to pick up literature about building barns, field shelters, and fences for your locale. And read these great books; they're our favorites and we refer to them all the time:

- Ekarius, Carol. *How to Build Animal Housing.*
- Clay, Jackie. *Build the Right Fencing for Horses; Storey Country Wisdom Bulletin A-193.*
- Damerow, Gail. *Fences for Pasture & Garden.*

Equines kept in poorly ventilated, tightly enclosed winter housing tend to suffer from respiratory ailments. Don't shut all the doors and windows; kept dry, donkeys can handle considerable subzero cold. If it's bitterly cold, blanket them instead of closing up or heating the barn. If you use a heat lamp for any reason — and they're very dangerous, so it isn't recommended — make certain that it's securely tied (don't hang it by its cord) and installed high enough to prevent bedding from igniting from the heat or curious donkeys from pulling it down. For the same reason, electrical wiring should be installed above chewable height and/or enclosed in a sturdy conduit. Windows should be well out of donkey reach or protected with strong, close-mesh wire (chicken wire isn't sufficient).

Gimme Water

Classic "horse troughs" come in a plethora of sizes, shapes, and materials. Because we live in the mid-South where algae is an ongoing problem and we need to dump, scrub, and bleach our tanks several times a week, we prefer midsize, heavy-duty fiberglass models with screw-out drains for their ease of cleaning. Standard metal tanks are also easy to maintain. Because heavy-duty rubber tanks grow underwater gardens of green scum in record time, though, we were disappointed with them.

One type of bargain basement horse trough we've used and endorse is a cast-off, old-style iron bathtub. The edges on these old tubs are nicely turned so that no one gets injured, and they're extra-easy to keep clean.

If you raise cattle, you probably have empty plastic mineral-lick tubs sitting around. If you do (or a neighbor or friend will give you some), put them to good use. The tall ones make sturdy, easily cleanable water troughs, and the smaller ones are first-class, portable grain feeders.

Another inexpensive summertime water container we use and that our equines seem to enjoy are

Reused mineral tubs make ideal feeders and water containers.

The Head of an Ass

- Asses' heads (representing the Greek storm giant, Typhon) were engraved on lead tablets that were used as curses to cause rivals in Greek and Roman chariot races to have accidents.
- Some ancient Romans slept in beds with donkey heads carved on the bedposts to assure that they'd have many children.
- Donkey skulls were posted around Portuguese villas to protect the inhabitants from evil influences.

secondhand turtle-shaped, green plastic sandboxes that we pick up at yard sales for a few dollars, tops. Besides being cheap, they're easy to clean and they hold enough water for two or three large donkeys for a day. Perhaps our donkeys and horses visualize themselves sipping cool, clear water from a mountain stream? For whatever reason, they bypass our conventional water containers and gravitate to the turtles every time.

Install water-filled tubs and buckets in the shade during the summer months. This helps inhibit algae growth, and since the water stays fresher, your donkeys will drink more water. When temperatures soar into the eighties or higher, freeze ice in gallon-size plastic milk bottles and submerge one in each trough or tub. Your donkeys will appreciate this treat. Refreeze them overnight and they'll be ready to use again by midmorning. Or freeze oversized ice cubes in plastic food tubs (we swear by Philadelphia Cream Cheese containers, and it gives us an excuse to eat more bagels) to drop into water containers throughout the day.

During the winter, prevent water supplies from freezing by installing bucket or stock-tank heaters. However, encase the cords in PVC pipe or garden hose split down the side and taped back together with duct tape; if you don't, your donkeys might gnaw through the cord.

Fencing Makes Good Neighbors

Several types of fences work well with donkeys: the most commonly used materials are board, woven wire, electric wire, and portable fencing. If horses share fenced areas, build to horse specifications; horses are flightier than donkeys and require taller, sturdier fencing.

Where predators are a problem, eschew fencing that allows dog-size animals to duck under or through rails, planks, or strands of wire to reach vulnerable individuals like Miniatures and newborn foals. Or, if your heart is set on fancy board fences, consider lining them with woven wire (or reinforce spaces between rails with strands of *hot* electrified wire) to keep undesirables outside where they belong.

ELECTRO-NET TEMPORARY FENCE

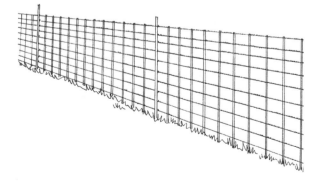

BOARD FENCE

Board Fences

Board fences, also referred to as post and plank or post and rail fences, are popular on many farms because they're attractive, highly visible, and relatively safe. This group includes fences constructed of treated or painted wooden planks nailed or screwed to the inside of wooden posts, split wooden rails that slide into holes in wooden posts, PVC plastic boards and posts, and wooden boards coated with vinyl. The only real disadvantage to board fencing is its cost, but with proper upkeep, these fences last 20 to 25 years.

Wooden planks are usually made of rough-cut oak or treated pine. Rough-cut oak lends a rustic appearance; it also has high tensile strength if animals lean on it, and most equines prefer not to chew oak planks. However, oak boards warp when freshly cut, and some spots (where the tree was weakened by natural processes) on some boards weather and rot faster than

PLAY IT SAFE

In 1999, the Equine Research Center in Guelph, Ontario, Canada, published the results of a fence safety survey in which 180 horse owners participated; the results, taking into consideration their more solid natures, applies to donkeys and mules as well. Seventy-three percent reported that no animals were injured in the past year. However, 27 percent reported fence-related injuries, and roughly half of them required veterinary treatment.

- Barbed wire, high tensile, and woven or welded wire accounted for 63 percent of injuries. Between 33 and 60 percent (depending on wire type) were serious.
- Diamond-mesh woven wire had only a 6 percent injury rate, and none required a veterinary call.
- Post and board fences accounted for 10 percent of the injuries; post and rail, 13 percent. Forty percent of the post and board injuries were serious, but most involved horses running into

fences or kicking them, resulting in splintering, and these aren't common behaviors in donkeys.
- Rigid polyvinylchloride (PVC) plank or rail fences were safest of all, with no injuries reported.
- Few injuries were reported with plastic and electrical tape, and the ones that were reported were minor.
- Most injuries occurred when a new animal was turned out with an established herd. Overcrowded lots, paddocks, and pastures contributed to horses running into fences.

others. Treated pine has a more finished look, it accepts paint (rough-cut pine can be stained but not easily painted), and the treatment it undergoes resists rotting and discourages equines from chewing — at least until the treatment wears off. On the other hand, pine isn't as strong as oak, so thicker boards are needed.

How Tall?

As a rule of thumb, fences containing equines should be withers-height to the tallest animal occupying the enclosure. However, while some sources recommend Miniature horse and donkey fences as low as 3 to 4 feet high, we consider 4½ feet the absolute minimum. Three feet might keep donkeys in, but it won't keep potential predators such as dog packs and coyotes out.

Vinyl Plank or Rail Fences

Solid PVC plastic fences cost more than wooden planks and rails but because they don't require painting (they're the same color throughout the material), they cost far less to maintain. Vinyl-coated plastic fencing is made of boards dipped in vinyl, so unlike solid PVC products, they can warp. White fences built from either type of vinyl fencing require washing with mildew-removing products at intervals, especially in the humid southern states.

FENCING OPTIONS

WHAT SIZE LUMBER?

| FENCE HEIGHT | PASTURES | | | RINGS AND PADDOCKS | |
	Line Posts	Corner Posts	Boards (based on rough-cut oak)	Posts	Boards (based on rough-cut oak)
4. 5"	4" × 4"; 7.5' long	6" × 6"; 7.5' long	1" × 6"; 14' or 16' long	6" × 6"; 8' long	2" × 6"; 12' or 14' long
5'	4" × 4"; 8' long	6" × 6"; 8' long	1" × 6"; 14' or 16' long	6" × 6"; 8' long	2" × 6"; 12' or 14' long
6'	5" × 5"; 9' long	6" × 6"; 9' long	1" × 8"; 14' or 16' long	6" × 6"; 9' long	2" × 8"; 12' or 14' long

SPACED HOW FAR APART?

(Distance in inches between boards, from top to bottom)

FENCE HEIGHT	THREE BOARDS	FOUR BOARDS	FIVE BOARDS
4.5'	12 - 12 -12	8 - 8 - 7 -7	—
5'	—	8 - 8 - 8 - 12	6 - 6 - 6 - 6 - 6
6'	—	7 - 7 - 7 - 7	6 - 6 - 6 - 6 - 8

Adapted from several sources but especially "Fencing Options for Horse Farm Management in Virginia," by Larry Lawrence (Virginia Cooperative Extension, 1999).

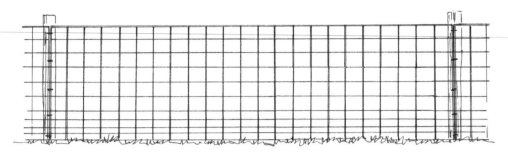

CATTLE PANEL FENCE

Cattle or Stock Panels

Cattle panels (sometimes called stock panels) are prefabricated lengths of sturdy mesh fence welded out of galvanized ¼-inch steel rods. Most cattle panels are 52 inches tall and built using 8-inch stays; horizontal wires are set closer together near the bottom of the panel to prevent small livestock from escaping. Cattle panels are usually sold in 16-foot lengths that can be trimmed to size using heavy bolt cutters.

Sheep panels are much like cattle panels except they're manufactured in 34- and 40-inch heights, and their horizontal wires are set closer together. Utility panels are the toughest of all; they're fabricated using 4-inch by 4-inch spacing and welded out of extra heavy-duty 4- or 6-gauge rods in a full 20-foot length. Utility panels come in 4- to 6-foot heights.

All are ideal for fabricating extra-stout pens and corrals and, where money is no object, long-lasting, maintenance-free perimeter fences around the farm. One bad thing about most cattle, sheep, and utility panels is that the raw end of each rod is very sharp. To make these panels user friendly, smooth each sharp end with a rasp to take off its razor-sharp edge.

Barbed Wire: Just Say NO

Barbed wire is the classic stockman's fence, but *don't use it*. While donkeys are far less likely than horses to injure themselves on barbed-wire fence, it can happen, and when it does it's nasty. If you already have barbed wire on your farm, especially if it was used to build cross fences that running equines could crash into, rip it out and replace it with something equine friendly. Equines and barbed wire don't mix.

WOVEN-WIRE FENCE

Woven-Wire Fences

Woven wire, also called wire mesh or field fence, is the fence of choice on many farms, ours included. It's made of horizontal lines of smooth wire held apart by vertical wires called stays. The spacing of the horizontal wires is usually closer near the bottom of the fence. The vertical stays in standard woven wire are 6 inches apart; "goat net" has 12-inch stays. Field fence's major failings are its price and the effort it takes to install it. However, correctly installed woven wire is the most secure form of affordable fencing, making it ideal for perimeter or boundary fences. Four-foot-high woven wire contains most equines, and installing one or two strands of electric wire above woven-wire fencing will keep most predators away from your herd, an important consideration for Miniature Donkey keepers.

When buying woven wire, read the tag; the numbers printed on it tell you how it's made. For instance, 10-47-6-9 fencing has 10 horizontal wires; it's 47 inches tall; there is a 6-inch spacing between stay wires; and the fence is made of 9-gauge wire. High-tensile woven wire costs more than standard woven-wire fencing but it's rust resistant, it sags less, and it's lighter in weight.

Nowadays most woven wire comes with galvanized (zinc) or aluminum coating. Both styles are further classified as Class I, II, and III wire; the higher the number, the thicker the coating and the more durable the fence. Class I galvanized woven wire will show signs

---

FENCE POSTS 101

If you live where Osage orange, Eastern, or black locust trees grow and you have more time and energy than money, you can cut your own wooden posts; untreated red cedar and black locust posts last 15 to 25 years, and untreated Osage orange will last a minimum of 25 to 30 years. Other suitable do-it-yourself fence post woods are untreated white oak, ironwood, honeylocust, catalpa, mulberry, and hickory; untreated, these last at least 15 years. Whichever species you choose, you must peel each post before setting it into the ground. It's easiest to peel them in the spring when rising sap causes bark to loosen.

The strength of wood posts increases as their top diameter becomes larger; a 4-inch post is twice as strong as a 3-inch post, a 5-inch post is twice as strong as a 4-inch post, and so on. Corner and gate posts should have a top diameter of at least 8 inches, brace posts 5 inches or more, and line posts can be anything from 2½ inches on up, but the bigger the posts, the stronger and more durable the fence.

Steel posts come in U bar, studded Y, punched channel, and studded T types, but they're all commonly referred to as T-posts. Although they lack the eye appeal of wood posts, T-posts are fireproof, longwearing, lighter in weight, and relatively easy to drive. They also ground the fence against lightning when the earth is wet. They do tend to bend if larger livestock leans against them or if you back into one with your tractor.

Equines have been known to impale themselves on the tops of T-posts, so it's best to cap them with special plastic toppers if you use them. Expect unbent T-posts to last 25 to 30 years.

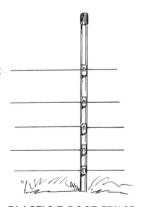

PLASTIC T-POST FENCE

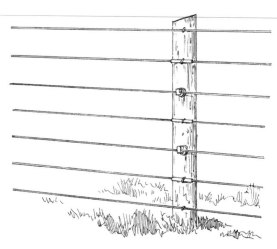

HIGH-TENSILE WIRE FENCE

of rusting in 8 to 10 years; Class III fencing begins to rust in 15 to 20 years. Aluminum-coated wire resists corrosion three to five times longer than galvanized wire with the same thickness of coating. Since a major part of fencing costs is installation, it's best to choose the longest-lasting wire you can afford.

Woven wire is sold in 20-rod (330-foot) rolls and is supported by wood or steel posts erected at 14- to 16-foot intervals. Wood posts come in treated and untreated varieties. Treated posts will last for 20 to 30 years; untreated ones will last two or more years, depending on the type of tree from which they're made.

BUILDING BETTER ELECTRIC FENCES

It pays to buy an adequate fencer. The box will tell you how many miles of fence it charges, but that's the greatest length of one strand of fencing operating under tip-top conditions. Think big. The more powerful your fencer, the fewer problems you'll have, so pick a model that packs a punch. Here are some pointers:

- **Read the instructions.** Mount solar fencers directly facing the sun; install ground rods as directed in the user's manual. Don't wing it. Unless you install your fencer correctly, it can't do the job that it was designed for.

- **Use quality insulators.** Sunlight degrades plastic; choose high-quality insulators, preferably a brand treated to resist damage done by ultraviolet (UV) light.

- **Don't skimp on wire.** The larger the wire, the more electricity it can carry.

- **Don't space wires too closely.** To get the most from your energizer, keep the wires at least 5 to 7 inches apart.

- **Train your donkeys to respect electric fences.** Electric fences are psychological rather than physical barriers; in order for animals to respect them, just once they need to get mightily zapped. Place untrained donkeys in a fairly small area equipped with a hefty fencer, and then entice them from the sidelines with a pail of tasty grain. Once most donkeys have been "bitten," they tend not to challenge electrified fences.

- **Buy a voltmeter and use it.** A good one costs less than $100, and it'll help you to keep your fences nice and hot. Check voltage every day. If the fence runs low on voltage or shorts out, you want to know it and correct the problem right away.

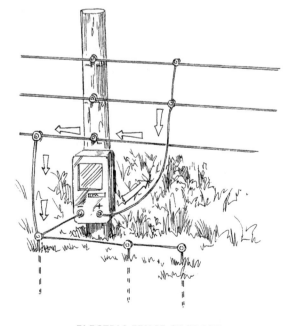

'Ear's a Tip

The Walking Ruler. Forget the metal tape measure the next time you build fences. When installing electric fencing, wear old jeans and use a felt marker to make fencing height marks on your pant legs; it will save you a great deal of time.

High-Tensile Nonelectrified Fence

High-tensile smooth fencing was once the horseman's fence of choice, but it's no longer recommended; too many serious injuries occur when horses hit tautly stretched wire while traveling at high speed. Because donkeys are wiser and far less likely to gallop blindly into fences, high-tensile wire fences might be an option on your farm, but think carefully before you commit. If there are horses in your future, you'd be wise to choose a different type of fence.

High-tensile wire fencing is installed using wood or T-posts and plastic insulators; it comes in 11- to 14-gauge models, and it has a breaking strength of roughly 1,800 pounds. High-tensile fencing is fairly inexpensive to buy and install, it's extremely durable, and it can be stretched tightly without breaking. Strong corner and end braces are needed, as well as the tensioners and strainers used to keep it bowstring-taut.

Electric Wire Fences

Electric fencing wire is sold in aluminum and steel varieties. Aluminum wire is the better of the two; it's rustproof and easy enough to work with that you can shape it with your bare hands, it conducts electricity much better than steel wire does, and it breaks if a silly horse slams into it.

Fence failures happen when the fencers (also called chargers or energizers) that power them aren't up to the job. Fencers are sold by voltage and the number of joules they put out (a joule is the amount of energy released with each pulse). One joule will power 6 miles of single-wire fence; a 4½-joule fencer will energize 20 to 60 acres, depending on the length of the fence and the number of wires that are used in its construction. Replace old-style "weed burner" fencers with modern low-impedance models that don't short out when damp vegetation touches a wire or spark grass fires during times of drought.

A four- or five-strand electric fence is adequate for most donkeys, depending on their size. To make a five-strand fence, install the first four wires 9 inches equidistant from one another, starting 9 inches from the ground, with the fifth wire a foot above the fourth.

Electric fences must be properly grounded or they lose their punch, yet according to some resources an estimated 80 percent of the electric fences in the United States are improperly grounded. A minimum of three 6- to 8-foot-long ground rods should be used with each fence charger, and it's important to follow the charger manufacturer's instructions when putting them in.

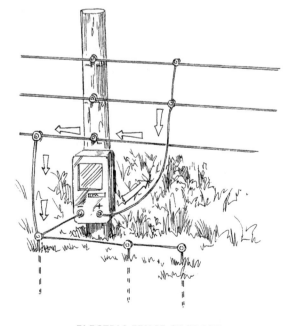

ELECTRIC FENCE CHARGER

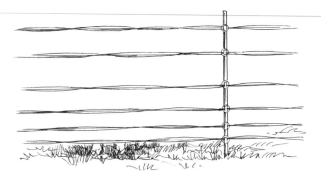

POLYTAPE FENCE

Portable Fencing

Polywire, polytape, and rope-style electric fencing used with step-in poly or fiberglass posts make fine interior fences and can be moved around with ease. Wide, flat tapes combine safety with high visibility. However, flat tapes whip around in the wind more than rope-style temporary fencing; to minimize whip, twist the tape once or twice between fence posts rather than installing it flat.

Net-style portable fencing is wildly popular with grass-based farmers and other folks interested in controlled grazing. There are two basic types: brands with built-in posts and the kind without. Most roll up onto easy-to-use reels, making moving these lightweight fences a breeze. It works well with Miniature Donkeys and others that really respect electric fence, but it might not be enough barrier for the big boys.

Gates

Gates should be as strong and safe as the fences they help constitute. Recommended: equine-safe tubular steel gates with smooth corners and welded cross pipes to minimize sharp-edged snags. Avoid gates with diagonal cross bracing that more easily traps heads, legs, and hooves.

Walk-through gates for handlers and animals should be at least 4 feet wide; drive-through gates should be at least 14 to 16 feet wide.

Avoid gaps between gates and support posts to prevent donkeys from getting hooves caught or foals from getting their heads trapped in a gap.

WALK-THROUGH GATE

Six Good Uses for Portable Net Fencing

1. To subdivide pastures for rotational grazing
2. To erect small pastures near the barn for special needs like confining jennies with new foals and providing limited grazing for animals in quarantine
3. To keep coyotes and marauding dogs away from your Miniature Donkeys
4. To keep livestock guardian dogs from wandering
5. To fence steep, rocky, or otherwise uneven land
6. To use as boundary fences on rented land

Noah's Donkey and the Devil

Noah and his family (his wife, Anak, Canaan, and Og excepted), together with a company of other believers, the number of whom some say was six, others ten, twelve, and even seventy-eight or eighty, half of them men and half women, including Jorham the elder, the preserver of the Arabic language, were saved, as well as the animals which Allah had caused to enter the ark. Among the latter was the ass, under whose tail Iblìs (the Devil) had hidden, disguised as a fly. This donkey, reluctant to enter the ark bearing the Evil One with him, was driven in by Noah with hard blows.

Presumably to pay the donkey out for this meanness, Iblìs whispered in his ear that all the females of his kind had been destroyed; whereupon the unfortunate beast made so terrible a noise of lamentation that the Evil One was scared and made haste to comfort him by adding, "But there is one left for you." At that the donkey's noise subsided in one long "Ah!" of relief. This is the origin of the donkey's braying.

— As told by J. E. Hanauer in his book
Folk-lore of the Holy Land, Moslem, Christian and Jewish, published in 1907

This photograph of llamas and pack donkeys in Argentina is dated 1940.

BRAY SAY: *While the stallions kick each other, the asses eat hay.*
— Belgian proverb

CHAPTER 7

FEEDING YOUR DONKEY

In Mexico, when two Indian farmers meet, they pass the time of day, inquire for each other's wife and children, and then always comes the question, "How is the burro?" Indeed, as the burro earns the daily bread for the people, this is natural enough.

— Theodore Ayrault Dodge, *Riders of Many Lands*, 1894

Ask half a dozen donkey keepers how to properly feed your donkeys and you're apt to get six divergent answers. However, most agree that the basis of a healthy donkey diet is forage: grass, browse, or hay. This chapter will explore the pros and cons of each.

Hay

Donkeys lacking access to a plentiful supply of pasture or browse should be fed free-choice quality hay. Hay is composed of grasses and/or legumes that have been air- and sun-dried until about 18 percent of their moisture remains.

There are several things to consider when selecting hay for your donkeys. For one thing, quality is more important than type. An example: Properly baled coastal Bermuda is higher in protein and other digestible nutrients than rained-on first-cutting alfalfa.

Hay should smell fresh, never sour or musty. There should be no sign of dust or mold. Donkeys who breathe

mold spores from hay can suffer respiratory ailments leading to permanent lung damage. Reject new bales that seem unusually heavy for their size or that feel warm to the touch; they weren't fully dry when baled. Dampness generates heat, which in turn causes hay to mold, or worse, triggers spontaneous combustion.

This donkey munches hay from a properly installed hay feeder.

Ask the seller to open several bales so that you can evaluate the hay inside. Most types of hay should be green inside. Hay that is yellow or brown inside has been rained on or sun-bleached in the field before baling. In either case the nutritional quality will be lower than that of properly put-up hay; however, don't worry about slight discoloration on the outside of the bale, especially with hay that was stacked in the sun. To minimize metabolic problems caused by changes in hay type or quality, buy as much hay as you can properly store at one time.

Buy tested hay or test the hay you bale or buy to determine its crude protein and TDN (Total Digestible Nutrients). Use a hay probe; you can buy one at farm stores or borrow one from your County Extension agent (he'll give you a list of forage-testing laboratories and a test kit, too). Sample three or four bales representative of all your hay by shoving the probe into the

How Much Is This Bale of Hay?

Most hay dealers sell hay by the ton instead of per bale. Purchasing hay by the ton allows the buyer to know precisely how much hay he is getting for his money, provided the hay is cured properly and accurately weighed.

To convert the price per ton to price per bale: Determine average weight of bales you are purchasing in pounds. Divide the price per ton by 2,000 and multiply the results times the average weight of the bales to determine the price per bale.

center of each bale to extract each sample. Put about a pound of combined sample material in a plastic bag, fill out the paperwork, and mail your sample to a certified laboratory. Depending on the mailing method used, expect the results back in two to seven days. A basic test should run about 15 to 25 dollars.

Establish terms before you finalize the purchase: prices, delivery terms, who is responsible for hay that doesn't meet your standards, and so forth. Don't depend on word of mouth or the classifieds when buying hay. For a wider choice of sellers, pick up a list of hay suppliers at your County Extension office or peruse hay lists online at the United States Department of Agriculture Farm Service Agency's HayNet Web site (*see* Resources).

Hay Quality Characteristics

If I were to name the best hay for donkeys I'd say timothy — which I haven't used in over 25 years because it doesn't grow where I live. That's true of virtually all types of grass hay; they're regionally specific, rather than being easily obtained throughout the United States. As a donkey owner, you need to ascertain what's obtainable where you live and then buy the best hay you can find based on those choices. How do you know

Feeding Rules of Thumb

1. Provide small portions two or three times a day.

2. Make changes slowly to allow your donkeys' digestive systems to acclimate to new feed.

3. Establish a schedule and adhere to it as closely as you can (feed at the same times every day).

4. Donkey diets should be forage-based. Don't feed concentrates unless your donkeys need them.

5. Feed according to your donkeys' needs, not on what they want to eat (given the opportunity, most any donkey will overeat to the point of obesity).

6. Discard dusty, moldy feed of all kinds.

7. Feed out of safe feeders designed for equines, never directly off the ground.

8. Make clean water available to your donkeys at all times.

if your hay is good quality? Below are some things to look for:

EXCELLENT HAY. Look for alfalfa cut in late bud to early bloom and grasses cut in the boot stage (when grass heads are still enclosed by the sheath of the uppermost leaf and no seed heads are showing). This hay is bright green, leafy, and free of dust, mustiness, and mold, medium green in color and high in protein, energy, minerals, and carotene but relatively low in fiber. For donkeys, legumes of this quality should be fed very, very judiciously (if at all).

GOOD HAY. This is legumes cut in 50 percent bloom or grass cut as it begins to head. This hay is reasonably soft, leafy, green, and free of dust, mold, and mustiness.

FAIR HAY. This is legumes or grasses harvested at full bloom. This hay is tinged green or yellow. It is stemmy, low in protein, energy, minerals, and carotene, and high in fiber. It may contain a moderate amount of dust; don't feed it inside an enclosed barn.

POOR HAY. This is any legume or grass hay cut after full bloom. It is stemmy, has few leaves, and is yellow or brown in color. It may be dusty, moldy, or musty. It's a poor buy; this hay is nutritionally bereft, and it may contain harmful amounts of dust or airborne mold. Buy a better grade of hay.

Hay Maturity at Harvest

There are a number of variables that can affect the quality of your hay. One thing to pay attention to is the maturity of your hay-of-choice when it's harvested. Below are some guidelines for the most commonly fed types of hay.

ALFALFA. Legume hays like alfalfa are generally too rich for donkeys and should be avoided whenever possible. However, because alfalfa is the only hay available in some parts of the United States, you may have to feed it. Keep in mind that alfalfa should be harvested in the bud stage (when there are buds at the tips of its stems) or slightly later in early bloom (when it has some purple flower petals and its stems are somewhat heavier than they were in bud stage). Alfalfa cut in full bloom or later may have seed pods in it, there will be fewer leaves, and its stems will be coarse and woody. Leafiness and soft, pliable stems are excellent indicators of quality alfalfa hay.

GRASSES. Orchardgrass, tall fescue, and reed canarygrass should be harvested in the boot (when grass heads are still enclosed by the sheath of the uppermost leaf and no seed heads are showing) to early heading (when seed heads are just beginning to emerge) stages. Timothy and brome should be cut in

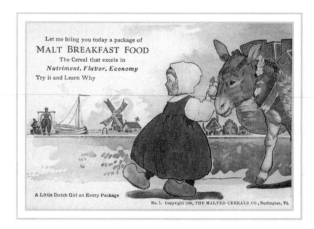

This trading card advertises Malt Breakfast Food with the slogan, "A Little Dutch Girl in Every Package," copyright 1906 by the Malted Cereals Company of Burlington, Vermont.

Feeding Don'ts

Never feed donkeys any of the following:

- Cattle feed containing urea (including many mineral licks marketed for cattle); urea is toxic to donkeys.
- Sheep or goat feed containing the coccidiostat Rumensin (this includes some but not all mineral licks labeled for sheep and goats); a single large feeding of Rumensin-laced feed can kill an equine, be it horse, pony, donkey, or mule.
- Fresh grass clippings; these can trigger colic and acute laminitis.
- Dog or cat kibble or any other especially high-protein animal feeds.

early bloom (when the plant's tiny flowers are beginning to form in the uppermost seed head) and when fully headed. Early-cut grass hay is greener than late-cut; plump brown seeds are indicators of less nutritious, fully matured grass hay.

MIXED HAY. Grass and legume mixes should be cut when the legume component reaches its ideal stage for harvest.

Hay Substitutions

We live in northernmost Arkansas, an area often hard-hit by drought. Sometimes quality hay trucked in from other states is available, and sometimes it isn't. In a pinch we've learned that it's easier (and better) to feed alternatives when hay brings top-dollar prices. When you can't find good baled hay, here are some options:

BAGGED HAY. When the going gets rough, pick up the phone and call the feed stores in your area. Chances are, they carry one or more types of bagged hay. Bagged hay is generally packaged in 40-pound, plastic-wrapped squares of compressed, chopped hay. To prevent waste, pour enough water over the hay to dampen it so that your donkeys are less likely to dribble it around.

BAGGED HAYLAGE. Haylage is a semiwilted hay product (grass or alfalfa dried to 55–65 percent dry matter as compared to 82–85 percent in hay), compressed and sealed in tough plastic wrapping. Donkeys love grass haylage, and, handled correctly, it's first-class donkey fodder (it's fed at the Donkey Sanctuary in England). It's important to remember that once bagged haylage is opened (or the wrapping is accidentally broken), mold spores begin to proliferate, and after three or four days any uneaten feed should be discarded. In the United States, the major manufacturer of bagged haylage for equines is Chaffhaye (*see* Resources).

> **BRAY SAY:** *Donkeys that bray eat last.*
> — German proverb

Storage

A lot of good hay is ruined through improper storage. Choose small bales stored indoors, forgoing bottom bales that harbor mold and top bales that are splotched with bird (and sometimes cat) droppings.

Large bales kept undercover are a better buy than field-stored bales, which in areas of heavy rain- or snowfall can represent up to 35 percent waste. If stored outside, they should be placed on pallets or poles to get them off the ground and covered with plastic tarps after they're fully cured (covering too soon or when they're damp from dew or being rained on results in spoiled hay). Don't store them close together, abutting each other from the sides (store them in a line, end-to-end), and don't stack them; both practices trap moisture between the bales and ultimately ruin a lot of hay.

HAY CUBES. Another option is to feed hay cubes. Their advantages are that they're readily available, and because hay cubes have adequate particle size to maintain normal digestive health and prevent wood chewing, they can be used to partially or totally replace baled hay. However, we've had a number of equines choke on cubes. If it's hay cubes or nothing, use them, but soak them for an hour or so before feeding to reduce them to a tasty mash.

Big Bales — or Not?

In some parts of the United States, it's becoming increasingly difficult to find standard-size, 45- to 100-pound "small square" bales of hay. It's cheaper, faster, and easier for producers to put up hay in big round bales ranging in size from about 400 to 1,200 pounds.

If you feed big bales, opt for a safe feeder like this one designed for horses.

Big round bales stored under cover (and not directly on the ground) can be fed to donkeys but with several caveats.

- ✘ Equines of all sorts waste an incredible amount of hay unless big bales are fed from a round-bale feeder, and, in our opinion, these feeders aren't necessarily safe. If you use one, at least once a week check for broken welds and other damage that your donkeys could injure themselves on (it happens more often than you think) and remove it altogether if young donkeys decide to play inside the ring.

- ✘ Make certain a bale is good *all the way through* before feeding it to equines of any kind. Big bales are rolled up in layers, and it's not unusual for bales to harbor moldy sections deep inside the bale.

- ✘ Remove the strings; they're strong and long, and all hay strings are dangerous. Hay strings and livestock don't mix.

> **BRAY SAY:** *It's time to shock the hay when the donkey blows his horn.*
> — Belgian proverb

- ✘ If you don't have enough donkeys to eat a big bale in a reasonable amount of time, don't put one where they can eat it. Instead, store the bale someplace under cover (where the donkeys can't reach it) and on a wooden pallet or old tires to get it off the ground, then unwind it as you need it, feeding just enough per day that your donkeys will eat however much you feed them.

Donkeys and the Law

- Archaic laws still on the books in Arizona, Georgia, and Brooklyn, New York, state that it's unlawful to keep a donkey in a bathtub (donkeys may be kept in bathtubs in Arkansas but alligators may not). It's also illegal to board a Pittsburgh, Pennsylvania, trolley in the company of a donkey or mule.

- On September 29, 2007, a donkey at an Algerian market ate the money of a man who came to buy him, making the buyer and the owner wonder who the animal belonged to. A local newspaper reported that the customer and the seller bargained too long and too enthusiastically to notice the donkey consuming the stack of banknotes meant as payment for him. Neither the Tizi Ouzou district court nor the city court was able to make a ruling that would satisfy both parties, so the case was referred to the Algerian Supreme Court.

- According to a BBC News story published on July 17, 2007, donkey owners in Limauru, Kenya, were up in arms over a town edict forcing them to diaper their donkeys. One donkey drover commented, "If we have to put nappies [diapers] on our donkeys, soon they will say our cows need them too." On the same note, donkey drivers on Hydra, an island in the Mediterranean where all motor vehicles are banned, are already required to fit their charges with manure-catchers or pay a $132 fine.

Grass and Browse

Pastured donkeys are generally happier (and healthier) than their stabled kin. Pastured, they respond to circadian rhythms that inspire feral donkeys to feed now, rest later, and then feed again. Their primary diet is grass and possibly browse: nature's perfect donkey foods. They roll when they want and loaf, race, or play-fight if they choose. In this more natural setting they behave — well, like donkeys. And for a donkey's body and mind, that's a very good thing.

However idyllic that image seems, your donkeys won't thrive on pasture unless they're properly cared for. They can't be simply turned out and checked on when it's convenient for you. They'll require a suitably sized, safely and securely fenced meadow of grass, forbs, and possibly browse, ready access to shelter, clean free-choice water, supplements such as salt, vitamins, and minerals (and sometimes additional feed), protection from biting flies and other insect pests, companionship, and daily monitoring, along with routine deworming and hoof and coat care.

All pastures are not created equal. Be certain that the one your donkeys graze provides enough forage for their needs. Pastures overgrown with unpalatable vegetation may seem ripe for grazing, but donkeys won't thrive when pastured there.

By the same token, lush spring pastures and meadows of legume hay are too rich for a donkey's metabolism. If you use them, your donkeys' grazing time must be limited and closely monitored, or the tubbers in your herd should be fitted with grazing muzzles.

Meadows must not be overgrazed. Depending on where you live and the condition of your pasture, it might take 1 acre or 30 to nourish a donkey through grazing season; ask your County Extension agent for local particulars.

Before committing your donkeys to pasture, be certain that all fencing and gates are safe. Walk every fence line, making repairs as needed. Replace equine-unfriendly materials such as saggy high-tensile or

barbed wire. If fences rely on electric current, make certain fence chargers are adequate and working.

Shelter from nature's occasional unpleasantness, be it searing heat, howling winds, rain, hail, early snow, or biting bugs, is essential. In very warm climates, natural shelter such as mature trees for shade and dense hedges or rock outcroppings for windbreak may be sufficient. However, all pastured donkeys appreciate (and most require) access to manmade shelter. Shelters should be sound and roomy enough to accommodate every individual in the herd. If horses are part of the herd dynamics, provide at least two exits so that cranky horses can't corner a passive donkey.

Pastured equines require free access to clean drinking water. Although thoughts of ponds and bubbling brooks spring to mind, these are not necessarily the

Gluttony Prevention

Grazing muzzles are nose baskets made of nylon webbing and neoprene with a small hole in the rubber bottom designed to prevent a horse, pony, or donkey from pigging out on lush grass. Best Friend Equine Supply (*see* Resources) manufactures them in two models: one with an attached headstall and one designed to be attached to a standard breakaway halter. They're an obese (or laminitic) donkey owner's best friend!

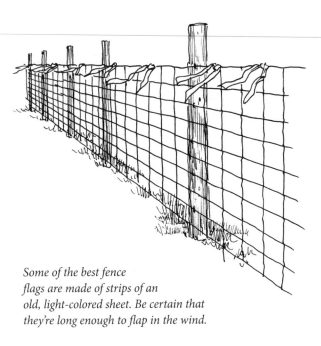

Some of the best fence flags are made of strips of an old, light-colored sheet. Be certain that they're long enough to flap in the wind.

best solutions. Natural water sources are often polluted, if not by tainted groundwater or chemical runoff, then by the feces of birds, deer, raccoons, opossums, and other wild creatures that expose your donkeys to a host of serious health problems. Even if your pasture contains a pond or stream, provide an automatic waterer or a frequently scrubbed, daily monitored watering trough. You'll find that most donkeys prefer it.

Lactating jennies, weanlings and yearlings, old geezers and other poor keepers often require supplementary feeding, and as spring's lush grazing gives way

to less nutritious summer growth, or when pastures are overgrazed, daily feeds of hay will be required to keep your donkey fit.

All donkeys need access to free-choice, equine-specific minerals, provided loose in a feeder or as a crumbly block. Hard salt blocks designed for cattle aren't sufficient. Pelleted vitamin supplements can be hand-fed or added to a tiny bit of grain fed from a pan.

Donkeys plagued by gnats, flies, and mosquitoes may camp in their shelters instead of grazing. Itchy fly bites and imbedded ticks cause them to scrub against trees, fences, and buildings, scraping their hides and injuring their eyes. Many of these pests spread disease. While appropriate shelter helps, depending on the insect load in your locale your pastured donkeys may require insect protection in the form of fly masks and sheets, pour-on, spray, or feed-through repellents, and, for ticks and bot fly eggs, daily "picking." Particularly bug-beset donkeys can be stabled in the daytime and turned out to graze at night.

Before They Go to Pasture

Let's suppose you've found the perfect pasture situation for your donkeys — what then?

1. First, make certain that you can catch them before turning them out to grass. Halters and pastured donkeys don't mix, and sometimes the best-mannered donkey, sans halter, refuses to be caught.
2. If any of your donkeys is young and active and the pasture is fenced with wire, knot a "fence flag" (a long strip of light-colored fabric or plastic tape) every few feet around the pasture's perimeter.
3. Show new donkeys around. Make sure that each donkey knows where his water, minerals, and shelter are. Walk the pasture fence line, leading him, so that he knows its boundaries.

Poisonous Plants

No matter where you live in the United States, there are poisonous plants growing in areas where your donkeys graze. This may or may not be a problem, depending on whether:

- ✗ *Your animals eat them.* Poisonous plants aren't necessarily attractive to donkeys. Pasture-wise animals of all sorts seem to know intuitively which plants they can safely consume. And many poisonous plants taste nasty, either acrid or extremely bitter, so donkeys won't eat these unless they're hungry and it's "eat them or starve."
- ✗ *They eat enough volume to matter.* Many "poisonous" plants are simply toxic. Unless they're eaten in massive quantities or over a length of time, they do no harm.
- ✗ *They consume the poisonous part of the plant.* In most cases, only a portion of a plant is poisonous, such as its roots or wilted leaves or seeds. Or the plant is poisonous only at certain stages of its growth and animals don't eat it at that time of the year.
- ✗ *They're immune to the compounds in a given plant.* Some poisons are species specific. Cattle thrive on acorns, sheep consume larkspur, and humans eat milkweed shoots, yet all of these plants can effectively poison a donkey.

PASTURE PROTOCOLS

Once or twice a day:

1. Remove each donkey's fly mask, fly sheet, turnout rug, or any other clothing and inspect him carefully, checking for cuts, scrapes, sunburn, rashes, bumps, hoof cracks, weepy eyes, and debris or stickers in his ears, hide, or tail swish. Lead him forward and back him up to evaluate his movement. Treat any injuries and pick out his hooves.

2. Spritz or pour on fly and tick repellent and pick off bot fly eggs and ticks.

3. Check the water supply. Be certain that automatic waters are functioning and tidy. Inspect water tanks, dumping and swabbing scummy or soiled ones before refilling. Do the same for loose mineral feeders.

4. Inspect electric fence chargers, then walk or ride the fence line, making necessary repairs.

5. Finally, do a quick sweep of shelters, inside and out. Mist with premise insecticide and zap wasp nests. Remove accumulated manure.

Weekly:

- Carefully inspect all shelters, combing them for hazards and mending any that you find.
- Strip soiled bedding and replace it with fresh.

Every 6 to 8 weeks:

- Trim hooves.
- Deworm your donkeys. Ask others using the pasture to deworm their equines, too. Otherwise, their animals will continue shedding worm eggs and your donkey will quickly become reinfested.

Since there are no known antidotes to most plant poisons, prevention is worth lots more than cure.

- ✗ Learn to recognize the poisonous plants growing in your locale. Ask your County Extension agent for pamphlets, invest in plant identification books, or visit poisonous plant Web sites online (type *horse poisonous plant* in the search box of your favorite search engine; if it poisons horses, it'll poison donkeys, too).
- ✗ Make sure that your donkeys have enough to eat. Avoid overgrazing; when pasture runs low, provide hay as substitute forage.
- ✗ Before moving your animals to new pasture, *check it out*. If you find poisonous plants growing there, remove them, fence your donkeys away from that area, or seek safer pasture elsewhere.

Nosebag Know-How

When you feed grain, use a nosebag. With nosebags, each animal in a herd consumes exactly what he needs, no less, no more, and because less assertive individuals can move away from the crowd and still eat dinner, fewer chasing injuries are likely to occur. There is no way for a wasteful donkey to dribble his feed when dining from a nosebag; when your donkey wears a nosebag, you know whether he's consumed his feed-through dewormer or medications; and nosebags make feeding away from home at shows and trail rides a breeze.

In years past, nosebags were crafted of heavy canvas or leather and were secured to a horse, mule, or donkey's head using a strong, adjustable, leather strap headstall. Newer choices include nylon mesh nosebags, sometimes with long-lasting hard plastic bottoms; these, as well as twill and canvas versions, can be purchased at saddle shops and from backcountry outfitters' catalogs.

Whatever materials they're made of, nosebags must be sturdy, well made, and washable. Check inside any nosebag you're thinking of buying: Are the edges finished? If not, they'll unravel when you wash the bag. Examine the headstall: Is it easily adjustable, and, if not, is it sized for the animal you'll use it on? And is the headstall made of sturdy material and well attached, but not so sturdy and well attached that it won't snap or rip away from the nosebag if your donkey gets it hooked on a snag?

Most important: Where is the nosebag patch located? The nosebag patch is a perforated leather insert set low and at the center of all good nosebags. If the lower edge of a nosebag's leather patch is set more than a few inches above the bottom of the bag (or worse, there is no patch inset at all), don't buy that bag! Many equines have drowned in nosebags. The perforated patch helps a nosebagged animal to breathe easier, but its main function is to let out water should the need arise. If, when wearing his nosebag, your donkey's nostrils are situated lower than those holes, he's in deep trouble should he dip his muzzle in a bucket or horse trough or stream, or even if he gets caught in a summer downpour. Sure,

---

Concentrates — Yes or No?

Working donkeys (packing, riding, driving), geriatric donkeys, some in late pregnancy and most lactating jennies, growing youngsters, and sick or underweight donkeys generally need grain in their diets, but most donkeys don't.

Unfortunately, little research has been conducted on the effects of feeding concentrates to North American donkeys. However, many veteran donkey owners who feed concentrates recommend feeding one-third of one pound of oats (whole, crimped, or rolled) or low-protein (10 percent) horse feed per one hundred pounds of donkey. Keep in mind that overweight donkeys are prone to a host of serious maladies, and it's easier (and safer) to keep a donkey's weight in check than to try to trim him down later on.

you'll waste a little grain when it sifts through those low-set holes, but better to lose some grain than your long-eared friend.

And no nosebagged horse, mule, or donkey should be left unsupervised, ever. If you can't keep constant watch, check on him every 5 or 10 minutes or so. Never forget that your donkeys are wearing bags. Once they've finished eating, most equines try to drink, even if they're wearing nosebags. And bags or no, once they've eaten, pastured animals drift out to "graze." Besides drowning, unsupervised animals can snag their nosebags on stall protrusions, trees, bushes, and worse. And if his nosebag is too big for him, a youngster can hook a foreleg inside of it while trying to graze. If the headstall doesn't snap, he's in trouble!

Another caveat: Be careful what you feed in a nosebag. Ventilation is limited inside that bag, so feeding dusty grain won't do. Powdered supplements should be mixed in well (when feeding from nosebags, pelleted supplements are best). If in doubt, mix a little water or vegetable oil with your donkey's grain before bagging him; it's messier but better for his lungs.

If you do feed messy additives like oils, expect to launder your nosebags often. Turn them inside out and stuff them into an old pillowcase, knot the end of the pillowcase, and drop it in the washer on the warm water setting. Hand washing works, too. Chuck your bags into a bucket of warm, slightly soapy water, one at a time and inside out; soak, slosh, and rinse thoroughly in two changes of warm, clean water. Gloss

A tourist poses with a donkey, which is far more interested in its dinner! Believed to be taken in Italy in 1958.

leather conditioner on your bags' leather headstalls if they have them. Don't machine-dry nosebags (some shrink); hang them on a clothesline, inside out.

Before nosebagging your donkey for the first time, be certain that the bag you're going to use fits him. A full-size model on a Miniature or Small Standard donkey spells disaster; sooner or later he'll get his foot inside that too-big bag (unfortunately, small nosebags are seldom available, so you'll probably have to make your own). Measured from top to bottom, a properly sized nosebag is long enough to contain the amount of grain you normally feed plus 8 to 12 inches clearance for a Standard donkey, and the top shouldn't be so floppy that it sags around your donkey's face.

Know how to adjust your donkey's nosebag before you hang it on him. Don't ratchet the headstall up so short that his muzzle is buried in feed.

For his first experience wearing a nosebag, it's best to put your donkey in a safe, semiconfined area; a round

Did You 'Ear?

According to Anthony Dent (writing in *Donkey: The Story of the Ass from East to West,* 1972), the word *donkey* is probably a derivative of the Flemish word *donnekijn,* meaning "small, dun-colored animal."

pen is ideal. Remove his halter if he's wearing one, and show him that there's grain in the nosebag. Let him nibble some grain from the bag while you're holding it. While he's nibbling, quietly slip the headstall around his head. With one hand on the headstall and the other on his neck to reassure him, gently swivel his head back and forth a few times so that he realizes that the bag is hanging on his nose, then stand back and let him get a feel for the thing; he'll soon learn to drop the bag and eat from it as if he were eating off the ground. Being the smart, sensible creatures that they are, most donkeys take to nosebags like cats to cream; however, a few sensitive individuals panic. If yours does, catch him and remove the nosebag; for the next few grain feeds, hold the bag and let him eat from it that way. After three or four feedings, fasten the headstall and try again.

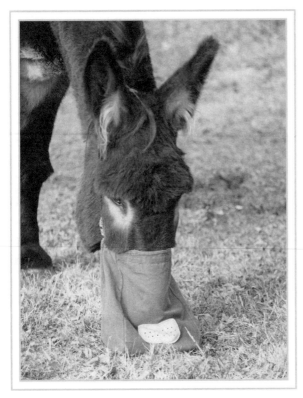

The author's donkey, Edward, dines from a homemade envelope-style nosebag.

Nosebags are a boon in lots of situations, but no one appreciates them more than she who must feed equines in a herd setting, especially when horses are involved. The same herd aggressors who mob a human carrying a bucket may overlook one carrying a more easily concealed nosebag. If they don't, at least they cut off their attack once the bag is on another individual's head. Once bagged, meeker herd members generally drift to the edges of the pack and since the herd tyrants can't see (or reach) their grain, they're usually left in peace. While dining from nosebags, most equines slip into a blissed-out trance, so unless they're aggressively harried, bullies usually won't bother them at all.

Some equines, especially youngsters, nibble and tug at another individual's nosebag. If they do — bag 'em! A handful of grain will keep interlopers occupied for a while, and when it's gone their own nosebags prevent them from worrying at another's.

Make-It-Yourself Nosebags

Where can you buy a nosebag, especially for Miniatures and Small Standard donkeys? That can be a problem. Not all saddle shops carry nosebags, nor does every mail order horse supply. Worse, not every bag on the market is well made (or safe). However, nosebags are easy to make, and, made my way, they cost mere pennies. No sewing machine? Not to worry; even the most stumble-thumbed donkey keeper can hand sew a suitable nosebag in an hour and a half or less.

Most commercial nosebags look like little fabric buckets: They're cylindrical in shape with round bottoms. The bag we use is modeled on a flat ethnic design based on a vintage nosebag that we discovered in an antiques shop. Unique in its simplicity, this flat nosebag is infinitely easier to make than standard bucket models, and because they're not as floppy, my donkeys don't step on and rip them the way they did their purchased nosebags.

FOR EACH NOSEBAG YOU'LL NEED:

- ✗ **A pair of heavy-duty, wide-at-the-thigh pants.** Measured just below the crotch, the legs must be more than wide enough to accommodate your donkey's face. Heavy denims work well, and canvas hunting pants or winter coverall legs make fine, sturdy nosebags. Secondhand pants, even pants with worn-out knees, work just dandy, and you can buy them at secondhand shops or yard sales for practically nothing.
- ✗ **A sturdy leather, nylon, or heavy cotton belt.** It should be ¾- to 1-inch wide and long enough to pass from the near side of the nosebag, up across the top of your donkey's head, and down to the bag's off side (you'll need a 36- to 38-inch belt or strap for an average Large Standard donkey). Adjustable straps from backpacks and waistbands from fanny pouches also work well.
- ✗ **A 3- by 5-inch piece of scrap leather.** Heavy cowhide or elk hide is ideal, but a thrift shop purse or worn-out work boot yields excellent leather, too.
- ✗ **Strong thread or cord.** Button thread (doubled), waxed dental floss, and artificial sinew all perform equally well.
- ✗ **A needle.** A #4 Glover's needle is best, but any sharp needle with an eye big enough to accommodate the thread you're using will work almost as well.
- ✗ **A revolving leather punch.**
- ✗ **A ballpoint pen, a scratch awl, or an extra-sharp nail.**
- ✗ **Sharp, sturdy scissors.** Leather shears work best.
- ✗ **Heavy-duty sewing pins.**

DIAGRAM 1

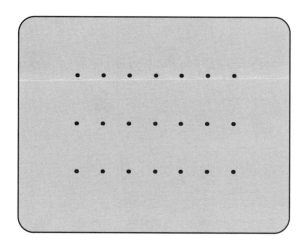

STEP 1

- ✗ Photocopy or trace the nosebag patch shown in diagram 1 and cut out the paper template.
- ✗ Trace the template's outline onto your scrap of leather and cut out the leather patch.
- ✗ Place the paper template over the patch and poke a ballpoint pen (or the tip of a sharp awl or nail) through the exact center of each hole, marking the leather beneath. Use your leather punch's largest tube to punch them out. Then, using the smallest tube, punch sewing holes about ⅛" apart around the patch's perimeter. A scrap of leather held under the patch while you punch will result in cleaner holes. Set aside the completed leather patch for now.

DIAGRAM 2

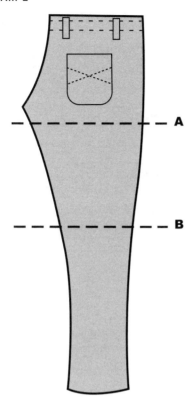

A

B

DIAGRAM 3

STEP 3

- ✗ Arrange the nosebag so that the old pants seams are at its sides.
- ✗ Cut your belt or strap in two approximately 5 inches from the buckle (if it's an adjustable strap, be sure the adjustment is on its longer segment). Hold the cut end of the 5-inch piece of belt in place 2 inches down on the inside of the left side of the nosebag. Sew it in place using a short, tight backstitch.
- ✗ Make several passes. Now do the same with the longer section of belt on the opposite side of the nosebag.

STEP 4

- ✗ Position the leather nosebag patch just above the front bottom seam of the nosebag (it can be set vertically or horizontally). Temporarily secure it at its sides, top, and bottom by sewing down through one perimeter hole at each location and up through its neighbor, then loosely knotting that stitch in place. Sew the patch in place using a backstitch through the perimeter holes, and then remove any temporary stitches.
- ✗ Turn the finished nosebag inside out and, leaving a ½-inch fabric allowance between the stitching and the hole you're making, snip away both layers of fabric covering the drain patch's central drain holes. *Don't omit this step.*
- ✗ Turn your nosebag right side out, and that's it!

- ✗ *Hint:* For an easy-to-make, lightweight nosebag, buy an appropriately sized, ready-made cotton tote bag at a rummage sale, snip off the handles, and omit step 2 altogether.

STEP 2

- ✗ Fold the pants in half as shown in diagram 2. Cut straight across both legs just below the crotch (point A). Decide how long the nosebag should be (14 to 18 inches is about right for a bag for a Large Standard donkey). Cut straight across at that length below your first cut (point B). Turn one nosebag piece inside out; stuff it inside the other piece so that their existing seams align.
- ✗ Turn this conjoined piece inside out. Pin all four thicknesses together at the narrower end and sew straight across using a tight backstitch. Remove the pins, and for best results, fold this seam once, then hem-stitch it down.
- ✗ Fold the top edge inward, twice, as shown in diagram 3. Pin it, and then securely hem-stitch it in place. Remove the pins and turn the nosebag right side out.

The Democratic Donkey

When Andrew Jackson ran for the presidency in 1828, his campaign slogan was "Let the people rule." His opponents labeled him a "jackass" for his populist platform. Jackson was not dismayed; he embraced the donkey for its tenaciousness and used it on his campaign posters. Later, during Jackson's term in the White House, he was widely depicted as a donkey for his legendary stubbornness in vetoing a recharter of the National Bank.

In 1879, Thomas Nast, the famous political cartoonist, chose a donkey to represent Lincoln's secretary of war, Edwin M. Stanton, in a cartoon published in *Harper's Weekly* (earlier, in 1874, a *Harper's Weekly* Nast cartoon linked the elephant to the Republican Party). The Democratic donkey so enchanted Nast that he continued using it to represent the Democratic Party, as have his successors.

Although the Democratic Party has never formally recognized the donkey as its official emblem, since Nast's era the two have been irrevocably linked.

Democratic donkeys are a favorite subspecialty among donkey memorabilia collectors, and each political election contributes additional merchandise to the mix. One rich source is eBay: on the day that I visited, I found bracelet charms and similar jewelry, apparel of all kinds, scores of lapel pins, Beanie babies, highly collectible dated Frankhoma coffee mugs, Jim Beam commemorative decanters, several styles of bobble-head Democratic donkeys — and a cute figurine of former President Clinton seated backward on a donkey's rump, playing his saxophone!

CHAPTER 8

GROOMING AND HOOF CARE

An ass is beautiful in the eyes of an ass.

— Turkish proverb

Grooming isn't done just to make your donkey look pretty. Daily grooming removes sweat, dirt, dead skin particles, and shed hair while unclogging pores, enhancing circulation, and soothing stiff muscles. And it makes a donkey feel good. That alone makes grooming worthwhile.

The Tools of the Trade

A grooming kit can be as plain or as fancy as you please. A few basic tools used correctly will do the trick. However, specialty items make grooming easier, quicker, and sometimes just more fun. You may want to add a few of them to the mix. You will need:

✗ One or more currycombs. Used to scrub away dirt and caked mud, currycombs can be made of soft or hard rubber or rigid plastic, but never metal. An exception is a fine-toothed loop of metal called a shedding blade, sometimes used to divest coarse-coated equines of their shedding winter woolies. If you have an old-fashioned

metal currycomb, use it to clean your brushes, not your donkey.

✗ A dandy brush. You'll use this stiff-bristled brush to sweep away dirt raised by the currycomb.

✗ A body brush. Body brushes have finer, softer, sometimes shorter bristles than a dandy brush,

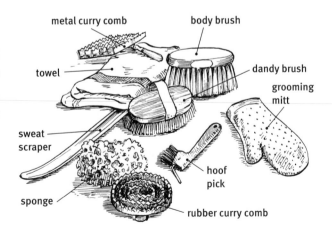

A nice selection of grooming tools makes coat care easier — and more fun!

and they are far more densely filled. Use it to remove fine dust particles and scurf (greasy, whitish dandruff flakes raised by your currycomb and dandy brush). It's soft enough to use on donkeys' heads and lower legs.

× A hoof pick. If you choose one backed with a stiff hoof brush, you won't have to buy a separate tool.

× A sponge, towels, washcloths, or disposable unscented baby wipes. You'll use these to clean around your donkey's eyes, nostrils, and mouth, and under his tail.

× A sturdy, reasonably tip-resistant container in which you can pack your grooming kit. A colorful plastic grooming tote is nice, but a sturdy plastic bucket or even a cast-off book bag or daypack works well, too.

At one time, everyone who crossed the border into Nogales or Tijuana had his or her picture taken on a burro. This postcard from 1918 is typical. What a sweet-looking couple, and the donkey looks so well cared for.

Optional items that you'll be glad you have:

× A grooming mitt. Sticky rubber mitts with pimply bumps or longer protrusions work wonders on loose hair and grime.

× A bot fly egg remover. A special knife, a commercial Styrofoam-block bot egg remover, fine sandpaper, or a safety razor all beat picking off teensy, stick-tight bot fly eggs with your fingernails.

× A sweat scraper. You'll use it to squeegee away water while bathing your donkey. The smooth side of a loop-style shedding blade makes a dandy sweat scraper, too.

Buying Brushes

Colorful synthetic-bristled brushes cost less than brushes constructed of natural fiber bristles but may work just as well for everyday grooming. They are, however, usually stiffer than their natural fiber counterparts and usually not constructed as well.

Natural fiber body brushes are usually made of horsehair or soft boar bristles drawn through a leather brush back and secured with wire that is covered with another leather backing. They're often fitted with leather hand straps. These cost a lot more than synthetic-bristle brushes, but if taken care of, they'll last for a very long time.

Nowadays some companies manufacture the same brushes in two or more sizes. Choose brushes that you like and that comfortably fit your hand.

Grooming Overview

Use your currycomb or mitt to loosen caked mud, shedding hair, and scurf by scrubbing in a circular or back-and-forth motion with and against the lay of your donkey's coat. The curry is also a massage tool, so dig in while currying heavily muscled areas, but go lighter on bonier parts. Don't use a currycomb on a donkey's lower legs or face unless it's an ultra-cushy curry like a rubber Grooma or a Jelly Scrubber. If using a hard plastic curry like the Sarvis comb (my personal favorite) or a traditional stiff rubber one, occasionally whack it on the floor, your boot, or the wall to clean it.

Next: the dandy brush. Use short, snapping strokes to dig clear down to your donkey's skin. Work with the lay of his coat. Every few strokes clean the brush by scraping its bristles with a plastic or metal currycomb. Press hard on less-sensitive body parts but go lightly on your donkey's head and lower legs.

If your donkey is particularly touchy you might want to substitute a softer implement like a body brush for the dandy brush; the body brush's soft, dense bristles make it ideal for grooming a donkey's more delicate parts. You can also use your body brush to remove fine dust and scurf from your donkey's coat and to condition and shine it. Press it in long, firm sweeps across his body. Currycomb or bang it clean every few strokes.

Next, moisten your grooming sponge, washcloth, or towel with plain water and gently clean around his eyes, nostrils, and lips; moisten it again and wipe his sheath (or her udder) and under his tail. Unscented baby wipes work well for this, too.

Finally, clean your donkey's hooves. Starting at his heel, dig the tip of your hoof pick under packed debris and pry it out. Be sure to pick accumulated debris out of the cleft of his frog and from the deep grooves on either side of it, then scrub the bottom of the hoof clean using the pick's built-in hoof brush. Alternatively, use a human's fingernail brush or vegetable scrubber in its place. *(For more details on cleaning hooves, see page 128.)*

Grooming Tips

Equines like routine. Establish a pattern and follow it every day. Finish with each tool before moving along to the next. Here are some other pointers:

- ✖ Although most donkeys enjoy being groomed, some don't appreciate certain procedures and might nip or kick. It's always safest to halter and tie a grumpy donkey when you groom him.
- ✖ It's wise to place your free hand someplace on your donkey's body so that you can feel him tense

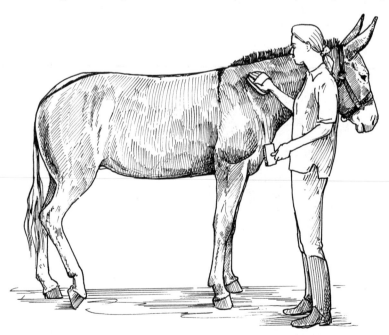

Most donkeys love to be groomed. It's a great way to bond with your long-eared friend.

Don't overdo bathing. Unless you show, a few times a year is usually sufficient.

if he gets annoyed. Also, with your free hand touching him, he always knows where you are and is never surprised.

✕ Wear boots or sturdy shoes. Should your donkey step down on your boot, don't try to jerk it free. Instead, shoulder him over and off your foot.

✕ When working on his legs, squat and stay alert; never sit or kneel under even the gentlest of equines.

✕ While brushing your donkey's face or ears, stay alert lest he accidentally whack you with his head.

✕ To safeguard your donkey's health, don't lend or borrow grooming tools.

Bathing

Bathing removes dust and oils from your donkey's coat, so you won't want to bathe him more often than necessary. If he's dark colored, not show-ring bound, and otherwise well groomed, you'll probably want to do it only a few times each year. If you show white donkeys, spots, or lighter roans and gray-duns, you'll need to bathe them more often.

When you bathe your donkey for an occasion, be sure to do it a day or so in advance and then plan on investing lots of pre-event grooming, especially body brushing and hand or towel rubbing, to restore his natural shine.

Bathe your donkey in a safe location and never on a day so cold that you wouldn't want to swim outdoors. Assemble the necessary tools in advance: a hose attached to a (preferably) warm-water source, horse shampoo (for light colors, choose one with whitening agents), conditioner or detangler and silicone coat spray, two buckets of warm water, a small towel or body sponge, a rubber curry, and a sweat scraper. Tie your donkey securely, using a quick-release strap or a slipknot in a tie rope you know won't bind when wet.

If your donkey isn't accustomed to being hosed with water, start slowly and begin low on his body by gently saturating his hooves, then his legs, gradually soaking

HOME GROOMING PRODUCTS

Some of the best grooming products money can buy don't come from the tack shop; you probably have them in your home right now. Just check your medicine chest, cupboards, and refrigerator for some of these products:

- Splash Listerine on your donkey's itchy spots to halt rubbing; mix it half-and-half with rubbing alcohol for a superlative dust-zapping coat wipe or spray; use it to freshen grooming brushes.

- When you open a new bottle of hoof black or hoof dressing, smear its threads with petroleum jelly, and the cap will screw off easily every time. Also, the faintest glaze of petroleum jelly makes a fine eye and muzzle highlighter when showing or photographing your donkey.

- A gloss of olive oil or baby oil is also an excellent highlighter. They make a great equine aftershave, too. Both are superlative hoof polish brighteners. Olive oil soothes mane, tail, and body itches and nourishes donkeys' skin. Baby oil massaged into recently roached manes eliminates that just-scalped look. Work a few drops of olive or baby oil into clean grooming brushes to soften and nourish their bristles. Use them as mane detanglers, too. When bathing dark-colored donkeys, bring out the shine by adding one-quarter cup of either oil to the final warm-water rinse.

- Lysol is an excellent grooming tool disinfectant.

- Heavy-duty hand cleansers like Goop and Gojo deftly remove pine pitch, chewing gum, and tar from donkeys' coats without drying or damaging them.

his entire body — mane and tail swish, too. Don't squirt his face; most donkeys hate it.

Then, using a large sponge or towel and shampoo-laced bucket of warm water, start at his back and work down, liberally soaping him one section at a time. Work the lather deep into his coat using your fingers or a rubber curry. Scrub stained leg or body markings with your rubber curry and full-strength shampoo to whiten them. Rinse from his topline down, making sure each segment is clean before moving on. Don't forget his mane and tail swish.

Be compassionate when washing your donkey's head. Unless you know he likes it, don't spray water in his face. Bucket-bathe him, taking care that soapy water doesn't drip into his eyes, ears, or mouth.

Move on to his tail. Dunk the swish in your second bucket of warm, soapy water and scrub. If it's white or light-colored, you may need to rinse and lather again.

If you like, follow with your favorite mane and tail conditioner applied according to directions.

Because soap residue is itchy, make sure that your donkey is rinsed squeaky clean. Give him a last, gentle, all-over body rinse, then squeegee away excess moisture using a sweat scraper. While damp, treat his coat with an all-over spritz of stain-repelling coat conditioner, then walk him or put him in his stall to dry.

Sheath Cleaning

Sheath cleaning seems a thankless job, but if you own a male donkey, you have to do it. Sebaceous glands lining male equines' sheaths secrete a dark, waxy substance called smegma. Smegma mingles with grime, grit, and shed skin debris, accumulating in the deep folds lining the sheath and in the urethral recess, a small pouch just inside his penis.

Although some donkeys squeak by without having their sheaths cleaned, ignoring the problem is risky business. Most jacks and geldings should have it done every six months and more often if they're elderly or their sheaths or penises have been damaged.

Suppose a donkey's sheath hasn't been cleaned on a regular basis? How would you know he needed a scrub?

✗ He might start rubbing his tail because his sheath was itchy and irritated.

✗ You'd probably notice dark, gummy smegma speckling his inside hind legs.

✗ When he lets down his penis, the pink-skinned parts of his penile shaft might be angry pink.

✗ The penis will probably be scaly or waxy looking.

If he isn't cleaned (and soon) and hard smegma (called a "bean") in his urethral pouch continues to accumulate, he'll find it increasingly harder to urinate. His penis might become inflamed. Bacterial infection could set in; then you'd find pus in the folds of his sheath, and his penis might ooze foul-smelling fluid.

Because the inside surface of the equine sheath is deeply folded (to allow the extension of the animal's dropped penis) and because, inflamed, it's capable of swelling to many times its usual size, the lining might expand to form a restrictive collar, trapping his penis inside or outside his sheath. With his penis trapped inside, he can urinate, but with difficulty. Urine might accumulate, causing still more infection and inflammation. Trapped outside, fluid can accumulate inside his penis, causing it to swell. Fluid seeping from his sheath and penis will make the organ increasingly raw and tender and its covering thick and dry. Eventually the skin dies and he might never be able to retract his penis again.

So it's important to keep him clean — but not *too* clean. Some breeders wash their jacks' penises before and after breeding, using antiseptics like chlorhexidine and Betadine. Clean is keen, but overuse of antiseptics destroys beneficial as well as harmful

Stain Chasers

When spotted or light-colored donkeys need a bit of stain removal, here are a few things to try:

- Sponge stains with undiluted whitening shampoo, then scrub hard using a clean dry sponge, followed by a wet terrycloth towel and a rinse.
- Glob on a thick paste made of Bon Ami cleanser, baby powder, baking soda, cornstarch, or powdered denture cleaner moistened with lemon juice or water. Scrub it in using a fingernail brush, and then let it dry. In a few hours or even the next day, loosen the crust with a damp sponge, rinse, and scour with a dry, rough-textured towel.
- Spot-clean stains with white vinegar, lemon juice, witch hazel, or rubbing alcohol. Alcohol can parch a donkey's skin, so use it judiciously. Spray or sponge on solutions, then scrub with a medium-stiff brush or clean dry towel.
- Prevent everyday stains by frequently spritzing your donkey's coat with stain-chasing silicone spray.

BRAY SAY: *"I remember when I was a boy I brayed as often as I had a fancy, without anyone hindering me, and so elegantly and naturally that when I brayed all the asses in the town would bray; but I was none the less for that the son of my parents who were greatly respected; and though I was envied because of the gift by more than one of the high and mighty ones of the town, I did not care two farthings for it; and that you may see I am telling the truth, wait a bit and listen, for this art, like swimming, once learnt is never forgotten"; and then, taking hold of his nose, he began to bray so vigorously that all the valleys around rang again.*

— Sancho Panza in Cervantes' *Don Quixote* (1605)

bacteria. Warm, soapy water followed by a clear-water rinse is sufficient.

Smegma accumulation is also a predisposing cause of squamous cell carcinoma, a malignant cancer of the penis and sheath. Sheath cleaning is *important*.

You don't need prepackaged sheath-cleaning products to clean your donkey's sheath; mild soap works just as well (not detergent, and not fragrance-laden body cleansers — just plain soap). Ivory is the traditional horseman's standby. Until 1991, Proctor and Gamble manufactured pure Liquid Ivory, but today's Liquid Ivory is an altogether different formula. Your best bet for an inexpensive, gentle sheath-washing solution is old-fashioned castile or Ivory bar soap worked into lather in your hands, but read the label before you buy. Proctor and Gamble markets two bar soaps. The one marked "Ivory Soap 99 44/100% Pure: It Floats" is the real McCoy. Or opt for the commercial alternative: packaged sheath cleaners for horses.

You'll also need a palm-size sponge, two buckets of warm (not hot) water, and, if your donkey will tolerate it, a warm-water hose without a nozzle. If he hasn't been cleaned for a while (or never), have petroleum jelly or olive, baby, or mineral oil standing by.

If you're squeamish, wear a close-fitting pair of latex gloves (floppy rubber gloves won't do). If you aren't especially fastidious, forget gloves; it's easier to feel what you need to and to grasp your donkey's penis without them. Though smegma is nobody's eau de cologne, it's not that terribly nasty, and you can scrub with antibacterial hand soap when you're finished.

If your donkey has never had his sheath cleaned or if he has and you know that he objects, recruit someone to help to restrain him. Herbal or homeopathic relaxants like B-Kalm, Quietex, or Calm and Cool could help. Or if your donkey has been unusually resistant to having his sheath cleansed before, ask your veterinarian to tranquilize him. Xylazine or acepromazine tranquilizers are a boon because as he relaxes, his penis will naturally drop.

Tie your donkey high and safely to an immovable hitching station, with your assistant standing by to restrain or reassure him as needed. Stand close to your animal's left side. Unless you're absolutely certain he won't cow-kick, stay well forward of his flank. If he kicks, be prepared to shoulder your weight against his barrel and shove him off balance to avert the blow. If your donkey objects, be patient but firm. While a few donkeys never learn to accept this process, most learn to enjoy or at least tolerate sheath cleaning.

HOW-TO
Cleaning the Sheath

1. Begin by soaking your donkey's sheath, inside and out, with a warm-water-saturated sponge. Liberally lather your hand or sponge with bar soap or squeeze a big dollop of commercial sheath-cleaning product into your palm, and then carefully enter the sheath. Work the lather or gel into the deep folds of the inner sheath, then withdraw and wait four or five minutes for the solution to soften and unstick the goo inside. Reenter the sheath and gently pick off any smegma that easily comes away, rinsing frequently to dislodge more debris.

Soak the sheath, inside and out, with a warm, wet sponge to loosen smegma.

2. Unless your donkey likes this process (some do), he's probably reeled his penis as far up into the nether regions of his sheath as it will go. Bolster your courage, reach in there, and grasp it as best you can. Don't tug; just hold on patiently. When he seems reasonably comfortable with this bizarre development, begin s-l-o-w-l-y inching his penis out of the sheath.

If a smegma bean is lodged inside the urethral pouch, gently remove it.

3. When it's extended as far as he'll allow, sluice it with warm soapy water, swabbing off loose scales and smegma as you go. Check the blind pouch above the urethral opening, located inside the tip of his penis. Gently tease out the smegma bean lodged there (if he has a bean). When everything seems slick and tidy, flood his sheath and penis with warm, clean water; only then let go of his penis. He'll quickly reel it in. If he'll allow it, finish by inserting a slow-flowing warm-water hose into his sheath to flush away traces of lingering residue.

Rinse penis thoroughly with warm, clean water.

If your donkey's sheath and penis were especially messy, you might not have gotten him as clean as you'd have liked. If so, reach back in his sheath and slather on a thick coating of petroleum jelly or olive, mineral, or baby oil to soften and loosen the smegma you missed. Clean his sheath again in two or three days.

Pliny the Elder on Asses

Part of Pliny the Elder's 37-volume *Natural History* (written about AD 77) was a medical encyclopedia. Some of the "cures" were pretty bizarre by modern standards. Since we're on the subject, here are a couple of his formulas involving donkey genitals:

- *The ashes of an ass's genitals, it is thought, will make the hair grow more thickly and prevent it from turning grey, the proper method of applying it being to shave the head and to pound the ashes in a leaden mortar with oil for application. Similar effects are attributed to the genitals of an ass's foal reduced to ashes and mixed with urine some nard, or fragrance, being added to render the mixture less offensive.*

- *To restore potency: The right testicle of an ass is taken in a proportionate quantity of wine or worn attached to the arm in a bracelet; or else the froth discharged by that animal after covering, collected in a piece of red cloth and enclosed in silver . . . Salpe recommends the genitals of this animal to be plunged seven times in boiling oil, and the corresponding parts to be well rubbed therewith.*

Winter turnout rugs designed for horses and ponies neatly fit donkeys of comparable size.

Now that your donkey is squeaky clean, occasionally monitor his sheath for smegma accumulation. When it begins to build and you can peel small clumps away with your fingers, it's time for another scrub.

When cleaning your donkeys' sheaths, check your jennies' udders, too. Nonlactating brood jennies or other jennies with large, slack udders sometimes amass impressive globs of smegma between their teats. If your jenny scratches her tail because of gooey smegma accu-

'Ear's a Tip

Cold-Weather Strategies. Donkeys require additional calories to generate body heat during cold spells. Know your donkey's needs and stockpile emergency feed to last for at least a week and preferably longer. Calculate bedding needs and add an adequate supply to your emergency cache. Make certain that your barn's first-aid kit is fully stocked, and stash away extra pharmaceuticals you'd need should your veterinarian be unable to reach you.

mulation, gently wash her udder with sheath-cleaning solution.

A final tip: Some folks delay the inevitable by smearing the insides of their animals' sheaths with olive, mineral, or baby oil once a month or so. Don't substitute petroleum jelly for routine maintenance; it collects too much dust and debris.

Freeze Factor

Donkeys evolved in the hot, arid regions of the world, so under normal circumstances they withstand summer's heat remarkably well. Subzero temperatures, freezing rain, and snow are another story.

Be prepared. When the weatherman promises an ice storm or two feet of snow, you won't have time (or possibly the resources) for last-minute fixes, so for your donkeys' comfort and safety, be ready before winter's first big storm.

Winterizing the Barn

When preparing a barn for winter, remember that adequate ventilation is vital. Don't button your barn up tightly — eliminating drafts is sufficient.

Snugly winterize heated areas such as tack rooms and wash racks by caulking and weather-stripping windows and doors and filling cracks. Service all heaters and well pumps. Check well-house insulation and apply heat tape where needed. Make repairs, give your barn a thorough cleaning, and check fences *before* first snow. Seasonal chores aren't fun when it's 10 degrees below zero and a gale force wind is gnawing at your frozen toes.

Cold-Weather Care

If your donkeys wear winter clothing (and they should if they venture out in a storm), hang out blankets to air and examine them to see if repairs or replacements are needed. It can be difficult to buy quality equine clothing locally during the height of snow season.

During the winter months, free access to water is essential. It takes six times as much eaten snow to generate an equal amount of water, and consuming cold substances lowers body heat. Warm water is best, especially for old, young, lactating, or debilitated donkeys. If you're watering only two or three animals, you could carry warm water from the house. However, if you top off frozen buckets with extremely hot water to help thaw ice, stand guard to prevent the recipient from drinking until enough ice melts to cool the water a little bit.

For larger groups drinking from communal water troughs, a tank heater or an automatic heated water bowl is a sound investment. Plug it into an extension cord equipped with a ground fault circuit interrupter (GFCI) to prevent shocks. When in use, check tank heaters often. A tripped GFCI could mean a frozen water supply, as can a heater fished out and dumped on the floor by equines (which is likely to happen if you keep young horses with your long-eared friends). If your animals play tank heater hockey, build a tank cover. Drill a central hole through a partial sheet of plywood and install this over the tank with an opening at either end, weighting it with rocks, bricks, or whatever is needed to keep it in place. When freeze-up arrives, you thread the heater's cord through the hole, safely out of reach of inquisitive muzzles. Even using a

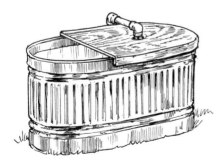

Playful youngsters can't dump their tank heater on the ground nor chew the cord and shock themselves with a lid and a covered tank heater cord installed.

GFCI-equipped model, you may want to fit an exposed cord with a length of PVC pipe to thwart chewing and prevent shorts and shocks.

Keep the tank clean. Scoop out organic matter daily. To clean the tank, bail water into buckets and dump them someplace that won't create a dangerous ice slick for your donkeys to navigate. If spills happen, put down plenty of bedding material to freeze in with and texturize the resulting ice. Ice slicks can also be sprinkled with rock salt, sand, or kitty litter for better traction.

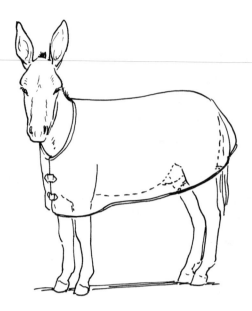

To measure for a blanket, measure your donkey from the center of his chest, almost to the center of his tailbone.

Leg straps around the hind legs help to keep a turnout rug in place.

Riding Out a Storm

If you board your donkey and must camp at the barn to care for him during a storm, stockpile enough convenience food, drinking water, paper products, and personal needs to last for the duration. Add flashlights with spare bulbs and batteries, cozy blankets or a comfy sleeping bag, a battery powered radio or TV, a cell phone, and amusements such as books or a favorite game.

Using Turnout Rugs

Donkeys aren't designed to live in cold, wet climates. If you live where phrases like Alberta clipper and Siberian express make sense or where winters equate with cold, torrential rains, consider blanketing your donkeys when they venture outdoors in the cold and damp. They'll be more comfortable, and they'll expend less energy on staying warm.

Choose the right type of donkey blanket. Fitted numbers called "stable blankets" won't do as outdoor wear. Because they have seams, stable blankets aren't fully waterproof. What you need are called "turnout blankets" or more traditionally "turnout rugs." Most turnout rugs are unfitted, one-piece garments, but if they're seamed down the center of the back, the seams are taped and sealed.

When buying a turnout rug, read the tags. Make certain that the one you choose is waterproof (not water-resistant) and breathable (never use foam-lined blankets; they don't breathe). A donkey wearing a nonbreathable blanket is like you wearing an old-fashioned plastic raincoat when it's hot outside. It's uncomfortable, and sweating inside a blanket is as just as bad as getting wet.

No one seems to manufacture rugs specifically with donkeys in mind; fortunately, it doesn't matter because garments made for horses and ponies fit donkeys of similar sizes and shapes. Use a seamstress' tape to measure for a turnout rug. Hold one end of the tape at the center of your donkey's chest, then measure along his side, back almost to the center of his tailbone; that's the size blanket he needs.

Choose a blanket suited for your climate. When we lived in central Minnesota, our donkeys wore turnout rugs rated for serious midwinter cold; here in Arkansas, a waterproof turnout sheet (an unlined blanket) when it rains does the trick.

Turnout rugs, by definition, come with stretchy elastic leg straps that help to keep the blanket in place. Choose a model with detachable leg straps; you can remove them to launder the rug and replace them if the stretch gives out before the garment does.

To blanket a donkey, fold the back half of the blanket forward over the front, then gently place the blanket on the donkey's back a little farther forward than it goes and slide it back into place. Smooth down the back half and fasten the chest strap. Next, fasten the surcingles that secure the rug around his middle, and only then should you fasten the leg straps. To prevent the straps from chafing your donkey's legs, secure one leg strap, then move to the other side of your donkey and run that leg strap through the first one before snapping it in place.

Remove your donkey's blanket once a day to make certain that nothing is rubbing his skin. Check his shoulders carefully, because that's where most blanket rubs occur (if he's sensitive, fit him with Lycra shoulder "underwear" to prevent this from happening). He'll appreciate being groomed then, too.

Keep in mind that good turnout rugs aren't cheap, so treat them well and make them last. Hang them up between wearings and keep them clean. Here are some additional tips to consider.

- ✖ Few home washers and drainage systems can cope with bulky, hair-encrusted turnout rugs. Instead of laundering them at home, brush off as much hair, mud, and manure as you can and run

them through front-loading commercial washers at your favorite laundromat. Bulky donkey clothing needs room to agitate, so don't overload the washers. If an item is truly filthy, you may need to send it through a second wash. When you're finished, swab the washer's interior with a damp cloth, and then run a complete cycle to flush away hair and debris.

- ✖ Before machine-washing turnout rugs, remove detachable leg straps and surcingles that could otherwise wrap around the machine's agitator and destroy the machine.

- ✖ Launder your donkey's outdoor clothing in warm water unless a care tag directs otherwise. Don't use harsh detergents; they can permanently damage synthetic fabrics, and detergent residue irritates many animals' skin. Choose pure soap flakes (read the label), pet shampoo, or a commercial horse garment cleaner such as Nature's Blend Horse Blanket Wash or Rambo Rug Wash.

- ✖ Never machine-dry turnout rugs; heat shrinks many fabrics and can damage some synthetics. Line dry or hang donkey togs on a fence or stall partition. Indoors, aim a fan at the garment to hasten drying.

- ✖ New Zealand–style rugs with canvas exteriors and turnout rugs made of especially bulky materials aren't washer friendly. To launder one of these items, hang it on a fence or stretch it out on a clean hard-surfaced floor; use a plastic curry comb, stiff brush, broom, or shop-vac to remove hair, manure, and muck, and then douse it with a hose or power washer and scrub, using any of the above-named cleaning solutions. Do the inside of the rug as well, rinse thoroughly, and air dry.

- ✖ An easy, inexpensive way to clean a bulky rug is to take it a car wash. Secure it to the building's floor-mat clips and power wash away. When the item is saturated, hand scrub it with your favorite

BRAY SAY: *There is no character, howsoever good and fine, but it can be destroyed by ridicule, howsoever poor and witless. Observe the ass, for instance; his character is about perfect, he is the choicest spirit among all the humbler animals, yet see what ridicule has brought him to. Instead of feeling complimented when we are called an ass, we are left in doubt.*

— Mark Twain, *Pudd'nhead Wilson*

soap solution, blanket wash, or animal shampoo. Power rinse, scrub the other side, and then rinse again. To save even more time and money, dump rugs into an empty horse tank or plastic manure basket, presoak them in lukewarm sudsy water, then power rinse them at the car wash.

✗ Restore a turnout rug's water-resistant finish by treating it with products like Nikwax Synthetic Rug Proof or Nikwax Canvas Rug Proof.

✗ To prevent it from mildewing, never store any type of equine clothing until it's bone dry. Store rugs or other bulky items in zippered bed-blanket bags, in heavy-duty trash bags with the tops sealed shut, or in covered plastic storage boxes; choose the latter if your blankets are stored in the barn or garage, where mice often gnaw through bags or cardboard boxes. Never add mothballs; mothball residue is toxic, it irritates donkeys' skin, and moths aren't attracted to blankets made of synthetic fibers (and nowadays most waterproof, breathable ones are) anyway. If your blankets have woolen linings, store them with plenty of mesh or cheesecloth bags of naturally moth-repellent herbs such as bergamot, hyssop, sage, or tansy. Cedar needles or shavings and eucalyptus leaves repel moths, too.

BRAY SAY: *Make yourself an ass and you will soon have every man's sack to carry.*

— Danish proverb

Cleaning Hooves

Buy donkeys with sound hooves, feed them right (to avoid laminitis), and keep their hooves clean and trimmed. If you do these things, hoof care is a simple process. The basis of proper hoof care is keeping your donkey's feet clean. To do this, pick out his hooves at least every few days (daily cleanings are better). Trained donkeys rarely mind picking up their feet; untrained donkeys are a completely different story. I'll show you how to clicker train a donkey to accept this important procedure in chapter 12. For now, let's assume that your donkey is already trained. To clean a donkey's hooves:

1. Unless you know that he'll stand patiently while you pick out his hooves, halter and tie your donkey or ask someone to hold him.

2. Holding a hoof pick in your right hand, start at the donkey's left front shoulder. Face your donkey's hindquarters and run your left hand from his shoulder down his foreleg to his pastern. Say "leg," "hoof," or "up" to tell him you want to pick up his foot. When he does, support the hoof in your left hand. (Don't hold his pastern; hold the hoof as close to the toe as you comfortably can.)

It's best to clean hooves every day.

3. Begin at the heel and work forward, prying out debris and taking care to remove anything stuck in the cleft of the frog or in the deep groves that border it.

4. When the bottom of the hoof is reasonably clean, use the scrub brush to finish the job. When you're finished, say "okay" and place the hoof back on the ground; don't just drop it.

5. Move to the left hind leg. Place your left hand on the donkey's hip and run it down the leg to the pastern and give your donkey his "pick up your foot" verbal cue. As you pick up the foot, step forward so that the donkey's leg is extended out slightly behind him. Support the hoof with your left hand, pick the hoof, and set it down as before.

6. Proceed to the donkey's right hind leg, then his right front. That's all there is to it!

Two things to keep in mind: Don't allow the donkey to lean on you (farriers hate that), and don't hold his leg too high or too far out to the side. Put yourself in your donkey's place; if it seems uncomfortable, don't do it.

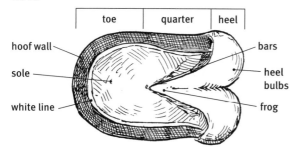

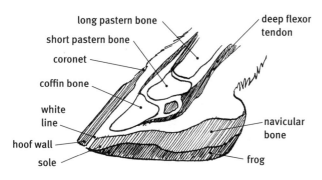

A good donkey farrier is a jewel of great price. When you find one, do your best to keep him.

Finding and Keeping a Top-Notch Farrier

Some donkey owners trim their donkeys' hooves themselves, but it takes skill, special tools, and a very good back to do it correctly. In most cases it's better to hire a farrier (a person who shoes equines and trims their hooves) to do the job for you.

To find a good one, ask horse and donkey owners in your locale whom they recommend. Ask why they recommend this person and if there are things about the farrier's service that they don't like. Also ask if there are farriers you should avoid. As is true of all kinds of tradesmen, some farriers are better than others. A good farrier:

✗ Takes pride in his work and does it well.

✗ Is patient and not prone to erupting in violent anger.

✗ Keeps appointments.

✗ Phones to let you know if he (or she) is running more than an hour or so late (this is often unavoidable, especially if you're scheduled for an afternoon appointment).

Another good way to find a competent farrier is to ask your vet. Vets and farriers work hand in hand in the treatment of serious hoof disease and when treating equines with lower leg injuries that require special shoeing. If you choose the farrier your vet recommends, you'll know that they're compatible and that the farrier knows his trade very well.

You might also find a certified farrier through the American Farrier's Association (AFA) (*see* Resources). Farriers jump through hoops to earn certification, so if you hire one, you're getting the best. There are three levels of American Farrier's Association certification:

INTERN CLASSIFICATION. This person has graduated from horseshoeing school or has apprenticed with a qualified farrier. He or she must pass a written test and a practical exam based on shoeing two feet (judged on hoof preparation; shoe preparation; and nailing, clinching, and finishing). He must also submit and discuss a shoe collection that demonstrates that he knows how to alter keg shoes for specific purposes. Earning an Intern Classification certificate doesn't imply that the holder of it is a qualified farrier (yet), but that he or she is working toward full certification.

CERTIFIED FARRIER. The certified farrier has all of the above qualifications plus a year or more of practical experience. He, too, must pass detailed written and practical tests and present a more specific horseshoe collection for the judges' scrutiny.

JOURNEYMAN CERTIFIED FARRIER. This is the American Farrier's Association's top rating. To earn it the candidate must have passed the Certified Farrier exam, have two years or more of practical experience, and pass extremely stringent written and practical shoeing exams.

Questions to Ask

Once you find a farrier you'd like to work with, arrange to meet him when he's working in your area to get a feel for his ability and ask a few pertinent questions:

1. *Do you work on donkeys?* Not all farriers do. Most donkeys are small (thus a strain on the farrier's back), and they have a reputation for being difficult to trim and shoe. Untrained donkeys are especially loath to have their hind feet picked up, and when it happens, they often ratchet the picked-up leg back and forth with surprising speed and force in an attempt to shake off the "thing" that snatched up their leg. Some farriers are happy to work with donkeys, others do it but charge extra, and some won't do donkeys on a bet.

2. *What are your rates?* Expect to pay a fair price for your farrier's services (allowing for regional differences, $25 to $50 is the current going rate to trim all four feet of a donkey). Also ask if the farrier charges mileage to come to your farm; many do if you have only one or two animals who require their services.

3. *When do you want to be paid?* Some farriers operate on a monthly billing system, but most expect payment when services are rendered.

4. *How do you schedule clients?* Should you call when you need him, or will he let you know when he's working in your neighborhood so that you can schedule an appointment? Or can you arrange for him to trim your donkeys at a mutually agreed upon set interval, say every six to eight weeks? If so, will he call to confirm? Should you?

5. *Does someone have to be present when you're working on my donkeys?* It's always a good idea to be around, but some farriers require it.

Do Donkeys Need Shoes?

The simple answer is no; donkeys have tough, resilient hooves that under most circumstances never need shoeing. However, there are always exceptions. If you ride or drive your donkey so hard or so often that his rate of hoof wear exceeds his rate of hoof growth, he'll need shoes. Also, if you ride on rough, extremely rocky terrain, he may need shoes to protect the soles of his feet.

The Original Eeyore

The original Eeyore was a stuffed toy donkey owned and loved by Christopher Robin Milne, the young son of English playwright Alan Alexander (A. A.) Milne, who immortalized Eeyore and several of Christopher Robin's other plush toys in two immensely popular children's books, *Winnie-the-Pooh* (1926) and *The House at Pooh Corner* (1928).

Illustrated by E. H. Shepard, the books became such beloved classics that after the author's death, Disney Studios purchased rights to produce several full-length feature cartoons and many cartoon shorts featuring Pooh, Eeyore, and their friends, as well as merchandising rights to Winnie-the-Pooh memorabilia. Disney's Eeyore is far less cynical than Milne's version and is even sometimes depicted smiling, something the original Eeyore would probably never do.

Milne describes Eeyore as an "old grey donkey," a quiet, gloomy guy who lives in a tumbledown stick house called the House at Pooh Corner, located in "Eeyore's Gloomy Place" at the southeast edge of the Hundred Acre Wood. Throughout the Pooh books, Eeyore soldiers on despite his melancholy outlook on life. He can write; he authored a poem and signs his name (eoR) to the farewell resolution given to Christopher Robin at the close of *The House at Pooh Corner*. He is good at the game of Poohsticks, and his favorite food is thistles.

The "real" Eeyore's birthday was December 25, 1921 — the day he was given to young Christopher Robin. However, for more than 40 years, fans in Austin, Texas, have celebrated Eeyore's Annual Birthday Party around May 1, with a gala celebration featuring live music, lots of good food, commemorative merchandise, and a maypole.

Christopher Robin's Eeyore is about 18 inches tall and 27 inches long. He now resides, along with Winnie-the-Pooh, Tigger, Piglet, and Kanga, in a climate-controlled showcase in the Central Children's Room at the Donnell Branch of the New York Public Library, where he and his friends are viewed by more than three-quarters of a million visitors per year.

Eeyore collectibles are always in hot demand. On a typical day an eBay visit turns up roughly 2,000 auctions and more than 5,000 Buy Now collectibles including (to name but a few) Eeyore Pez dispensers, a huge array of clothing for all ages, hundreds of figurines, books, watches, bracelet charms and other jewelry, collectors' pins, coffee mugs and dishware, and more toys than you can possibly imagine.

Did You 'Ear?

- The ancient Egyptian word for donkey was *eeyor*. The name came from the sound of a donkey's bray.

- Eeyore in Swedish is *Lor*.

The Care and Feeding of Farriers

Once you find a farrier you like, you need to do your best to keep him. If you treat your farrier right, he's able to do a better job on your donkeys' hooves and he'll be happy to keep coming back to your farm.

Most important is to teach your donkeys to stand quietly and to pick up and hold up their feet. Farriers are not trainers — that's your job — and if your farrier has to wrestle your donkeys to trim their feet, expect to pay extra for the service. Other things to do include:

- Before calling for an appointment, prepare an accurate list of precisely what you want to have done. If you add more work when the farrier arrives, he may not have time to do it all. If he does it anyway, he'll run late the rest of the day.
- Have your donkeys ready to trim when he arrives. That means that they should be clean (especially legs, bellies, and any other parts that come in contact with the farrier), wearing halters, and in an area where they're easy to catch.
- Provide a safe, uncluttered, well-lit area under cover where your farrier can work.
- Arrange a sitter for the children and pen the barn dogs in an empty stall. The farrier doesn't want anyone to get hurt, including himself.
- Offer to hold your donkeys while he works. If he prefers to have them tied, provide a sturdy tying ring set at withers height or taller.
- Offer a mug of coffee or cocoa in the winter or a glass of soda or ice water when it's hot. If the farrier is there over lunch or supper, offer him a sandwich.
- And please, have your payment ready when the work is done; don't make him ask for it.

Can This Hoof Be Saved?

Sooner or later you're bound to meet a neglected donkey with overgrown, misshapen, "sultan's slipper" feet. Donkeys' tough hooves don't chip off and wear down the way horses' do, so left untrimmed they can grow to alarming lengths, curling up at the toe as they grow.

Donkeys with hooves like this are sometimes foundered (they suffer from advanced, chronic laminitis). Even with skillful, long-term trimming, donkeys with hooves like these are rarely sound enough to ride or drive, but they still make loving pets.

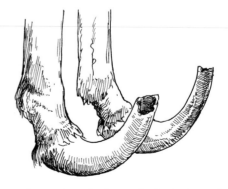

This poor donkey has a bad case of sultan's slipper feet.

Turning into an Ass

In *The Adventures of Pinocchio,* written by Carlo Collodi in 1883 (and made into a movie by Walt Disney in 1940), the hero is turned into a donkey, along with many boys who do nothing but play all day. Here is how the novel describes their transformation:

"They had hardly finished speaking, when both of them fell on all fours and began running and jumping around the room. As they ran, their arms turned into legs, their faces lengthened into snouts and their backs became covered with long gray hairs. This was humiliation enough, but the most horrible moment was the one in which the two poor creatures felt their tails appear. Overcome with shame and grief, they tried to cry and bemoan their fate. But what is done can't be undone! Instead of moans and cries, they burst forth into loud donkey brays, which sounded very much like, 'Haw! Haw! Haw!'"

CHAPTER 9

PARASITE CONTROL

The ass knows well in whose face it brays.

— Dutch proverb

All donkeys have gastrointestinal worms and sometimes lungworms, too. Donkeys also host external pests that can make their lives unpleasant — itchy, painful things like ticks and flies and lice. It's not that hard to keep these pests in check, but it takes diligence on a donkey owner's part to do it correctly. Here are some things that you should know.

Internal Parasites

Internal parasites (most of us simply call them worms) cause a variety of problems for our donkey friends. The simplest ones include weight loss, dull coat, poor appetite, mild colic, and an itchy backside. If those aren't addressed, more serious symptoms surface, such as persistent diarrhea, anemia, susceptibility to infections, nonhealing sores, coughing, and significant or recurrent colic.

If things get really out of hand parasites can cause pneumonia; emaciation; debilitating diarrhea and colic; and gut emergencies such as torsion, telescoping bowel, and gut perforations. Some of these emergencies can result in death. However, the milder signs are the ones

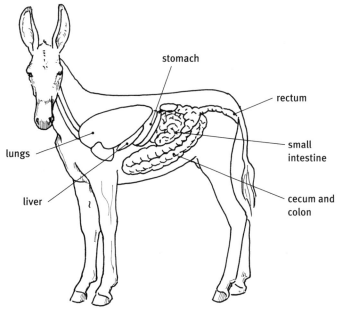

133

most likely to occur with well-kept donkeys like yours. Below is a quick overview of some common parasite irritations:

- ✖ Worms that live in a donkey's intestines irritate the gut; this eventually leads to diarrhea and colic.
- ✖ Pinworms lay their eggs around the donkey's rectum, which leads to intense itching.
- ✖ Tapeworms can accumulate in a small area and attach to the gut lining by way of strong mouthpieces; this irritates the gut lining, and a lot of worms attached in one area cause blockages and sometimes death.
- ✖ Bots attach to and damage the lining of the stomach.
- ✖ Parasites that travel through the tissues cause swelling and inflammation.
- ✖ Bloodworms can cause swelling of the arteries they navigate, leading to blocked blood vessels.
- ✖ Ascarid larvae damage the lung and windpipe, causing coughing and pneumonia; adults live in the gut and sop up nutrients from their host's feed.
- ✖ Small strongyles chomp chunks of tissue from the lining of the intestine and feed on blood; this leads to diarrhea and anemia.
- ✖ Tapeworms trigger life-threatening colic.

Sound scary? It is. But before we discuss what you can do about it, let's take a closer look at these invading critters.

Did You 'Ear?

According to Tibetan herdsmen interviewed by Sir Charles Bell in *The People of Tibet* (1928), when a donkey inadvertently dines on poisonous aconite, the cure involves slitting his ears and pinching his bottom.

A Rogues' Gallery of Parasites

Read the following descriptions to get acquainted with the eleven most common parasites and the symptoms they are likely to cause in your donkeys.

ASCARIDS (roundworms). These large white worms are as big around as a pencil and up to 14 inches long. By the time a donkey is four or five years old, he's usually built up immunity to this worm; however, ascarids can be a major problem for foals beginning as early as two to three months of age. Migration to the lungs leads to coughing, fever, and pneumonia. Larger worms can clog the intestine, causing impaction colic and death. Because ascarid eggs have a thick, resilient shell, they can survive in the environment for years.

Symptoms: Dry cough, nasal discharge, loss of appetite, poor appearance, diarrhea, weight loss, potbellied appearance, lethargy, and colic.

BOTS. During the summer and autumn months, bot flies lay tiny yellow nits (eggs) on the underjaw, shoulders, legs, and necks of donkeys and other livestock. When the host's warm, moist mouth encounters eggs, the eggs hatch and the larvae burrow into the animal's gums and tongue. In time they migrate to the stomach, where they attach to the stomach wall, causing inflammation and ulceration. Severe infestations lead to digestive problems, infections, and permanent damage to the stomach. After 8 to 10 months in the stomach, bot larvae detach and exit via the host's manure.

BOTS

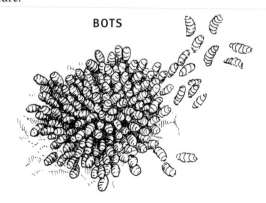

Symptoms: Colic and other signs of gastrointestinal distress; larvae in the mouth can cause ulcers on the tongue and gums.

HAIRWORMS. As their name implies, hairworms are extremely thin; they measure only ⅓ inch in length. Hairworms damage villi, tiny, fingerlike projections attached to the stomach lining, thus decreasing the host's ability to digest food. Severely damaged villi and underlying blood vessels can cause bleeding into the intestines; bleeding may lead to anemia. Foals are especially vulnerable to hairworms, and can get them by eating infected grass. Hairworms are most likely to be a problem for equines kept with sheep and goats.

Symptoms: Dark, smelly diarrhea and lack of condition.

INTESTINAL THREADWORMS. These hairlike parasites are about ⅓ inch in length. Donkeys can be infected in three ways:

1. Through swallowing larvae in the environment
2. By larvae passing through the donkey's skin (in which case they migrate to the lungs, where they cause bleeding and respiratory problems, and then up the windpipe where they are coughed up and swallowed)
3. When foals nurse infected milk from their dams

Symptoms: Diarrhea, respiratory distress, poor appetite, anemia, weakness, and failure to thrive.

LUNGWORMS. These slender worms are up to 2¾ inches in length. Once ingested, larvae pass through the walls of the intestine and into the circulatory system, where they are carried to the lungs. Surprisingly, lungworm infestations have little negative effect on donkeys, nor do donkeys typically show symptoms of lungworm infestation. In horses, lungworms cause severe coughing, difficulty breathing, and weight loss due to lack of appetite; horse foals can die from lungworm infection.

NECK THREADWORMS. Threadworm larvae enter the donkey's body via the bite of an infected fly. They occur in the ligaments of the neck and forelegs. Adult worms are long and coiled; females are up to 12 inches long and males about 2 to 3 inches. Infection causes soreness and swelling in the neck, as well as the suspensory ligaments and flexor tendons of the forelegs. Microfilaria sometimes infiltrate the lens of the eye and cause swelling of the cornea and sometimes blindness.

Symptoms: Skin lesions, hair loss, itching, swelling of the fetlocks, and lameness.

PINWORMS. Donkeys become infected with pinworms by ingesting eggs in contaminated food or water; they occur in donkeys of all ages but are more common in younger animals. Female pinworms are 3 to 6 inches long; males are considerably smaller, about ⅓ inch long or less. Pinworms occur in the large intestine, colon, and rectum. While they do minimal damage to the digestive tract, pinworms cause intense itching of the anal area.

Symptoms: Suspect pinworms if your donkey rubs the root of his tail, sometimes abrading his skin.

LARGE-MOUTH STOMACH WORMS. Maggots ingest stomach worm larvae as they feed; about three weeks later, the maggots develop into infected adult flies. Flies deposit the larvae in skin lesions and on the donkey's nostrils, lips, and other naturally moist areas. When the donkey ingests these larvae, worms mature inside his stomach. If larvae remain in the wound, they sometimes create weepy, expanding wounds called "summer sores" in horses and "jack sores" in asses. These sores frequently become chronic, nonhealing wounds that easily become infected. Larvae deposited around the eyes can cause conjunctivitis leading to blindness. Larvae in the stomach mature into 1-inch adults; their eggs are passed in manure.

Symptoms: Internal stomach worms cause diarrhea and sudden weight loss. Jack sores, as well as previously clean wounds that become enlarged and covered with a reddish yellow tissue, indicate larval deposits under the skin.

LARGE STRONGYLES. These are the most destructive and damaging of all internal parasites; they range from ¾ to 2 inches in length. One type, *Strongylus vulgaris,* passes through the donkey's bloodstream, damaging arteries and creating the potential for blood clots and aneurysms. *Strongylus equinus* produces cysts in the liver, pancreas, and intestines. *Strongylus edentatus* causes swelling of the liver and the abdominal lining. Adult worms attach themselves to the cecum and large colon to suck blood, and this can lead to bleeding ulcers. Infection can also lead to weakening of blood vessel walls; severely damaged walls can burst, causing immediate death.

Symptoms: Fever, poor appetite, weight loss, colic, lethargy, a rough hair coat, and intermittent bouts of diarrhea and constipation.

SMALL STRONGYLES. Much less harmful than large strongyles, these parasites live within the intestinal wall, potentially causing colic and intestinal ulcers. It is not uncommon for animals to have large numbers of small strongyles, which adversely affect gut motility. They range in size from ¼ to 2 inches long. Twenty-nine species of small strongyles are found in the United States.

Symptoms: Chronic mild colic, weight loss, diarrhea, and loss of appetite.

TAPEWORM. Eggs develop in an intermediate host, the oribatid mite; when swallowed, depending on the species ingested (there are three in North America), the worms mature within the host donkey's stomach or small intestine. Rather than being composed of long chains of segments as is true in tapeworms infecting other species, equine tapeworms are pumpkin-seed-shaped parasites roughly 1 inch long and ½ inch wide. They have four suckers that enable them to attach to their host's intestinal lining, where they absorb vital nutrients. No studies have been done on tapeworms in donkeys, but according to a nationwide survey of veterinarians conducted in 2002 and 2003, from 2 percent of the horses in California to 98 percent in Minnesota are infected with tapeworms. Heavy tapeworm infestations are associated with colic and intestinal blockages.

Symptoms: Tapeworms generally cause no outward symptoms, making it especially important to deworm using a product containing praziquantel, one of the few chemical agents effective against equine tapeworms.

Fecal Exams

To discover how many parasites your donkeys might have and which are involved, have EPG (eggs per gram) fecal tests run on your animals.

You'll need a fresh fecal ball or two from each donkey you want tested. The key word is *fresh,* so stand by with a labeled plastic sandwich bag turned inside out on your hand; when the donkey delivers, pick up a dropping (preferably one from the top of the heap that didn't touch the ground), turn the bag right side out around the dropping, and there you are!

Worms are very rarely visible in droppings, so after processing the manure, your vet will view it under a microscope to identify what kind of eggs are present. Then he'll count the number of eggs per species that he finds in one gram of prepared sample. By counting the types and numbers of eggs present in the fecal sample, your vet can recommend the right deworming agents. Counts of fecal eggs per gram indicate the degree of parasite infestation on a farm or within a herd. The fecal exam is a cost-effective follow-up to deworming to determine whether the dewormer has worked, so it's a good practice to have another fecal egg count run about two weeks after deworming.

Lay people can learn to run their own fecal counts as well. It takes some special equipment and a little practice, but you can do it. I learned via the Fias Co Farm Web site (*see* Resources).

Husbandry Hints

There are a number of things you can do to help lower parasite contamination on your farm. Consider implementing these options.

1. Dispense hay and grain from proper feeders instead of off the ground.
2. Provide a clean water supply free of manure contamination.
3. Remove manure from stalls, field shelters, pens, paddocks, and pastures at least once a week. Spread it on cropland or other ungrazed areas, compost it, or sell it as fresh garden enhancer, but don't pile it where your donkeys can reach it.
4. Mow and chain harrow pastures to break up meadow muffins and expose parasite eggs and larvae to the elements.
5. Rotate pastures. When you move equines to fresh grass, allow other livestock (such as sheep, goats, llamas, alpacas, or cattle, none of which are susceptible to the same species of internal parasites that horses are, nor are equines susceptible to theirs) to graze the field they're leaving, thereby interrupting the life cycles of any equine-specific parasites dwelling there.

BANISHING BOTS

Bot fly egg removal is a chore, but for your donkey's well-being, removing them before they hatch is infinitely better than letting nature run its course. Here's what you need to know before you begin:

- The common bot fly cements 87 percent of her pale yellow eggs on donkeys' forelegs, each near the tip of a single hair shaft. They need sudden heat and moisture to hatch; the perfect vehicle is a donkey's warm, moist tongue.

- The throat bot deposits her darker yellow eggs deep in the hair under a donkey's jaws; they hatch in a week without stimulation.

- Hairy and beelike, female bot flies dart about, hovering, depositing eggs. Bot flies can't bite or sting; in fact, during their short adult lives they don't feed at all. After a four- to seven-day egg-laying frenzy, female bot flies die, each leaving 500 to 1,000 eggs in her wake.

- Handpicking common bot fly eggs is a tedious job that takes forever. Save yourself the aggravation and use some type of removal tool. Contour-curved bot egg knives scrape them off with ease, but Styrofoam-like bot blocks (Farnam's Slick 'n Easy Grooming Block is a good one), fine-grit sandpaper, and safety razors work well, too.

- It's harder to spot throat bot fly eggs deep in a donkey's underjaw hair, but if you suspect they're in there, look. Handpick them or smear the area with petroleum jelly to suffocate existing larva and deter adult flies.

- If you can't remove eggs promptly, it's not enough to scrub with hot water, kerosene, or vinegar (as some people do). Spray them with a commercial, equine-friendly larvacide so they don't hatch. A bonus: Larvacides discourage further egg-laying forays.

- Plump, rounded eggs contain larvae; deflated ones have already hatched. Act quickly, and zap them while it counts!

6. To prevent overgrazing and reduce fecal contamination, avoid overstocking pastures.

7. Remove bot fly eggs from donkeys' hair coats to prevent ingestion; do it every day.

8. Rotate deworming agents, not just brand names, to prevent chemical resistance.

9. Set up a deworming schedule and stick to it.

Factors such as local climate, the season, soil conditions, and the number, age, and type of equines using a facility all need to be factored in when formulating a parasite-control program. Consult your equine vet or County Extension agent for management tips particular to your locale.

Demystifying Dewormers

I can't say this often enough: It's best to discuss deworming and dewormer with your vet, rather than attempting to wing it alone. This is especially important now that some parasites are becoming resistant to certain deworming agents; if you do the job incorrectly, you will not only give your donkey something that does very little to reduce his worm load — you can also add to resistance buildup on your farm. That said, let's look at the basics.

There are dozens of brands of dewormers on the market, but all rely on one or more of these chemical agents:

IVERMECTIN. Effective against large strongyles, small strongyles, pinworms, ascarids, hairworms, large-mouth stomach worms, bots, lungworms, and intestinal threadworms. It's a very safe product. Ivermectin is available in paste and feed-through products. Many horse owners prefer to dose their animals orally using injectable Ivermectin products for cattle; it's more economical than using horse products, but because this is off-label usage, if you want to try it you need to discuss it with a vet. Ivermectin products labeled for horses include generic-brand ivermectin pastes, Zimecterin,

Equimectrin, Equell, Rotectin, and IverCare, as well as IverEase (an in-the-feed product).

MOXIDECTIN. Controls large strongyles, small strongyles, pinworms, ascarids, hairworms, large-mouth stomach worms, and bots. This gel dewormer melts in a donkey's mouth and thus prevents spit-out and waste. It should not be used in foals under six months of age, debilitated animals, or those suspected of having large parasite burdens. Moxidectin is presently available only in one form: Quest Gel.

FEBENDAZOLE. Works against large strongyles, small strongyles, pinworms, and ascarids. It's the active ingredient in Safe-Guard Paste. Safe-Guard is an extremely safe product; worm resistance to this deworming agent is a problem in some areas.

PIPERAZINE. A very mild agent somewhat effective against large strongyles, small strongyles, large-mouth stomach worms, and pinworms. It's available as applesauce-flavored Farnam Alfalfa Pellets in-the-feed dewormer.

OXIBENDAZOLE. Effective against large strongyles, small strongyles, pinworms, and threadworms. It's very safe. Buy it as Anthelcide EQ Paste.

PRAZIQUANTEL. Controls tapeworms and roundworms. It's sold as Equimax, Zimecterin Gold, ComboCare, and Quest Plus Gel (with moxidectin).

PYRANTEL PAMOATE and **PYRANTEL TARTRATE.** This combination, available only in pastes and pour-on, controls large strongyles, small strongyles, pinworms, and ascarids; it's also effective against tapeworms when double-dosed. Dewormers containing pyrantel pamoate and tartrate include generic Pyrantel, Exodus Paste, Rotectin P, TapeCare Plus, and Strongid P pastes as well as Liqui-Care P and Primex Equine

Liquid Wormer (both can be dosed orally or poured on feed).

PYRANTEL TARTRATE. The active ingredient in dewormers, designed to be added to concentrates once or twice a day. Daily dewormers include Strongid C, Strongid C 2X, Continuex, Strongyle, Pellet-Care P, and Equi Aid CW.

Get the Most Out of Deworming

There's no sense deworming unless you do it correctly. Dewormer smeared on your shirt or plopped on the ground does no one any good. And underdosing leads to chemical-resistant parasites — a growing concern all over the United States. Here's what you need to know before you begin.

- ✗ Always read dewormer packaging and follow the instructions.
- ✗ Know the weight of each animal to ensure that he gets the correct dose.
- ✗ Never underdose as this could lead to resistance.
- ✗ Deworm all equines in the herd on the same day with the same active ingredient.
- ✗ Have fecal counts run on all newcomers and deworm them while they're in quarantine; have fecals run again two weeks later and deworm again based on their results.
- ✗ Deworm donkeys 48 hours before moving them to clean pasture.
- ✗ Keep accurate records of your deworming.

HOW-TO
Paste Deworm a Donkey

Deworming your donkey needn't be stressful (or wasteful) if you follow this easy protocol.

1. Check the syringe (the tube of paste or gel deworming product) to make certain that everything is in working order, you've set the knurled knob correctly, and that the knob doesn't slip. Sometimes they do, and you don't want to accidentally give a Miniature Donkey an entire tube of paste.
2. Clean everything out of the donkey's mouth, especially lingering grass or hay. Dewormer paste gets hung up on these particles and makes it easy for the donkey to spit out the paste. If

'Ear's a Tip

Play It Safe. Lice dropped or pulled from the host die in a few days, but because discarded eggs may continue to hatch over a two- to three-week period during warm weather, premises should be disinfected before being used for clean stock.

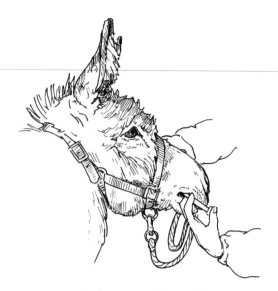

Place the syringe in the corner of the mouth and insert it as far as you can.

you don't want to remove stuff from the donkey's mouth, wait until he finishes eating it. Don't omit this step!

3. When the donkey's mouth is clear, hold the syringe in your left hand while standing on the donkey's left side, facing the same direction that he is facing, or in front of him.

4. If he's a Miniature and you're tall, reach across his neck, grasp him under the jaw, and lift his head. If he's taller than you are, reach under his neck and place your right hand on the bridge of his nose.

5. Steady the donkey with your right hand while you place the syringe in the corner of his mouth (there are no teeth there) and insert it as far as you can. Point it upward because you want to squirt the paste up onto the base of the tongue, not under it.

6. Quickly depress the plunger. Stay calm.

7. Hold the donkey's head up, preferably with his nose higher than the rest of his head, until he noticeably swallows. Do not be tricked. Equines can hold paste in their mouths, waiting to spit it out, for much longer than you might think.

8. Dispose of empty tubes responsibly. Dogs love to chew them up and certain deworming agents are toxic to some breeds of dogs. If a partial tube remains, clean the tube, recap it, and store the remainder in the refrigerator.

Lice

Lice spend their entire lives on their host. Both immature and adult stages suck blood and feed on skin. Horse lice are host specific, so they infest only horses, ponies, zebras, donkeys, hinnies, and mules.

Louse-infested donkeys have dull, matted coats. They rub themselves against any sort of solid object, resulting in raw skin or loss of hair. The constant crawling and piercing or biting of the skin causes nervousness in hosts. In severe cases, loss of blood to sucking lice leads to serious anemia.

There are two types of lice:

1. Sucking lice: these pierce the host's skin and suck blood

2. Biting lice: these have chewing mouthparts and feed on particles of hair and scabs

Lice are generally spread via direct contact, often when infested animals join an existing herd. Populations vary seasonally. Most sucking and biting lice proliferate during autumn and reach peak numbers in late winter or early spring. Summer infestations are fairly rare. Wintertime infestations are usually the most severe.

HORSE-SUCKING LOUSE

HORSE-BITING LOUSE

To see if a donkey has lice, part the hair on his face, neck, ears, topline, the base of his tail, and his tail swish to examine his skin (use supplementary lighting if you do this indoors). Biting lice are active, so you'll probably see them moving through his hair; sucking lice move more slowly, and you'll usually find them attached and drawing blood.

Treatment, generally with an over-the-counter residual pesticide designed for livestock, is needed whenever an animal scratches and rubs to excess. Louse control is difficult because pesticides kill lice but not their eggs. Because eggs of most species hatch 8 to 12 days after pesticide application, retreatment is necessary two or three weeks following the first application.

Louse treatments come in dusts, wettable dusts, and permethrin-based pour-ons. Because the percentage of active ingredient in commercial pour-on formulations varies from 1 to 10 percent (our vet recommends a 2 percent product for donkeys), it's important that the formulation is approved for the animals being treated. A wipe-on formulation of permethrin is available for lice control on horses and their kin, including donkeys.

Mites

Several types of tiny mites cause skin conditions, collectively called mange, in equines. Itch or mange mites feed on the surface skin or burrow into it, making minute, winding tunnels from ⅛ to 1 inch long. Fluid discharged at the mouth of each tunnel dries and forms scabs. Mites also secrete a toxin that causes intense itching. Infested donkeys rub and scratch themselves raw. Frequent grooming is important for early detection.

Confirmation of a mite infestation is usually by microscopic examination of skin scrapings or excised mites, so get your vet involved early on. Mites spread from donkey to donkey by direct contact or by the use of common grooming tools and tack. Infestations are highly contagious; if one donkey has mites, your vet may advise you to treat your whole herd.

Mosquitoes

Of the 2,500 or so mosquito species identified worldwide, roughly 200 of them are found in North America. Mosquitoes carry serious equine diseases such as equine infectious anemia; eastern, western, and Venezuelan encephalitis; and West Nile virus.

Like most other biting flies, both sexes feed on nectar but only the female drinks blood. She can sense a likely meal up to 110 feet away. Mosquitoes are attracted to carbon dioxide and lactic acid emissions, dark colors, motion, and floral scents. Remember that when you're out among the donkeys at dusk; if you wear dark colors and walk around doused in floral scents, you're going to become mosquito bait!

Mosquitoes require blood protein to produce eggs, which they lay in marshes, ditches, puddles, slow-moving rivers and creeks, lakes and ponds, ornamental pools, stock tanks and horse troughs, buckets, barrels, and junk containing pooled water, like discarded feed pans and old tires. They also live in tall grass and near weed-sheltered stagnant water.

Most North American mosquito species dine an hour or two before and after dawn and dusk, although a few are strictly day or nighttime feeders.

To best control mosquitoes, eliminate their breeding grounds. If something holds water and you're not using

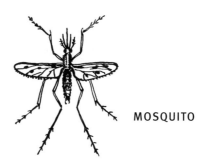

MOSQUITO

it, get rid of it; dump and refill pet dishes, wading pools, birdbaths, livestock watering buckets, horse troughs, and stock tanks every two days; drain pools, puddles, ditches, swamps, and other organic water holders or treat them with biological larvicides.

Horse- and Deerflies

Horseflies (*Tabanus* spp.) are the Humvees of the insect realm, stout bodied and as much as 1½ inches in length. Deerflies (*Chrysops* spp.) are smaller; most are the size of common houseflies. Only female horse- and deerflies feed on blood. They harvest blood the way most biting flies do, by slashing a host's skin with their sharp mouthpieces, then lapping pooled blood.

Most prefer to feed on equines (including donkeys), cattle, and deer. Of the two, deerflies are more likely to bite humans.

Deer- and horseflies are day feeders. They're most bothersome on warm, sunny days and are attracted by moving objects, warmth, and carbon dioxide emissions.

SKEETER BEATERS

It's hard to enjoy riding or working with your donkey when hordes of mosquitoes spoil the fun. You can certainly coat your donkey and yourself with commercial mosquito sprays, but here are some other solutions:

- Investigate the Cashel Company's array of sturdy, comfortable, fine-mesh fly-netting products for equines and equestrians, including a nice line of donkey- and mule-friendly gear (*see* Resources).

- According to the *Annals of Internal Medicine* (June 1998), plants whose essential oils repel mosquitoes for up to two hours include citronella, eucalyptus, lavender, pine, cedar, verbena, geranium, peppermint, cinnamon, thyme, basil, garlic, and allspice. Other sources add bergamot, lemongrass, rosemary, chamomile, tea tree, neem, and cloves. To blend an effective botanical wipe-on, combine in a glass jar 1 ounce of mosquito-repellent oils (in any combination that pleases you) with 16 ounces of apricot kernel, almond, jojoba, olive, or soybean oil. Shake vigorously each time you use it. For a spray, substitute 16 ounces of rubbing alcohol or vodka for the oil.

- A strong tea (infusion) brewed of equal parts chamomile and elder leaves used as a wipe will repel mosquitoes for about 20 minutes.

- One cup of Avon Skin-So-Soft combined with 2 cups of white vinegar, 1 tablespoon of eucalyptus oil, and 1 cup of water makes a dandy mosquito spray. Or add 2 ounces of Skin-So-Soft to 1 ounce of citronella oil, 1 cup of apple cider vinegar, and 1 cup of water.

- Latifah's Fly Spray repels mosquitoes and stable flies, too. Latifah is my Nubian dairy goat; she complains bitterly (and sometimes the milk pail flies across the stall) if anything lands on her while she's fastened in the milking stanchion. We don't want chemicals near our raw milk, so this is what I spritz on Latifah at milking time. Simply mix one 30-ounce bottle of Ultra Ivory Concentrated Dishwashing Liquid with 5 quarts of white vinegar; the recipe is the essence of simplicity, and it works!

Permethrin-based insecticides offer short-term relief to equines and other livestock and can be applied to clothing but not to human skin. Other chemical repellents seldom work.

Because both deer- and horseflies are voracious but flighty feeders, often skipping from host to host to complete a meal, they spread many diseases, including Potomac horse fever, encephalitis in its many forms, and equine infectious anemia. Under sustained attack, donkeys cease grazing and huddle together for protection, sometimes resulting in considerable weight loss.

Standard pheromone-baited flytraps don't attract deer- or horseflies, but *Michael Plumb's Horse Journal*, the *Consumer's Guide* of the horse world, gives Newman Enterprises' innovative visual-attractant HorsePal Horsefly Trap (*see* Resources) two thumbs up for efficiency.

Blackflies

Blackflies, also called buffalo gnats, are the nasty wee cousins of the horsefly. They're found worldwide, more than 1,000 species strong, in boreal to tropical climates.

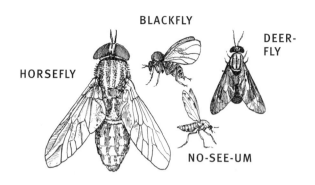

A variety of biting flies may try to be your donkey's stable companions.

Like most biting flies, both sexes feed on nectar, but the female also drinks blood. Her bite is painfully out of proportion to her size. She slashes and sucks pooled blood like the tabanids, injecting an anticoagulant that triggers mild to severe allergic reactions in most humans. The swelling and itch that follows lasts up to two weeks or more. Hordes of blackflies pose a serious threat. Megabitten hosts sometimes die from acute toxemia or anaphylactic shock.

Many blackflies feed on birds, others on mammals, and a few prefer human blood. Most blackflies are daytime feeders, and they rarely venture indoors. Biting varieties target donkeys' ears. DEET-based repellents are only minimally effective.

No-see-ums

No-see-ums, also called sand flies, sand gnats, punkies, and biting midges, are among the world's tiniest biting flies. Most are dark gray or black with spotted wings. Only females suck blood, using the biting flies' standard slash-and-suck mode to obtain it. Bites are initially largely painless, but within 8 to 12 hours, tissues swell and an intense itch sets in.

Most no-see-ums feed at dawn and twilight, from early spring through midsummer. A few species are daytime biters, especially on damp, cloudy days. Both

This postcard from 1912 is one of at least 20 different "tail-o-phone" cards commonly seen on eBay. The title reads, "Drop a nickel, please." When was the last time you made a phone call for a nickel?

types frequent salt marshes, sandy barrens, riverbanks, and lakes. Chemicals won't repel them.

These tiny terrors are highly attracted to dogs and livestock, particularly to their ears and lower legs. Gloss painfully chewed areas with a smear of petroleum jelly and keep animals indoors during prime feeding time.

Ticks

Here in the United States, ticks are responsible for the spread of nine serious human diseases: Lyme disease, Rocky Mountain spotted fever, ehrlichiosis, babesiosis, tularemia, tick-borne relapsing fever, southern tick-associated rash illness (STARI), Colorado tick fever,

ARISTOTLE ON ASSES

Aristotle, an ancient Greek philosopher who lived from 384 to 322 BC, was a student of Plato and teacher of Alexander the Great. Aristotle wrote on hundreds of topics, including zoology. His multivolume *History of Animals (Historia Animalium),* written around 343 BC, describes countless species of fish, birds, mammals, and other animals and their anatomies, including asses. Here are some entries from a couple of the volumes.

The ass suffers chiefly from one particular disease which they call "melis." It arises first in the head, and a clammy humour runs down the nostrils, thick and red; if it stays in the head, the animal may recover, but if it descends into the lungs the animal will die. Of all animals of its kind, it is the least capable of enduring extreme cold.

• • •

The ass of both sexes is capable of breeding, and sheds its first teeth at the age of two and a half years; it sheds its second teeth within six months, its third within another six months, and the fourth after the like interval. These fourth teeth are termed the gnomons or age-indicators.

• • •

A she-ass has been known to conceive when a year old, and the foal to be reared. After intercourse with the male it will discharge the genital sperm unless it be hindered, and for this reason it is usually beaten after such intercourse and chased about. It casts its young in the twelfth month. It usually bears but one foal, and that is its natural number, occasionally however it bears twins.

• • •

The she-ass has milk in the tenth month of pregnancy. Seven days after casting a foal the she-ass submits to the male, and is almost sure to conceive if put to the male on this particular day. The she-ass will refuse to cast her foal with any one looking on or in the daylight and just before foaling she has to be led away into a dark place. If the she-ass has had young before the shedding of the index-teeth, she will bear all her life through; but if not, then she will neither conceive nor bear for the rest of her days. The ass lives for more than thirty years, and the she-ass lives longer than the male.

• • •

When there is a cross between a horse and a she-ass, or a jackass and a mare, there is much greater chance of a miscarriage than where the commerce is normal. The period for gestation in the case of a cross depends on the male, and is just what it would have been if the male had had commerce with a female of his own kind. In regard to size, looks, and vigour, the foal is more apt to resemble the mother than the sire. If such hybrid connexion be continued without intermittence, the female will soon go sterile; and for this reason trainers always allow of intervals between breeding times.

WHERE THE TICKS ARE

Deer Tick/Black-Legged Tick *(Ixodes scapularis)*
Carries: Lyme disease, ehrlichiosis
Location: All states east of the Mississippi; all provinces east of Manitoba; central United States from Minnesota to Texas

Western Black-Legged Tick *(Ixodes pacificus)*
Carries: Lyme disease, ehrlichiosis, babesiosis, leptospirosis
Location: British Columbia, all Pacific Coast states, Nevada, and Arizona

Rocky Mountain Wood Tick *(Dermacentor andersoni)*
Carries: Rocky Mountain spotted fever, tularemia, Colorado tick fever, tick paralysis, Q fever
Location: From eastern Washington, Oregon, and California to the Central Plains; from southwestern Canada to northern Arizona and New Mexico

Brown Dog Tick *(Rhipicephalus sanguineus)*
Carries: Ehrlichiosis, babesiosis (especially canine versions), Q fever
Location: This soft tick lives indoors, where high infestations are possible, throughout the United States and worldwide

Lone Star Tick *(Amblyomma americanum)*
Carries: Rocky Mountain spotted fever, tick paralysis, tularemia, possibly Lyme and leptospirosis, Q fever
Location: Southeastern United States; west to Texas and north to New Jersey

American Dog Tick *(Dermacentor variabilis)*
Also called "wood ticks" in the upper Midwest
Carries: Rocky Mountain spotted fever, tularemia, ehrlichiosis, babesiosis, leptospirosis, Colorado tick fever, bovine anaplasmosis, possibly Potomac horse fever
Location: The eastern and central United States and Canada; all Pacific Coast states, plus Idaho and Montana

TICK AREAS IN THE UNITED STATES AND SOUTHERN CANADA

- Rocky Mountain Wood Tick
- American Dog Tick
- Deer Tick, Western Black-Legged Tick
- Lone Star Tick

Brown Dog Tick: everywhere

and tick paralysis. Donkeys also suffer from Lyme disease, ehrlichiosis, and tick paralysis.

There are two kinds of ticks: hard ticks (Ixodidae family) and soft ticks (Argasidae family). Ticks neither jump nor fly, and their bites cause little, if any, initial discomfort, although many become itchy a day or so later. All ticks will attach and suck blood, but only the female engorges to many times her unfed size.

After feeding, engorged females fall off their hosts and lay their eggs — as many as 20,000 in certain species — in soil. Eggs hatch into larvae ("seed ticks"), then climb on vegetation to seek a host (usually small mammals like mice); after feeding they molt, become nymphs, feed again, and then molt again to finally emerge as adult ticks.

Hard Ticks

Hard ticks have hard plates on their backs; their mouthpieces are visible from above. When they bite, they secrete cementlike saliva that glues them in place. They can go several months without feeding.

Hard ticks are attracted by heat, vibration, shadow, and carbon dioxide emissions. They locate a host through a process called "questing" whereby they perch on vegetation waiting until something happens by; when the tick senses any of these sensations, it extends its front legs and snags the passing host as it brushes past.

In her 6- to 8-day life as an adult, an engorged female hard tick is capable of expanding to two hundred times her unfed weight.

SOFT TICK   HARD TICK

Soft Ticks

Soft ticks have soft, leathery bodies; their mouthpieces aren't visible from above (they're located on the underside). A few species seek hosts by questing, but most are nest dwellers preferring established burrows and nests (including dog beds). Females engorge to 5 to 10 times their size in a few hours, and they look like inflated balloons.

Tick Removal

If you find a tick attached to one of your donkeys (or yourself), don't hesitate — get it off. The longer a tick is attached, the more likely it is to transmit any disease it's carrying. Here are some removal tips:

- ✖ Don't touch a match or cigarette to the tick's backside and don't swab it with gasoline, cooking oil, fingernail polish remover, or petroleum jelly. These ploys can cause it to inject additional toxins, and most of them plain don't work.
- ✖ Because a tick's mouthpiece is barbed, not spiraled, you needn't "unscrew" it during removal.
- ✖ Use needle-nose tweezers, a hemostat, or a commercial tick remover (buy them at pet shops and outdoors stores) to grasp a tick's mouthpiece as close to the host as possible, and then pull slowly and steadily straight back. Be careful; squeezing the tick will cause it to inject additional toxins. Check to make certain that its head is intact. If all or part of its mouthpiece remains imbedded in your donkey's skin, watch closely over the next few days to make certain that the bite doesn't become infected.
- ✖ Dispatch the tick by dropping it into a container of alcohol (soapy water will do in a pinch). If the tick is engorged, simply place it on a hard surface (a rock or board) and step on it.
- ✖ After removal, disinfect the bite with alcohol and dab on a spot of antibiotic ointment.

✘ Monitor the bite site. Watch for abscess formation or any sign of rash.

If your donkey is host to a great many ticks (I once picked 138 Minnesota wood ticks off a standard-size jack, accumulated on pasture overnight), large-scale picking may be the only logical solution. Some suggestions:

✘ Halter and tie him as he might not approve of this process. Then arm yourself with gloves and a widemouthed container of alcohol. Using your gloved fingers, gently grasp each tick near its

head and gently pull straight back, then immerse the tick in alcohol to kill it.

✘ Many tick species mate while the female is feeding. After removing an engorged female, check closely for much tinier males attached in the same location.

✘ Expect to find ticks attached under your donkeys' jaws, on chests, in "armpits" and groins, and especially in and around their manes and tails, and always check between a jenny's teats.

DONKEY DREAMS

What does it mean if you dream about donkeys? Here are some notations from *10,000 Dream Interpretations*, by Gustavus Hindman Miller (1901).

To dream of a donkey braying in your face denotes that you are about to be publicly insulted by a lewd and unscrupulous person.

• • •

To hear the distant braying filling space with melancholy, you will receive wealth and release from unpleasant bonds by the death of some person close to you.

• • •

If you see yourself riding on a donkey, you will visit foreign lands and make many explorations into places difficult of passage.

• • •

To see others riding donkeys, denotes a meager inheritance for them and a toiling life.

• • •

To drive a donkey, signifies that all your energies and pluck will be brought into play against a desperate effort on the part of enemies to overthrow you. If you are in love, evil women will cause you trouble.

• • •

If you are kicked by this little animal, it shows that you are carrying on illicit connections, from which you will suffer much anxiety from fear of betrayal.

• • •

To see children riding and driving donkeys, signifies health and obedience for them.

If you lead one by a halter, you will be master of every situation, and lead women into your way of seeing things by flattery.

• • •

To fall or be thrown from one, denotes ill luck and disappointment in secular affairs. Lovers will quarrel and separate.

• • •

To dream of drinking the milk of a donkey, denotes that whimsical desires will be gratified, even to the displacement of important duties.

• • •

If you see in your dreams a strange donkey among your stock, or on your premises, you will inherit some valuable effects.

• • •

To dream of coming into the possession of a donkey by present, or buying, you will attain to enviable heights in the business or social world, and if single, will contract a congenial marriage.

• • •

To dream of a white donkey, denotes an assured and lasting fortune, which will enable you to pursue the pleasures or studies that lie nearest your heart.

Tick Control on the Farm

The best way to control ticks is by altering their habitat, so mow your pastures and keep barn and outbuilding areas weed and litter free. Discourage visits by rodents, raccoons, and other small mammals (they are the favorite host for many tick species nymphs) by removing attractive food sources such as outdoor pet food dishes and spilled wild bird feed, and eliminating convenient cover like rock heaps, wood ricks, and junk piles.

Also consider adopting a flock of free-roaming guinea fowl. Folks who keep these speckled tick-pickers swear by their efficiency — and their wacky antics will keep you in stitches.

DONKEY TAILS

The Man with the Donk

One of the most heroic donkey men in history was John Simpson Kirkpatrick, who enlisted in the Australian and New Zealand Army Corps (ANZAC) as John Simpson. With the help of several donkeys, Private Simpson saved scores of wounded soldiers at ANZAC Cove, on the Gallipoli peninsula of Turkey, during World War I.

Jack Simpson, variously known by his ANZAC mates as "Murphy" or simply "the man with the donk," was born in 1892 in England. Young Simpson had an affinity for donkeys. Each summer he worked at Murphy's Fair from 7:30 A.M. to 9:00 P.M., giving donkey rides to children for a penny a ride. After work he rode one of the donkeys two miles to home. It's said that the donkeys responded well to his kind, easy manner and that he was deeply fond of them.

After his father died, Simpson sailed to Australia, where he jumped ship and pursued various jobs as miner, stevedore, ferryman, and stoker, always sending money home to his family. At the outbreak of World War I, Simpson joined the army and was sent to Egypt to be trained as a stretcher bearer.

On April 25, 1915, at dawn, he and his mates waded ashore at ANZAC Cove. Of the 1,500 men in the first wave, 745 were dead or seriously injured by day's end. By dawn of the next day, the ANZACs were holding onto a 500-acre chunk of land adjoining the cove. The Turks held surrounding higher ground and looked down on the ANZAC position from nearly every angle.

Simpson, finding himself without other recourse, spied several donkeys who were previously landed and abandoned, so he requisitioned them for his own use. It's unclear how many donkeys there were, but Simpson referred to them at various times as Duffy, Murphy, Abdul, and Queen Elizabeth.

Simpson fashioned a headstall and lead from bandages, and acting as an independent unit, he and his donkey began carrying wounded from the most deadly zone at Quinn's Post, where the opposing trenches were sometimes as little as 15 feet away. They passed through 1½ miles of sniper fire and shrapnel to the beach for treatment and evacuation; on the return trip they carried water for wounded soldiers. He started on some days as early as 6:30 A.M. and continued working as late as 3:00 A.M. the next morning, making 12 to 15 trips.

Simpson cared deeply for his donkeys, whenever possible leaving them under cover while he col-

Simpson stamps issued in Australia in 1965.

lected wounded. To provide them with fodder, he linked up with the 21st Kohat Indian Mountain Artillery, as they had plenty of feed for the mules who hauled their armament. He slept and ate with the Indian troops, who idolized him and called him *Bahadur,* meaning "brave one."

Simpson and his donkeys continued their heroic work until May 19, 1915, the day a Turkish machine gunner opened fire on Simpson, killing him instantly. Private John Simpson Kirkpatrick is buried in Hell Spit Cemetery near ANZAC Cove. He was awarded a number of service medals, and at least five statues in Australia have been erected in Simpson's (and his donkeys') honor. They've been pictured on stamps and banknotes as well.

On May 19, 1997, the Royal Society for the Prevention of Cruelty to Animals awarded their Purple Cross to "Simpson's Donkey, Murphy." The accompanying certificate further states, "and for all the donkeys used by John Simpson Kirkpatrick for the exceptional work they performed on behalf of humans while under continual fire at Gallipoli during World War I."

He (Private Simpson) had a small donkey which he used, to carry all cases unable to walk. Private Simpson and his little beast earned the admiration of everyone at the upper end of the valley. They worked all day and night throughout the whole period since the landing, and the help rendered to the wounded was invaluable.

— Colonel John Monash,
Commander of the 4th Brigade at Gallipoli

If ever a man deserves the Victoria Cross it was Simpson. I often remember . . . that cheerful soul calmly walking down the gully with a Red Cross armlet tied round the donkey's head. That gully was under direct fire from the enemy almost all the time.

— Chaplin-Colonel George Green,
who officiated at Simpson's burial service

When the enfilading fire down the valley was at its worst and orders were posted that the ambulance men must not go out, the Man and the Donkey continued placidly their work. At times they held trenches of hundreds of men spellbound, just to see them at their work.

— E.C. Buley, *Glorious Deeds of Australasians in the Great War,* 1916

CHAPTER 10

HEALTH CARE

*Don't call your brother an ass,
for you are next of kin.*

— Maltese proverb

The trick to having healthy donkeys is to start with healthy animals and to keep them that way. It's fairly simple to keep your herd in the pink if you nip health problems in the bud. It's a good idea to monitor your donkeys at least once a day (more often is better). Count noses and make certain that everyone appears to be healthy and uninjured. Address sickness or injuries immediately; don't wait to see if they get better by themselves. If you don't know what's wrong or you're not positive you know how to treat what ails an animal, don't wing it: *get help!*

Disclaimer

I am not a vet. Because it's illegal to prescribe without a license, I've made no attempt to include specific treatment protocols in these pages. Your first source of such information should *always* be a knowledgeable veterinarian.

Health Care Pointers

First things first — take care of the basics. Follow the guidelines below, and you will be laying the groundwork for a healthy donkey herd.

Maintain your donkeys in clean, dry surroundings. Provide adequate drainage in corrals, barn lots, and shelters, shovel manure out of structures on an ongoing basis, and provide adequate, draft-free shelter from the elements. Don't have too many animals for your facilities, taking into consideration all of the types of animals that you pasture or house together. Overcrowding, especially in housing areas, leads to stress.

Police your pastures, paddocks, and pens to remove accidents waiting to happen (protruding nails, pulled-down wire, sharp edges on metal buildings, hornets' nests, and so on). Set up a weekly or monthly schedule and follow it. Control rats and mice in feed areas. Don't allow barn cats, dogs, or poultry to defecate in hay and other feeds.

Packs of dogs can easily kill miniature equines, especially foals; so if you keep Miniature Donkeys, provide adequate predator protection. Maintain predator-proof fencing (this is difficult, and in some locales nearly impossible), house your Mini donkeys close to human habitation from twilight through midmorning, and/or add livestock guardian animals to each herd's mix.

Try not to stress your animals unnecessarily during handling, hauling, or weaning. Stress is a killer; avoid it whenever you can. Assemble a first-class first-aid kit and learn how to use it. Keep the kit where you can find it in an emergency. As you use an item, replace it. Do an inventory every year; check expiration dates and discard outdated products.

Closely monitor pregnant jennies and be there at foaling time (no exceptions). Know how to assist if necessary, and keep a well-stocked foaling kit on hand.

Choose donkey-friendly feeds; don't improvise. If you choose to feed your donkeys concentrates but don't know how to formulate a safe, nourishing mix, hire a nutritionist to do it for you or feed commercially bagged equine grains. Feed quality hay; discard dusty or moldy bales. Make certain that timid individuals aren't being pushed away from hay bunks; they need their fair share, too. Make all feed changes gradually, allowing your donkeys time to get used to new types or quantities of feed.

Provide copious supplies of fresh, clean water. If you wouldn't drink it, don't expect your animals to drink it either. Add tank and bucket heaters in the winter; place watering facilities in the shade during the hot summer months. You want your donkeys to drink as much water as they can.

BRAY SAY: *When the sheikh's ass dies, everyone goes to the funeral; when the sheikh himself dies, no one goes.*

— Saudi Arabian proverb

Quarantine all incoming animals, and thoroughly disinfect your quarantine area after they leave. Maintain a hospital area for sick animals (use your quarantine pen if it isn't already occupied); don't expect them to tough it out in the main herd.

Have dead animals, especially aborted fetuses, tested to determine cause of death. Remove dead animals and birthing tissues immediately. Don't allow dogs to eat them. If the materials aren't going to be necropsied or tested for pathogens, dispose of them properly by commercial removal, burning, burial, or composting.

Vaccinate for tetanus no matter where you live. In addition, discuss a vaccination program with your vet and add any other immunizations recommended for equines in your locale.

Trim hooves on a regular basis to maintain soundness and prevent hoof deformities.

A livestock guardian dog is a valuable addition to any Miniature Donkey herd.

BUILD A BETTER FIRST-AID KIT

We keep our farm-based first-aid kits (for our sheep, llamas, and goats as well as our horses and donkeys) in three five-gallon plastic food service buckets fitted with snug lids. On the top and both sides of each we've affixed big "Red Cross" symbols using red duct tape so that the buckets are easy to spot when we need them. We keep the buckets in the house in a walk-in closet; they're returned to their place immediately after each use — no exceptions.

In large letters, one bucket is marked "Ruminants" (for the sheep, goats, and llamas), and the one holding our horse and donkey supplies is labeled "Equines." These are organized using resealable plastic bags; each bag is labeled (in big, black letters using a felt-tipped marker) according to its contents' basic uses.

One bag contains wound cleanup and bandaging materials sized for the species indicated on the bucket:
- Gauze sponges
- Telfa pads
- VetWrap self-stick disposable bandage
- A roll of 2½-inch-wide sterile gauze bandage
- 1-inch- and 2-inch-wide rolls of adhesive tape
- A partial roll of duct tape with ¾ to 1 inch thickness of tape left on it
- Two heavy-duty sanitary napkins (they can't be beat for applying pressure wraps to staunch bleeding)
- A sandwich bag of flour or cornstarch (it works better than commercial blood-stopper powder)
- A small bottle of Betadine Scrub
- A bottle of regular Betadine
- A 12-ounce bottle of generic saline solution

Another plastic resealable bag holds hardware:
- Blunt-tipped bandage scissors
- A hemostat (we prefer it to tweezers)

- A flashlight (the flat kind that you can hold between your teeth)
- A stethoscope
- A digital thermometer in a hard-shell case (add an extra thermometer or at least extra batteries in case the ones in the primary thermometer go dead)

A third bag contains basic medicines such as:
- Bach Rescue Remedy
- Our wound treatments of choice (triple antibiotic ointment, emu oil, homemade herbal salve, and Schreiner's Herbal Solution)
- Topical antibiotic eye ointment
- A full tube of probiotic paste

The third bucket (unlabeled) houses emergency equipment shared by all of the animals. It holds a selection of lead ropes and a half dozen halters ranging from our smallest ruminant halter (an alpaca halter that fits Boer kids and our miniature sheep) up to a huge one that fits our Thoroughbred mare; the horse halters are hand-tied rope versions that can be crammed around other supplies to conserve space. When we need something in an emergency, we carry the equipment bucket to the site and simply dump everything on the ground so that we can pick out whatever we need. The bucket also contains a fencing tool and a small length of aluminum electric fence wire for making impromptu fence repairs if needed.

We also store over-the-counter and prescription drugs we'd need in an emergency in separate, easy-to-grab-when-we-need-them plastic baskets in our pharmaceutical refrigerator (a dorm-size model from the used-a-bit store). A separate, scaled-down first-aid kit is kept stowed behind the seat of the truck (see Build a Traveling First-Aid Kit on page 65).

To Be Your Own Vet (or Not)

Unfortunately, it's difficult to find donkey-wise vets in parts of North America. Because of this, some adventurous souls tend to become their own vets. This is not without peril for many good reasons. Here are two:

- ✘ Quite a few equine maladies look alike. Treating for one when your donkey has another could be catastrophic for your long-eared friend.
- ✘ Some of the pharmaceuticals you'd need in order to treat your own animals are prescription drugs, and many of them are extra-label for donkeys. Most vets won't hand them out without seeing your animals; others dispense prescription drugs only to established clients. One way or another, you'll have to get them from a vet.

The most workable solution: Find a vet who's qualified to treat your donkeys (see Finding a Veterinarian on page 52) and establish a relationship, but also learn to address minor problems and routine veterinary procedures yourself.

Wound Care 101

While donkeys aren't as accident prone as their horsy cousins, they are sometimes cut, bruised, bitten, and abraded. Although a vet should treat serious injuries of any kind, it's usually all right to handle minor bangs and boo-boos yourself.

However, always call the vet if:

- ✘ An injury is bleeding profusely
- ✘ The injury is on or close to a tendon or joint
- ✘ A wound is badly contaminated by dirt or other debris
- ✘ You've just discovered the wound and it's already infected
- ✘ It's a puncture wound (these are always best treated by a vet)

Keep a good supply of saline solution on hand for cleaning cuts, bites, or abrasions. If you're out, flush wounds using lots of cool water from a hose. Then apply a simple, mild disinfectant like dilute Betadine solution to kill bacteria left on the wound. Be gentle; don't scrub.

Here's where things get tricky. What to put on this nicely cleaned wound? In many cases, nothing. Clean, open wounds heal better (and faster) than similar injuries coated with lots of gunk. When we do dress a wound, we prefer holistic liquid dressings like Schreiner's Herbal Solution (see Resources) and emu oil, and sometimes we will gloss a wound with a thin coating of homemade herbal ointment if pesky flies won't leave a cut or abrasion alone.

Checking Vital Signs

When contacting your vet or mentor, be ready to provide your sick donkey's vital signs (temperature, heart rate, and respiration) and to describe his symptoms in detail.

TEMPERATURE. The first thing your vet or mentor will ask is, "What is your donkey's temperature?" Elevated temperatures are usually associated with infections or dehydration; a very subnormal temperature can indicate that a donkey is suffering from hypothermia or that he is dying. Unusually high or low temperatures

> ### To the Rescue!
>
> Vet-to-vet consultation is only one of many services offered by the Donkey Sanctuary of Sidmouth in Devon, England. The Donkey Sanctuary's veterinarians have more experience treating donkeys than any other veterinary practice in the world, and they are happy to liaise with vets in the United Kingdom and abroad. If your donkey is injured or sick and your veterinarian doesn't know how to proceed, ask him to contact the Donkey Sanctuary for further information (see Resources.)

are red-flag warnings; these animals need treatment right now.

To take a donkey's temperature you'll need a rectal thermometer. Veterinary models are best, but a digital rectal thermometer designed for humans will also work. Traditional veterinary thermometers are made of glass and have a ring on the end to which you can attach a string. Add an alligator clamp to one end of a length of cord and knot the other to the thermometer; you can clip the clamp to your patient's body hair or tail before inserting the thermometer. That way you will neither lose the thermometer inside the donkey (yes, it can happen) nor drop and break it.

Glass thermometers must be shaken down after every use: hold the thermometer firmly and shake it in a slinging motion to force the mercury back down into the bulb. However, a digital model is the best choice for working with any kind of livestock, including donkeys. Why? It is faster, it beeps when done, and it needn't be shaken down (simply press a button and it's reset).

To take a donkey's temperature, follow these three easy steps:

1. Restrain the donkey. Halter and tie him or recruit a helper to speak to and steady him as you proceed.
2. Lubricate the business end of the thermometer with K-Y Jelly, Vaseline, or mineral oil and insert it about 2 inches into the donkey's rectum.
3. Hold a glass thermometer in place for at least two minutes or a digital model until it beeps.

Check the pulse in the artery under the lower jaw. Count the beats for 15 seconds and multiply by four.

After recording the reading, shake down the mercury (in glass models), clean the thermometer with an alcohol wipe, and return it to its case. Always store your thermometer at room temperature.

HEART RATE. There are several spots where you can take your donkey's pulse, such as behind his left elbow, under his tail at the tailbone, and under his jaw. The easiest way to check a donkey's heart rate is with a stethoscope; simply place it, listen, and count the number of beats per minute. If you don't have a stethoscope you can do it manually instead. To take the pulse under a donkey's jaw, support his head with one hand on his muzzle and use the fingertips of your other hand to

DONKEY VITALS

	ADULTS	YOUNG DONKEYS
Temperature °F	98.8° (97.2–100°)	99.6° (97.8–102.1°)
Temperature °C	37.1° (36.2–37.8°)	37.6° (36.6–37.8°)
Heart rate (beats per minute)	44 (36–68)	60 (44–80)
Respiration (breaths per minute)	20 (12–44)	28 (16–48)

From *The Professional Handbook of the Donkey, Third Edition*, compiled for the Donkey Sanctuary by Dr. Elisabeth D. Svendsen, M.B.E. (*see* Resources).

locate the artery, then press lightly. Count the number of pulses in 15 seconds, and multiply that number by four.

RESPIRATION. Simply watch your donkey's rib cage as it moves and count the number of breaths he takes in 15 seconds and multiply that number by four.

Keep in mind that external conditions can affect your readings. Donkeys' temperatures rise slightly as the day progresses and may be up to a full degree higher on hot, sultry days. Extreme heat and fear or anger elevate pulse and respiration. Take these caveats into consideration before you react; slightly elevated readings are sometimes the norm.

22 Classic Donkey Maladies

Equines are prone to too many illnesses to list them all here, but these are either important ones or the ones you're most likely to see. To learn more, buy a good book or two on equine veterinary medicine; they're worth their weight in gold when you need them.

Choke

A donkey chokes when he has an object (nearly always a wad of poorly chewed food) stuck in his throat. It's a serious matter but not the extreme medical emergency it is when humans choke; this is because the blockage is in his esophagus instead of his windpipe and he can still breathe. Choke generally occurs when a famished or greedy donkey gobbles his food; equines are particularly likely to choke on pelleted concentrates, unsoaked or improperly soaked beet pulp, dry hay cubes, and bulky treats like whole apples or large chunks of carrot.

Symptoms: A donkey suffering from choke may seem distressed, but unlike horses, choking donkeys usually don't panic. Suspect choke if your donkey lolls around, head down and not eating or drinking, while copious amounts of slimy, green goo stream from his nose.

Treatment: Call your veterinarian; don't delay! Then, to prevent future attacks:

* Switch to nonpelleted grains or eliminate concentrates altogether
* Soak dry feeds such as beet pulp and hay cubes before feeding
* Slice or break treats into manageable chunks
* Place obstructions such as fist-size rocks in your donkey's grain pan so that he has to shove them out of his way or eat around them and thus stop gobbling his food

Colic

Colic is defined as severe abdominal pain; it's a clinical sign or a symptom rather than a disease. There are several types of colic and many causes; all are serious and require prompt veterinary intervention. Colic is a

---

Colic Kills!

According to a study sponsored by the Donkey Sanctuary and conducted by veterinary researchers at the University of Liverpool using 4,596 Donkey Sanctuary residents, 93 percent of 807 cases investigated between January 1, 2000, and March 31, 2005, culminated in death. Most (54.8 percent) represented suspected or confirmed impactions of the gastrointestinal tract, with an overall instance of 5.9 colic episodes per 100 donkeys per year. Researchers found that elderly donkeys, donkeys consuming too much feed, thin donkeys, donkeys suffering from musculoskeletal problems or dental disease, and donkeys who previously colicked were at increased risk. They also found that donkey colic lasted significantly longer than colic in horses, sometimes going on for days instead of hours.

major cause of premature death among equines.

Symptoms: Whereas most colicky horses show dramatic signs of abdominal pain (pawing, looking back or nipping at their sides, stomping, and rolling), donkeys, being far more stoic than horses, are likely to keep to themselves, refuse food, and suffer in relative silence. Always suspect colic when a generally outgoing donkey seems unduly depressed, stops eating, and stops making manure. Other symptoms include:

- ✖ Increased respiration
- ✖ Excessive sweating
- ✖ Sometimes (but not always) lack of gut sounds
- ✖ Bright red mucous membranes (check the donkey's gums or inside of an eyelid)

Treatment: If colic lasts longer than 30 minutes, call your vet! In the meanwhile (and until the vet arrives), allow the donkey to rest but prevent any horselike rolling (this can contribute to twisted gut). If your vet doesn't live close by, discuss colic treatment protocols before you need him. He may prescribe a dose or two of Banamine, a powerful prescription painkiller and anti-inflammatory in paste or injectable form, to have on hand to administer while he's in transit.

Prevention: It's better to prevent colic than to try to cure it. Unfortunately, different things cause different types of colic. The most common type, spasmodic colic, is caused by gas buildup in the digestive tract. Things apt to trigger gas-related colic are:

- ✖ Eating spoiled feed or drinking stagnant water
- ✖ Overeating on rich grass, legume hay, or grain
- ✖ Sudden changes in the diet or feeding routine
- ✖ Stress as a contributing factor

Donkeys who ingest sand and dirt while grazing or eating feed directly off the ground may suffer from sand impaction colic; donkeys unable to properly chew their food, as well as particularly wormy donkeys, suffer impactions, too. Other contributing factors include:

- ✖ Feeding excess concentrates to donkeys who don't need it while not providing enough drinking water. This is apt to occur when water

sources freeze during the cold winter months.

- ✖ Jennies sometimes suffer a brief bout of colic after foaling, due to painful contractions of the uterus; foaling colic passes quickly and doesn't normally require veterinary intervention.

Cushing's Disease

Cushing's disease (also called hyperadrenocorticism) is a disease of the endocrine system caused by an abnormality of the pituitary gland, which then causes the body to produce excessive amounts of cortisol — the body's natural steroid hormone. Elderly equines are far more prone to Cushing's than younger ones, but it's not strictly a disease of old age.

Symptoms:

- ✖ Increased thirst, appetite, and/or urination
- ✖ A thick, wavy hair coat in the summertime (complete failure to shed the winter coat)
- ✖ A pot-bellied appearance with loss of topline muscling
- ✖ Chronic laminitis
- ✖ Depression, lethargy

Cushing's disease is fairly common among elderly donkeys.

- A compromised immune system leading to increased susceptibility to respiratory disease, abscessed hooves, skin infections, and periodontal disease

Treatment: There is no cure for Cushing's disease, but several effective drugs are used in the treatment of the disease; your vet will prescribe the best one for your situation. Because stress contributes to the problem, do your best to help your Cushing's donkey to lead an easy, low-key life.

Encephalitis

There are at least six viruses that cause encephalitis (sleeping sickness) in equines:

- Eastern equine encephalitis (EEE)
- Western equine encephalitis (WEE)
- Venezuelan equine encephalitis (VEE)
- West Nile virus (WNV)
- The neurological form of equine herpes virus (EHV-1)
- Rabies

EEE, WEE, VEE, and WNV are transmitted via mosquitoes, which usually pick up the respective viruses from birds and can pass it on to horses, humans, and other vertebrate species. There are effective vaccines available for all four diseases, so if you live where mosquitoes are endemic, use them!

Symptoms:
- Depression
- Lack of muscular coordination
- Weakness
- Circling
- Head tilt
- Paralysis
- Muscle tremors
- Convulsions

Treatment: Because these are viruses, treatment is generally unsatisfactory; most equines who contract encephalitis die.

Equine Infectious Anemia (EIA)

Equine infectious anemia, also known as swamp fever, is a deadly disease caused by a retrovirus and transmitted by bloodsucking insects (especially deerflies and horseflies), saliva, milk, and body secretions; jennies and mares can also transmit the disease to their unborn foals via the placenta. You'll probably never know (or know of) a donkey, mule, or horse infected with equine infectious anemia, yet you should know about it because it's the disease that the Coggins test is designed to detect.

Symptoms: High fever, anemia, weakness, swelling of the lower abdomen and legs, weak pulse, irregular heartbeat, and sudden death.

Treatment: There is no treatment and no vaccine to prevent equine infectious anemia. The United States doesn't have an eradication program because the disease is so rare.

Equine Influenza

Equine influenza (horse flu) is caused by several varieties of influenza virus that are endemic in equines. The disease has a nearly 100 percent infection rate in unvaccinated equines not previously exposed to the virus; the incubation period is one to five days. Effective flu vaccines are readily available.

Symptoms: Fever, dry cough, runny nose, depression, reluctance to eat or drink.

Treatment: With rest and good nursing, most animals recover spontaneously in two to three weeks.

Fescue Toxicosis

Fescue is a common, perennial pasture grass grown throughout much of North America, where it's also widely baled as hay. It's often infected with an endophytic fungus called *Acremonium coenophialum,* which causes catastrophic foaling problems in jennies and mares who eat it.

Symptoms: Fescue toxicosis causes significantly increased gestation lengths, arrested milk production, retained placentas, premature separation of the placenta, increased placental weights and thickness, and the birth of dead or weak foals.

Treatment: Jennies and mares should be removed from fescue pasture and stop eating fescue-laced hay at least 60 days prior to foaling. If that isn't an option, discuss the problem with your vet; a prescription product called Equidone (domperidone oral paste), administered for at least 20 days prior to foaling, eliminates the symptoms of fescue toxicosis in most instances.

Hyperlipidemia

This potentially fatal condition is triggered by a negative energy balance (starvation) or a stressful event that causes an enormous mobilization of fat from the tissues to the liver. The liver is completely overcome by the fat overload and fails, usually resulting in death. Donkeys, miniature horses, and small-breed ponies, particularly obese ones, are all very prone to developing hyperlipidemia if they stop eating for more than a day or two. Jennies and mares (especially if they're pregnant or lactating) are more likely than geldings or jacks to develop hyperlipidemia. Never, for any reason, drastically cut back a donkey's feed in an effort to "diet" him!

Symptoms:
- ✘ Initially: anorexia, lethargy, weakness, and depression
- ✘ Followed by: jaundice, ventral edema, head pressing and circling, and other indicators of liver and kidney failure

Treatment: If your donkey stops eating and you suspect hyperlipidemia, call your vet without delay.

Laminitis

Laminitis is inflammation of the sensitive laminae (layers) of the hooves; chronic laminitis is often referred

This donkey is standing in an extreme laminitic stance. Not all donkeys afflicted with laminitis do so.

to as "founder." There are two sorts of laminae in an equine's hoof:
- ✘ Sensitive laminae attached to the coffin bone
- ✘ Insensitive laminae attached to the inner hoof wall

The function of the laminae is to bind to each other and keep the coffin bone suspended within the hoof capsule, allowing for normal hoof-wall growth over the coffin bone. "Founder" means to send to the bottom (like a ship); this is precisely what happens when the coffin bone sinks down through the hoof capsule due to an afflicted animal's body weight tearing through weakened laminae. Obese donkeys are exceedingly prone to developing laminitis, as are donkeys on overly rich diets or who have gorged on lush, spring grass or rich feed. It is essential for donkey owners to learn to recognize the earliest stages of laminitis (*see* Resources).

Early symptoms: Heat in the hoof wall; increased digital pulses in the pastern; lying out flat, reluctant to rise; hesitant gait ("walking on eggshells"); standing in a sawhorse stance with the front feet stretched out in front to alleviate pressure on the toes and with the hind feet camped farther forward under the body than normal to bear more weight.

Mud Fever

Mud fever (also known as scratches, dew poisoning, greasy heel, and dermatophilosis) is caused by a fungus-like actinomycetes (an organism that behaves like both bacteria and fungus) called *Dermatophilus congolensis*. Some authorities claim that it's present in soil and others say that it isn't, but several conditions must be present for it to take hold and proliferate: The organism has to be present on the animal's skin; the skin must be kept moist; the skin must be damaged in some way (a cut, scrape, or fly bite) for the organism to invade the epidermis layer of the skin (in this case, the pastern, coronary band, and heel areas on a donkey or other equid's lower legs). Equines with unpigmented skin on their legs (areas overlain by white markings) are more prone to mud fever than those with pigmented skin.

Symptoms: Thick, crusty, and inflamed skin. There is usually swelling of the pastern, and the donkey may be lame.

Treatment: Soak crusty areas, then carefully remove the scabs with your gloved fingers or a soft-bristled nail brush (dispose of scabs in an appropriate manner; burn them, bag and place them in the trash, or flush them down the toilet, but don't just drop them on the ground). Allow the area to dry, then apply Betadine or Nolvasan and continue applying it once a day for one full week.

Pneumonia

Pneumonia, a serious and potentially fatal inflammation or infection of the lungs, is caused by any of a host of bacteria, fungi, and viruses that gain a toehold due to environmental factors such as stress, aspiration of drench material, damage caused by lungworm infestation, drafts or being hauled in open trailers, and dusty feed, bedding, or surroundings. Some forms are contagious; others are not.

Symptoms:

- ✘ Loss of appetite, depression
- ✘ Rapid respiration and labored breathing
- ✘ Standing with forelegs braced wide and neck extended
- ✘ Thick, yellow nasal discharge
- ✘ Congestion, coughing, an audible "rattle" in the chest

Treatment: If you suspect pneumonia, call your vet without delay!

Potomac Horse Fever

Potomac horse fever (equine monocytic ehrlichiosis) is a serious disease of equines caused by exposure to a tiny bacterial organism called *Ehrlichia* sp.

Symptoms: Fever, depression, poor appetite, and in most cases, violent diarrhea. Some sufferers develop laminitis, and pregnant jennies and mares may abort.

Treatment: Potomac horse fever is currently treated with tetracycline antibiotics, but the success rate varies depending on the severity of the infection and the timing of the treatment. If you live in an area where equines have been diagnosed with Potomac horse fever, vaccinate your donkeys; the vaccine doesn't necessarily prevent Potomac horse fever, but if a vaccinated animal becomes sick, it reduces the severity of the disease.

Rain Rot

Rain rot, also referred to as rain scald or streptothricosis, is caused by the same actinomycetes (*Dermatophilus congolensis*) organisms that cause mud fever.

Symptoms: In most cases rain rot manifests as small, crusty scabs or slightly raised, matted tufts of hair. The crusts and tufts are easily (though painfully) lifted to reveal pus and raw, pink skin. Lesions generally appear in clumps across the infected donkey's neck, back, rump, and legs, although occasionally infections are far more extensive.

Treatment: Bathe the donkey using antifungal or antibacterial shampoo (they work about equally as well), following instructions to the letter. Soak crusty areas, remove crusts and tufts of hair (as with mud fever, dispose of them properly; don't just drop them on the ground), and allow the areas to dry, then apply Betadine or Nolvasan and continue applying it once a day for one full week.

Rhinopneumonitis

Two subtypes of equine herpes virus EHV-1 cause rhinopneumonitis, also called rhino, rhinovirus, or (more colorfully) "the snots." Rhino causes respiratory disease in equines and abortion in pregnant jennies and mares. Rhino often makes the rounds of young equines when they're brought together in groups. Two types of rhino vaccines are available; one is a modified live-virus vaccine, and the other contains a killed product. Both reduce the severity of, but don't fully prevent, respiratory infections; they are, however, effective against subtype one, the strain that induces abortion in pregnant equines. Jennies and mares should be vaccinated when five, seven, and nine months pregnant, and jennies who haven't foaled by the end of the eleventh month should be vaccinated again.

Symptoms: Moderate depression, fever, watery eyes, nasal discharge, and rapid breathing during the fever stage.

Treatment: Given rest, uncomplicated cases resolve themselves (without treatment) in three to seven days.

Ringworm

Contrary to popular belief, ringworm (also known as dermatophytosis) is not a worm but a fungal infection.

Symptoms: Ringworm lesions form round, dime- to quarter-size crusty patches that, when removed, leave scaly red skin and hair that lifts out in clumps. An infected donkey may have two or three small patches or be covered with lesions from head to toe. Ringworm is highly contagious, and it easily spreads from one species to another (humans included); it's transmitted though both direct (e.g., animal to animal) and indirect means.

Treatment: While self-limiting (ringworm usually resolves itself in six weeks to three months), you must treat it quickly and aggressively lest it spread. The best way is by bathing the infected donkey with an iodine-based shampoo formulated specifically for equine fungus problems (tack shops carry it or you can get it from your vet). Wear disposable gloves! Also, disinfect whatever the infected donkey touches using diluted chlorine bleach solution (¼ cup per gallon of water).

Sarcoid

A sarcoid is a type of nonfatal cancer limited to skin and the tissues directly under the skin. It often appears suddenly and grows quickly. Donkeys are particularly prone to developing sarcoids. The Donkey Sanctuary has prepared an excellent leaflet on sarcoids ("Sarcoid Fact Sheet for Owners") downloadable from their Web site (*see* Resources). The diagram of sarcoid sites (see below), showing the frequency of distribution of sarcoid lesions, is adapted from that resource.

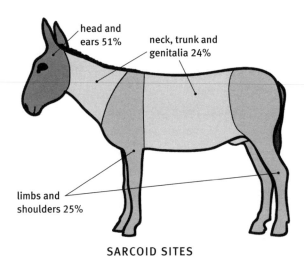

SARCOID SITES

head and ears 51%

neck, trunk and genitalia 24%

limbs and shoulders 25%

Symptoms: There are six main types of sarcoids.

✘ Occult: gray, hairless, often circular or roughly circular areas usually appearing on the face, armpit, inside thigh and groin

✘ Verrucose: gray, scabby, or warty in appearance, and sometimes have small shotlike, solid nodules within them; most common on the face, groin, sheath, and armpit

✘ Nodular: solid nodules of variable size and are common under the skin of the eyelid, armpit, inside thigh, and groin

✘ Fibroblastic: fleshy masses, sometimes with bloody surfaces, anyplace on the body

✘ Mixed: a combination of two or more forms

✘ Malevolent: cords of tumor tissue interspersed with nodules and ulcerating fibroblastic lesions; these are found mostly on the face and elbows

Treatment: Sarcoids can be treated in a number of ways ranging from banding with rubber bands to surgical removal to freezing to chemotherapy or radiation. Treatment is rarely 100 percent successful.

Strangles

Strangles is one of the oldest known equine diseases. It's a highly contagious and serious infection of equines caused by the bacterium *Streptococcus equi.* Strangles is most common in animals less than five years of age and especially in groups of weanling foals or yearlings. Foals under five months of age are usually protected by colostrum-derived passive immunity from their mothers. Recovered animals shed *S. equi* from their noses and in their saliva for up to six weeks following infection.

Symptoms:

✘ High fever, poor appetite, a soft cough, and depression

✘ Thin, watery nasal discharge that quickly becomes thick and yellow

✘ The upper respiratory lymph nodes, particularly the ones between the jawbones, become enlarged

✘ Abcesses form on the lymph nodes; these usually rupture and drain copious amounts of nasty, yellow pus

Treatment: Isolate the patient in comfortable quarters and provide good-quality feed. Your vet may prescribe phenylbutazone (bute) to reduce fever, pain, and swelling. Encourage swollen lymph nodes to rupture and drain externally by applying hot packs to the swollen areas. About 15 percent of sufferers develop complications, often of a serious nature; the rest recover with only supportive care.

Prevention: The best way to prevent strangles outbreaks is to quarantine all equines new to your farm; once you've weathered a strangles storm, you'll probably never make another exception to that rule. A vaccination against strangles is readily available; it doesn't fully protect against the disease, but it lessens its incidence and severity. Ask your vet for details.

Sunburn

Painful sunburns are often a part of life for pink-skinned donkeys kept on pasture.

Symptoms: Crusty patches atop peeling, sore skin.

Treatment: Keep pink-skinned donkeys indoors during the day and turn them out to graze at night. If that isn't feasible, apply sunscreen products made for horses or humans or colorful zinc oxide products to exposed pink skin. Another option: Fit a susceptible donkey with a fly mask that covers his face clear down to his nose (*see* Resources).

Sweet Itch

Sweet itch, also called summer sores or summer eczema, is properly referred to as summer seasonal recurrent dermatitis (SSRD). It's caused by allergic reactions to the saliva of culicoides flies, better known as ear gnats, midges, and no-see-ums.

Symptoms: Fly bites form itchy blisters that cause the donkey to rub himself fiercely, in turn creating areas of significant hair loss, crusting, and scaling on the head, neck, shoulders, back, and tail. Lesions generally heal over winter but recur again the following year.

Treatment: Some veterinarians recommend deworming with ivermectin products at 30-day intervals, feeling that ivermectin taken internally discourages culicoides flies. Your veterinarian may wish to pursue steroid treatments; however, the Donkey Sanctuary recommends against this as they feel that steroids contribute to laminitis problems in donkeys.

Prevention: Culicoides flies breed around ponds and marshes but rarely travel more than a few hundred yards from their breeding areas, so moving susceptible donkeys at least one-quarter of a mile from such areas should dramatically reduce fly exposure. These flies are most active at dawn and dusk, so stabling your donkey an hour or two each side of sunrise and sunset generally helps a lot as well. Use fly repellents beginning early in the spring before fly season starts for the year; pyrethrin pour-ons are especially effective.

Tetanus

Tetanus occurs when wounds are infected by the bacterium *Colostridium tetani*. These bacteria thrive in anaerobic (airless) conditions such as found in deep puncture wounds, fresh umbilical cords, and wounds caused by recent castration. Unless treated very early and aggressively, tetanus is nearly always fatal.

Symptoms:

- ✘ Early on: stiff gaits and anxiety
- ✘ Later: a rigid rocking-horse stance, drooling, inability to open the mouth (hence the common name "lockjaw"), head drawn hard to one side, tail and ear rigidity, seizures, then death

Treatment: If you suspect tetanus, call your vet without delay!

Prevention: All equines, no exceptions, should be vaccinated against tetanus; tetanus is a terrible way to die.

West Nile Virus

West Nile virus (WNV) belongs to the Flaviviridae family, part of the Japanese encephalitis (JE) complex of viruses; it causes encephalitis or meningitis (infections of the brain and the spinal cord or their protective coverings). West Nile virus emerged as a distinct virus about 1,000 years ago and occurs in both tropical and temperate regions of the world. The disease primarily infects birds but has also been known to infect asses, horses, zebras, humans, dogs, cats, bats, chipmunks, skunks, squirrels, and domestic rabbits.

The principal transmitter of West Nile virus is the northern house mosquito (*Culex pipiens*). Mosquitoes are exposed to the virus when they feed on infected birds. Once the mosquito is infected, it may transmit the virus to people or other animals when it bites them.

Symptoms:

- ✘ Loss of appetite, depression
- ✘ Fever
- ✘ Weakness or paralysis of the hind limbs
- ✘ Muzzle twitching
- ✘ Impaired vision
- ✘ Lack of coordination and aimless wandering
- ✘ Convulsions

- ✗ Inability to swallow
- ✗ Circling
- ✗ Hyperexcitability
- ✗ Coma

Treatment: Several preventive vaccines can be used, but there is currently no effective treatment for West Nile virus. Horses vaccinated against eastern, western, and Venezuelan equine encephalitis are not automatically protected against West Nile virus.

White Muscle Disease

White muscle disease, also called nutritional muscular dystrophy, is caused by a serious deficiency of the essential trace mineral selenium. Most of the land east of the Mississippi and much of the Pacific Northwest is selenium deficient; these are the areas where white muscle disease is most likely to occur.

You'll rarely see white muscle disease listed in a lexicon of equine ailments. However, we learned the hard way that foals born in selenium-deficient areas are often afflicted with this malady. Supplementing late-term mares and jennies with selenium can eliminate unnecessary breeding problems and foaling dystocia — and save you from heartache.

Find out whether you are located in a selenium-deficient area before treating animals for white muscle disease. Your best source of information regarding conditions in your immediate locale (especially in areas where selenium levels vary widely from farm to farm) is your County Extension agent.

Symptoms:
- ✗ Foals: stillborn foals, weak foals unable to stand or suckle, tremors, stiff joints, neurological problems
- ✗ Adults: infertility, abortion, foaling dystocia, retained placenta

Treatment: Injections of Bo-Se or Mu-Se (prescription supplements that you can get from your vet) often dramatically reverse symptoms, especially in neonatal foals.

Prevention: All animals, especially breeding stock, raised in selenium-deficient areas should be fed selenium-fortified feeds, have free access to selenium-added minerals, or be given Bo-Se shots under a vet's direction. To prevent foaling problems and protect unborn foals, jennies should be injected with Bo-Se four to six weeks prior to foaling.

Medicating Your Donkey

There are lots of relatively easy ways to "pill" a donkey; but before you try them, be sure to ask your vet these questions:

- ✗ Can this medication be powdered or must it be administered whole? Don't assume that you can grind your donkey's pills and simply add them to his feed; coated pills sometimes have to be delivered to his digestive system intact.
- ✗ If powdering is acceptable, are there safety protocols to follow? Some animal medications can

---

Using Health Care Products

You'll want to get the most out of the vaccines, medicines, and dewormers you pay for, so use them to your best advantage.

When you buy a new product, first things first: Read the label. Check for storage instructions and store it accordingly. Some drugs need to be refrigerated, others kept out of direct light; some should be shaken before use and aren't effective if that step isn't taken. Note any warnings: Certain pharmaceuticals cause abortion if used on pregnant animals.

Reread the label prior to using any product. If a product has been stored incorrectly, discard it. Also check its expiration date before using a drug or vaccine and pitch it if it's outdated, and don't buy the large, economy size if you can't use it up before it goes bad.

be toxic to humans when inhaled or absorbed through bare skin, especially over prolonged periods of time. It's always wise to wear latex gloves while processing medication, but depending on the potential toxicity of the medication involved, your vet may suggest more stringent measures.

Open Wide . . .

The easiest, most efficient way to pill any equine is by hand. To do this:

1. Stand alongside your donkey, facing him. Insert your fingers into the bars of his mouth.
2. When he opens his mouth, reach in and grasp a fistful of tongue. Hold tight.
3. Coax the tongue down and out the side of his mouth. As long as you maintain traction, his mouth will gape open.
4. Taking care not to snag yourself on his teeth, place the pill as far back on his tongue as you can.
5. Now quickly release the tongue. This carries the pill to the back of his throat where he'll gulp it down.

Not crazy about sticking your arm inside that gaping mouth? Neither was the inventor of the balling gun. This plunger-fitted tubular implement is often used to propel pills down large animals' throats. Metal or plastic balling guns can be purchased from veterinary supply catalogs. To use one:

1. Draw back the plunger and insert a pill in the barrel, holding it so that the pill doesn't slide out.
2. Elevate your donkey's head and — with or without extracting his tongue — insert the implement into his mouth up to the base of the tongue and zap the pill down his throat.

Here are some other pointers regarding the balling gun.

✘ Never use horse-size boluses (large-size livestock pills) to dose foals, Miniature or small Standard donkeys, or donkeys with a history of suffering from choke. Don't break the boluses in half; ask your vet for smaller tablets or grind and administer them in feed or by syringe.

✘ Lightly glazing a pill with vegetable oil or lubricating gel just prior to dosing will help it to slide down.

✘ Never insert a balling gun farther than the base of your donkey's tongue. Ramming it into his throat can cause serious, permanent damage.

Grind It Up

If you'd rather feed your donkey his medication or dose it from a syringe, first you'll have to grind up his pills.

How Much Is . . . ?

1 milliliter (1 mL) = 15 drops = 1 cubic centimeter (1 cc)

1 teaspoon (1 tsp) = 4.75 grams (4.75 gm) = 5 cubic centimeters (5 cc)

1 tablespoon (1 tbsp) = ½ ounce (½ oz) = 15 cubic centimeters (15 cc)

2 tablespoons (2 tbsp) = 1 ounce (1 oz) = 30 cubic centimeters (30 cc)

1 pint (1 pt) = 16 ounces (16 oz) = 480 cubic centimeters (480 cc)

Here are some pointers:

- To efficiently reduce pills to granules or powder, use a mortar and pestle (find them at kitchen supply or import shops and from food co-ops that sell herbs) or a small electric coffee grinder. Whichever you chose, reserve your pill powdering apparatus for veterinary use only, or scrub it in scalding, soapy water before processing food for human consumption.
- To rough-crush huge, hard pills before grinding, tuck them into a small paper envelope, stick that inside a plastic sandwich bag, and whack the pills with a hammer. Crush smaller, softer pills between the bowls of two spoons.
- Powder only enough pills for a single dose and process them just before dosing. Unless you're certain that a medication is nontoxic, wear gloves, and avoid breathing the powder if you can.

Down the Hatch

Anyone who's medicated any kind of equine knows how frustratingly fussy some of them are about eating doctored feed. Still, yours can probably be conned into doing it. The trick lies in discovering a carrier food he likes.

- Does he crave carrots or apples? Blend his powdered medication with mashed, cooked carrots or with applesauce and stir the mix into his grain.
- If he has a sweet tooth, add powdered pills to crushed after-dinner mints, pancake or corn syrup, molasses (or molasses whipped with brown sugar), or honey.
- Powdered, fruit-flavored Jell-O or Kool-Aid, half an envelope per medicated feeding, tempts many donkeys. So does yogurt in fruity flavors.
- If you feed your donkey sweet goodies, try this messy but effective trick: Stir powdered medication into a glob of canned cake frosting and let him lick it off your hand.

Does your donkey adore hand-fed treats? Here's a trick that works nearly every time. Quarter an apple and set three of the quarters aside. Cut a deep slash in the fourth quarter, pack it with crushed or powdered pill, and press the edges shut. Show your donkey the apple; make a big deal out of it. Feed him two of the plain quarters, one at a time as fast as he'll eat them, and then slip him the doctored piece, rapidly chased by the third unadulterated quarter.

Here are some other pointers for successfully dosing your donkeys with medicated feed:

- Unless you're positive that your donkey will eat doctored feed, add a dollop of medication/carrier mixture to only a moderate amount of grain (don't overwhelm him). After he cleans up the spiked feed, you can give him the rest of his ration.
- Allow him three hours to consume medication-laced feed. Then if a significant amount remains, remove it and try a different carrier.
- If you're not certain that his medicated grain is being consumed instead of scattered, feed him using a nose bag.
- To make adulterated feed more attractive, allow your donkey a flake or two of roughage and free access to water, but don't let him fill up on hay until after he's eaten his grain.
- Don't whip up huge batches of medicated carrier, planning to store part of it in the refrigerator for later. Grind and blend each dose immediately before feeding.
- Thoroughly blend spiked carrier into your donkey's feed; don't just drizzle it on top.
- Most medicated carriers blend well with sweet feed, and the grain's sweetness helps to mask bitter flavors.
- Plan ahead. Experiment with carrier foods and know which appeal to your donkey. To train him to accept a specific carrier, gradually add a dab more at each feeding until he readily accepts its flavor.

Syringe It Down

If your donkey turns up his nose at medicated feed, try dosing him using a syringe. Grind his pills and fold them into no more than two ounces of runny, adhesive carrier. Good ones include:

- **✗** Finely puréed baby foods like carrots or apple-sauce
- **✗** Fat-free yogurt (it's stickier than the low-fat kind); plain, vanilla, lemon, and coffee flavors work especially well
- **✗** Sugary syrups or molasses
- **✗** Smooth peanut butter liberally thinned with vegetable oil

  Load the spiked carrier into:

- **✗** A catheter-tip irrigation syringe (get one from your vet)
- **✗** A large volume, single-dose paste deworming syringe (scrub it first in steaming, sudsy water)
- **✗** A drenching syringe, if the carrier is runny enough *(see page 165)*

Administer the dose as though you were deworming your donkey with paste dewormer *(see page 139)*, taking care to squirt the medication well back on his tongue. As soon as you've emptied the syringe, grasp your donkey's jaw and elevate his head. Hold it up until he swallows.

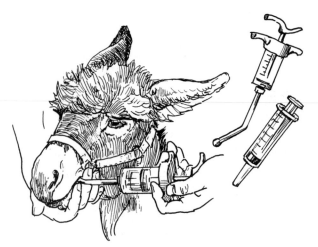

The correct tools make any job easier — in this case for syringing ground medications into a reluctant donkey.

If your donkey resists paste deworming and you know he'd hate being dosed with a syringe, practice while he's well, using a particularly tasty carrier. You'll perfect your technique and he might decide that being dosed is actually quite okay.

Don't Wait, Vaccinate!

Not everyone routinely vaccinates their equines, but they should. At the minimum, vaccinate for tetanus. Depending on where you live and what sort of "bugs" haunt your locale, your vet will probably recommend other vaccinations as well.

Tetanus Toxoid vs. Antitoxin

Toxoid is the vaccine to choose when you're seeking long-term immunity for your donkeys. Toxoid is given in stages: an initial injection, followed three weeks later by a second shot, followed by boosters once or twice a year. Immunity, however, is not immediate, so if an unvaccinated animal is already exposed to danger, anti-toxin is injected to impart immediate immunity.

Foals acquire immunity from disease via their vacci-nated dams' colostrum, so it's wise to boost a pregnant jenny's immunity with toxoid vaccines five or six weeks prior to foaling. This will protect the foals until their own immune systems kick in, at around five months of age. This is a good time to start the foals on a series of tetanus toxoid injections.

To be absolutely certain that they're protected, teta-nus antitoxin is routinely administered when castrating jacks or when any donkey suffers a deep puncture wound. The protection imparted by antitoxin is short-lasting, a week or two at most. If the donkey's wound isn't healed at the end of seven to ten days, he will require revaccination with antitoxin.

Administering Shots

Every animal owner on a budget quickly becomes adept at giving shots. Here are the steps we follow.

1. Select the correct vaccine or other injectable and reread the label. Don't omit this very important step.

2. Use the smallest disposable syringe that will do the job; they're easier to handle than big, bulky syringes, especially for women with small hands. Disposable syringes are inexpensive and readily available; choose disposables over reusable syringes and avoid the work of sterilizing them between uses. Never try to sterilize disposable syringes — boiling compromises their integrity, so it simply isn't wise to reuse them.

3. Choose the correct needles for the job. Most antibiotics and vaccines should be injected intramuscularly (IM), within a major muscle mass, using a 1- to 1½-inch needle; 18 and 20 gauge are easiest to use. A few antibiotics are very thick, and when injected, their carriers make these injections sting; for these, some people prefer 16-gauge needles so that they can inject the fluid quickly before the patient objects.

4. Select enough needles to do the job. You'll need a new needle for each animal, plus a transfer needle to stick through the rubber cap on each product. Using a new needle each time is less painful for the donkey than using a reused, dull needle, and it virtually eliminates the possibility of transmitting disease via contaminated needles.

5. Secure the donkey. Recruit a helper or tie the donkey using a halter and sturdy lead rope.

6. Insert a new, sterile transfer needle through the cap of each pharmaceutical bottle. As you draw each shot, attach the syringe to the transfer needle and draw the vaccine or drug into it, then detach the syringe and attach the needle you'll use to inject the pharmaceutical into your donkey. Never poke a used needle into the cap to draw vaccine or drugs!

7. If you're drawing 3 cc of fluid from the bottle, inject 4 cc of air (to avoid the considerable hassle of drawing fluid from a vacuum), then pull

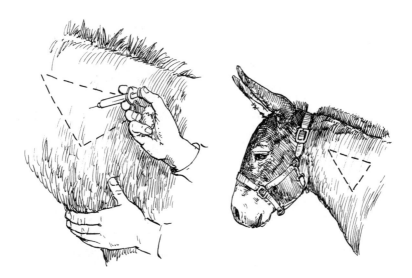

Intramuscular shots are best given into the natural triangular hollow on the donkey's neck.

a tiny bit more than 3 cc of fluid into the syringe. Attach the needle that you'll use to inject the donkey, and then press out the excess fluid to remove any bubbles created as you drew out the vaccine or drug.

8. Select the best injection site. You want to inject the vaccine or drug where it will work well but won't injure the donkey. The preferred site for SQ (subcutaneous injections; injected just under the skin) is into the loose skin of the "armpit." Intramuscular injections are usually given into the thick muscles of the animal's neck.

9. When injecting a relatively large volume of fluid, break the dose into smaller increments and inject it into more than one injection site (for example, no more than 10 cc of penicillin should be injected into a single injection site). If you're unsure whether this is necessary with the product you're using, ask your vet.

10. Swab the area with alcohol (prepackaged alcohol swabs are easy to use). Never inject anything into damp, mud- or manure-encrusted skin.

11. To give a subcutaneous injection, pinch up a tent of skin and slide the needle into it, parallel to the donkey's body. Take care not to push the needle through the tented skin and out the opposite side, or to prick the muscle mass below it. Slowly depress the plunger, withdraw the needle, and rub the injection site to help distribute the drug or vaccine.

12. When giving an intramuscular injection, quickly but smoothly insert the needle deep

DON'T LEAVE HOME WITHOUT IT

Epinephrine, also called adrenaline or epi, is a naturally occurring hormone and neurotransmitter manufactured by the adrenal glands. It was first isolated and identified in 1895 by Napoleon Cybulski, a Polish physiologist, and artificially synthesized in 1904 by Friedrich Stolz. It's widely used to counteract the effects of anaphylactic shock, a serious and rapid allergic reaction that, if severe enough, can kill.

Any time you give an injection to an animal, no matter the product or amount injected, you must be prepared to immediately administer epinephrine to counteract an unexpected anaphylactic reaction. If your donkey goes into anaphylaxis (indicated by glassy eyes, increased salivation, sudden-onset labored breathing, disorientation, trembling, staggering, or collapse), you won't have time to race to the house to grab the epinephrine, and you might not even have time to fill

a syringe. You have to be ready to inject the epi right then.

Many folks who give their own shots keep a dose of epinephrine drawn up in a syringe in the refrigerator. Keep it in an airtight container (we use a clean glass jar with a tight-fitting lid); it will keep as long as the expiration date on the epinephrine bottle. Take it with you every time you give a shot. It may save the life of a valuable or much-loved friend. Standard dosage is 1 cc per 100 pounds; don't overdose as it causes the heart to race.

Previously available over the counter, epinephrine is now a prescription drug and available only through a vet. An alternative some animal owners favor is over-the-counter Primatene Mist sprayed under the animal's tongue. Every 14 or 15 squirts of Primatene Mist contains the same amount of epinephrine as a 1-cc dose.

into muscle mass, then aspirate (pull back on) the plunger ¼ inch to see if you hit a vein. If blood rushes into the syringe, pull the needle out, taking great care not to inject any drug or vaccine as you do, and try another injection site.

13. After injecting one product, you can use the same needle and syringe to inject the next vaccine or drug into the same donkey — but only if you use a new transfer needle.

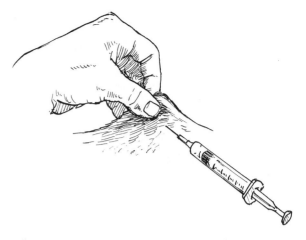

Subcutaneous shots should be injected into a pinch of tented skin.

HOW-TO
Drenching a Donkey

Liquid medicines and some dewormers are given orally as drenches. Drenches can be administered using catheter-tip syringes (not the kind you use to give shots; *see page 166*) and even turkey basters, but the most efficient way to drench is with a dose syringe.

To drench your donkey:

1. Restrain him and elevate his head using one hand under his chin — not too high but enough to let gravity help you a little bit.

2. Insert the nozzle of the syringe between the donkey's back teeth and his cheek; this way he's less likely to aspirate part of the drench.

3. Slowly depress the plunger, giving him ample time to swallow.

When giving semisolid drenches such as paste-type horse dewormers or gelled medications, simply deposit the substance as far back on his tongue as you possibly can. In either case, keep the donkey's nose slightly elevated until he visibly swallows. And be careful — if you stick your fingers between his back teeth you're likely to be accidentally bitten.

Antibiotics: Good or Bad?

Nowadays most everyone knows that antibiotic overuse is a real and rapidly expanding problem. Many people, realizing this, think, "I'm not going to use antibiotics on my animals." However, in some cases avoiding antibiotics simply isn't feasible. If your vet says you have to use them for one reason or another, keep these tips in mind.

Follow his directions to the letter. Use precisely the recommended dosages and complete the series as directed. Otherwise weak pathogens die but "super germs" survive to proliferate.

Because antibiotics destroy good bacteria as well as bad, always follow antibiotic treatment with oral probiotics such as Probios or Fastrack (the leading brand among horse folk) to restore the patient's digestive system to good health.

Queen Victoria's Donkeys

Queen Victoria ruled England for 63 years, longer than any other British monarch in history. Victoria loved donkeys. Throughout her long life, the queen's pets consisted almost entirely of dogs, horses, and donkeys.

According to *The Education of a Royal Princess*, published in 1837, the donkey that her uncle gave her when she was a child was "the greatest treasure she then possessed in the world: the King had never seen it, and with infantine simplicity she believed that she could not pay her Royal Uncle a greater compliment than to visit him on her favorite, 'Dickey.' . . . When the weather was favourable she was constantly to be seen twice a day upon the sands at Ramsgate, in the morning on her donkey, and in the afternoon on foot, always attended by her governess, and one or two men-servants, and sometimes attended by her mother and sister."

Later, as an aging monarch, Victoria frequently visited the French Riviera. On one such visit she decided, as it was getting hard for her to get around on foot, she needed a donkey. One day when she was out in her carriage, she spied a donkey that she fancied; it was hitched to a peasant's cart. Victoria haggled with the peasant herself. When he told her that he paid 100 francs for the donkey, she doubled the price and the little ass became hers. She named him Jacquot, and he served her for many years.

Jacquot was eventually replaced with a pure white donkey, Tewfik, that Lord Wolseley brought her from Egypt. In July 1899, Lord Kitchener was charged with finding "the finest white female donkey in Egypt" to be his mate.

An undated, anonymous book titled *Queen Victoria* describes the queen's donkeys thus: "Jenny, a white donkey, twenty-five years of age, which has been with the Queen since it was a foal. Tewfik, a white Egyptian ass, bought in Cairo by Lord Wolseley. . . . A strong handsome donkey called Jacquot, with a white nose and knotted tail. This donkey draws the queen's chair (a little four-wheeled carriage with rubber tyres and a low step), and has accompanied her to Florence. A gray donkey, the son of the Egyptian Tewfik, carries the Queen's grandchildren."

CHAPTER 11

THE HOLISTIC DONKEY

A living donkey is better than a dead doctor.

— German proverb

A few short decades ago, not many horse and longear owners would consider treating their four-legged friends with "that weird health-nut stuff." Nowadays, complementary therapies have come of age. People like you and me discuss energy modalities as we spoon up chili at the café, old-school horsemen feed herbs and homeopathic calming pastes to their animals, and we can even buy Bach Rescue Remedy off the shelf at Wal-Mart. Why? Because complementary modalities represent a holistic means for us to help our veterinarians to heal our animals in need. And they work!

Herbs

Donkeys respond exceptionally well to herbal medicine. In fact, when exposed to native pasture they harvest their own wild herbs. Who hasn't noticed a donkey nibbling dandelions or thoughtfully savoring a yellow dock leaf? But the fact is, most modern equines lack access to the sort of plant diversity they need to forage healing herbs for themselves. That's where we humans come in.

There are hundreds of herbal products specially formulated for horses sold online, through catalogs,

and at local retail outlets; these work great for donkeys, too. Many are imported from Great Britain and Europe, where feeding herbs to equines is a long-standing tradition.

You can also create herbal remedies at home. Neighborhood food co-ops retail hundreds of bulk herbs to choose from, or you can grow or wildcraft your own. But before dashing off a check for a commercial

Disclaimer

Do-it-yourself alternative and complementary therapies should be used only to treat minor ailments and injuries, to augment conventional veterinary treatment, and in some cases to address behavioral quirks. Life-threatening injuries and illnesses can be treated with these modalities but only under the direction of a holistic veterinarian. If none is available, don't experiment — consult your usual vet. The information in this chapter is for educational purposes only. It is not intended to medically prescribe or diagnose in any manner.

Traveling photographers often carried costumes for children to wear while posing on animals. This boy makes a super soldier! This postcard was sent to Philadelphia, PA, in 1910.

herbal feed additive or setting out to the meadow to gather your own herbs, you must know precisely what you're doing.

Getting Started with Herbs

Until you are very experienced, never substitute herbs for professional veterinary treatment. Even then, never attempt to treat life-threatening injuries or illnesses; call a vet!

There are a number of things to be aware of when working with herbs. First and foremost, realize that donkeys are individuals and that what's right for one isn't necessarily even safe for another. Always monitor your donkeys after feeding or otherwise administering any herb, even externally. If an animal suffers an allergic reaction (such as shivers, hives, or labored breathing), even a slight one, immediately discontinue using that herb or product and wait several days before trying another.

Unless you rely on responsibly prepackaged, well-labeled herbal products designed specifically for equines, carefully educate yourself before you experi-

ment. Pregnant jennies can abort after ingesting devil's claw, parsley, thyme, tansy, sage, wormwood, or rosemary; some popular herbs, like the natural tranquilizer valerian, test positive in competition drug testing; and a handful of herbs interact, for better or worse, with certain conventional drugs. If in doubt, consult an equine health practitioner before trying an unfamiliar herb or herbal product.

Harvesting wild herbs is a time-honored tradition, but for beginners there are pitfalls. To the uninitiated eye, yarrow looks a lot like poison hemlock, and a tiny dose of poison hemlock can quickly kill a full-grown horse or donkey. Before gathering wild herbs, take wildcrafting lessons from a state or national park naturalist or an experienced herbalist. Don't risk endangering your donkey's life.

Whether using commercial products or creating your own, dose responsibly. In this, herbs are like conventional medicine: Just because a little is good, a lot more is not necessarily better. Medicinal herbs contain powerful active constituents, some of which are toxic in huge doses or when administered over a lengthy period of time. When using commercial herbal products, read their labels and follow directions. And if crafting your own, extensively research each herb and dose accordingly.

Don't expect overnight miracles. Unlike conventional drugs, herbs work relatively slowly. When using internal applications such as feed additives, teas, or tinctures, don't expect measurable results for two to four weeks (or in some cases, even longer). External applications often perform faster, but still not overnight.

Time-Tested Favorites

Herbs do work, sometimes remarkably well. Here are three that have been used effectively for centuries:

COMFREY. Extensive European studies show that its main constituent, allentoin, effectively promotes the growth of connective tissue, bone, and cartilage, and

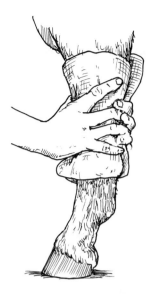

Comfrey poultice is a classic treatment for wounds, bruises, and swellings of all types.

it's easily absorbed through the skin. Fresh or dried comfrey leaves can be used in poultices, fomentations, salves, and oils to treat wounds, bruises, sores, minor burns, and insect bites. Modern science confirms what herbalists have claimed for hundreds of years: Comfrey poultices effectively ease pain, reduce inflammation, and support the healing of minor fractures; comfrey salve or oil rubbed into arthritic joints, muscle strains, and bowed tendons bring healing relief; in emergency situations, powdered comfrey leaves and flowers quickly staunch bleeding.

YARROW. This time-honored healing herb has been proved to contain more than 100 healing constituents. Dried yarrow herb in steam inhalations effectively soothes coughs and other respiratory distress; fed, it's a fine digestive tonic; and its proven anti-inflammatory properties make yarrow a safe, effective additive to an arthritic animal's feed. Or smear yarrow-laced salve on slow-healing wounds, sores, and rashes; you could also treat them using a healing yarrow poultice.

NETTLES. These itchy scourges of the fencerow are actually a rich natural source of essential vitamins and minerals, especially vitamins A and C, iron, silica, potassium, manganese, and sulfur. Fed fresh (but wilted to remove the "sting"), dried and crumbled, or tinctured in alcohol, nourishing nettles increase brood jennies' milk production, put gloss on any donkey's coat, and add clout to an anemic donkey's diet. Nettle salves and oils or nettle tea applied as a soak soothe and heal sores, scrapes, wounds, bites, and rashes.

While there are very few books about herbs specifically for equines, libraries and booksellers stock a wide selection of helpful herbal books for humans,

---

Feast of the Ass

In the days before the average churchgoer could read, medieval churches held theatrical spectacles based on scriptural history to impress religious truths upon their congregations. One of these pageants, originally known as the Festum Asinorum (Latin) or Fête de l'âne (French) and now as the Feast of the Ass, commemorates the Holy Family's flight into Egypt. It was (and is occasionally even now) celebrated on January 14.

Charles du Fresne, sieur Du Cange (1610–1688), French philologist and historian, describes the French Medieval Feast of the Ass in his *Glossary of Medieval and Late Latin* (Paris, 1678). The pageant begins, Du Cange tells us, with a solemn procession through the streets of the city. The principal players are a beautiful young girl cradling an infant in her arms and the splendidly decorated donkey on which she rides. The donkey and his burden are escorted to the city's principal church and placed near the high altar. Then follows a special mass in which, in place of the usual responses ("amen"), the congregation brays like an ass.

At the conclusion of the Mass, the officiating priest brays three times instead of reciting the usual "Ite, Missa est"; in turn, instead of replying "Deo Gratias," the congregation cries, "Hinham, hinham, hinham" (hee-haw, hee-haw, hee-haw).

as well as a dozen or so good ones about herbs for pets. Most herbs affect all species in a similar manner, but there are notable exceptions. When ingested by equines (especially in large doses or over a long period of time), St. John's wort, that popular over-the-counter herb used to treat nervousness, depression, and insomnia in humans, can cause phototoxicity leading to serious dermatitis, itching, and even death; other safe-for-humans but toxic-for-equines herbs include ground ivy, horsetail, horseradish, and fireweed. So if after investigating any herb in references designed for humans or companion animals, you're still not certain it's safe, cross-reference it in an equine-specific source or consult a holistic veterinarian before using it on your donkey.

Did You 'Ear?

During Europe's Middle Ages a poultice of donkey dung was widely used for treating eye problems, and donkey urine was used as a disinfectant. At various times donkey bile has been used to restore eyesight and to cure jaundice.

The location of Castle Cawdor near Inverness (made famous in *Macbeth*), was chosen when the first Thane of Cawdor, William Calder, had a dream that the castle should be built where his donkey lay down to rest. The donkey lay down next to a holly tree; the remains can still be seen on the castle grounds.

In *The Quest of the Holy Grail*, Hector has a dream in which Lancelot is wreathed in holly and riding a donkey. When asked to do so, a hermit interprets the dream, saying that the holly gown is a hair shirt of penitence, and that the donkey is a symbol of humility.

Using Herbal Products

For most of us, the easiest and safest way to administer herbs, especially internally, is to buy professionally processed individual herbs and herbal blends. Although a few liquid herbal solutions are available, most feed additives are packaged as powdered or crumbled plant matter. These arrive sealed in heavy cardboard or plastic kegs, which should be stored in a cool, dry place, especially after opening.

Herbs can be administered in a number of ways:

ADDED TO FEED. Stir the herbs into your donkey's grain, possibly feeding from a nose bag so that you know that they've been consumed. Some donkeys refuse herbs, especially when bitter or pungent plants are part of the mix. If yours turns up his nose, make his herb-laced grain more palatable by adding a dollop of molasses, honey, carrot-flavored baby food, or applesauce.

INFUSION. Another solution is to brew his daily herbs into a strong tea (also called an infusion). To make one, drop the suggested amount of dried herb into a glass jar with a tight-fitting lid, top it with the specified amount of boiling water, tightly cap the jar, and steep the herbs at room temperature for four hours. When the time's up, strain the liquid through cheesecloth. Use an infusion to moisten your donkey's hay or grain or add the infusion to his drinking water. If that fails, rely on liquid feed additives or tinctured herbs.

TINCTURES. These concentrated herbal medicines are concocted by steeping fresh or dried plant matter in alcohol for six weeks or longer. The decanted liquid is sold in brown or amber glass dropper bottles. Most donkeys accept tinctures stirred into grain. If yours doesn't, administer it in water as a drench (*see chapter 10*).

POULTICES. Herbs can be applied externally as homemade poultices, concocted with dried or fresh herbs. The simplest poultice is a smashed or chewed fresh herb such as yellow dock or wild daisy applied to a bee sting, or milkweed sap dabbed on a wart. For a full-scale poultice:

- Smash fresh herbs into a pulpy mass and heat them briefly in a glass or enamel pan on the stove or in a glass bowl in the microwave.
- If using dried herbs, or fresh ones if they aren't juicy enough, moisten the herbs with water, cider vinegar, or a mixture of bread and milk before heating.
- Spread the resulting goo on half of a wet, hot cloth and double the other half over it to contain the pasty pulp.
- Apply the poultice to the bruise, bang, sting, or bump; secure it if you can, otherwise hold it in place until it cools.

FOMENTATIONS. These are liquid poultices. To make one for your donkey:

- Reheat a quart of steeped, strong tea or infusion; you needn't strain it.
- Dip a towel into the hot infusion (it should be as hot as your donkey will tolerate) and apply it to the injury.
- Hold the towel in place until it cools; repeat until the infusion is gone.

COMMERCIAL PRODUCTS. Also available are commercially prepared herbal shampoos and conditioners, hoof dressings, insect repellents, liniments and gels, lotions and oils, itch creams, salves and ointments, eye drops, and more.

Give herbs a try. Whatever your needs, for home doctoring minor problems, maintaining your donkey's health, or (with your veterinarian's guidance) supporting his recovery from serious injury or illness, herbs can be an effective alternative to drugs.

Flower Essences

The flower essence movement began in 1930 when Dr. Edward Bach, a prominent English physician, walked away from his thriving practice to pioneer a brand-new healing system. Bach sensed vast, untapped healing energies coursing through flowers, energies

> ## Musical Jawbone
>
> The *quijada de burro* is a unique South American musical instrument made of the cleaned and dried lower mandible (jawbone) of an ass. While the percussion sections of many Latin American groups boast a quijada or two, this unusual percussion instrument takes center stage in Afro-Peruvian Landó bands alongside melodic guitars and other unorthodox instruments like the *cajita* (a small, lidded box used to take up collections in Catholic churches, played by clapping the lid open and closed and beating on one side with an open hand) and *cajón* (a wooden fruit box straddled by the player, who beats out a rhythm with his hands).
>
> Crafting a quijada is the essence of simplicity. Securely gluing the first molar in place leaves the remaining molars free to vibrate and buzz in their sockets when the player strikes the side of the jawbone with his palm, his fist, or a stick. If you want one but don't have a donkey jawbone to spare, you can buy a genuine quijada on eBay.

that could be extracted through a process he'd devised. Between 1930 and 1936 he settled in various parts of rural Britain, testing blossoms and perfecting his new medicines. In November of 1936, Edward Bach died peacefully in his sleep, but not before creating and carefully proving 38 flower essences (called flower remedies in his native Britain) and a five-flower blend that he called Rescue Remedy.

Dr. Bach's flower remedies are still manufactured in England, while hundreds of other essences are manufactured and distributed by at least 38 other companies throughout the United States, Canada, Australia, England, Scotland, Wales, France, Germany, and the Netherlands.

Flower essences are natural, nontoxic, self-chosen, self-administered, over-the-counter remedies. It's impossible to overdose, and there are no side effects.

What isn't simple is explaining why they work. It is said that water has the ability to hold an etheric imprint of a flower's vital life force energy, that each plant's blossoms emit a different energy, and that this life force, the so-called active ingredient in flower essences, has the ability to alter faulty mental and emotional states as well as mind-linked physical symptoms.

Do they work? Try this: Buy a bottle of Rescue Remedy and for the next few weeks, whenever you're stressed, place two drops under your tongue. See how you feel 15 or 20 minutes later. Rescue Remedy works so well that it is often the only flower essence people use. It is a combination of the essences of Cherry Plum, Clematis, Impatiens, Rock Rose, and Star of Bethlehem.

When do you use the other essences, and for what purpose? When to Use Bach Flower Essences (see below) lists 28 commonly used remedies and what donkey behavior they are meant to address. Only behavior modification uses are listed in this chart. Essences treat the same quirks, foibles, behavior problems, and psychosomatic illnesses displayed by any animal, including humans. Before treating physical injuries or ailments using flower essences, please consult a veterinarian familiar with alternative therapies.

WHEN TO USE BACH FLOWER ESSENCES

All essences are used in the same dosages: 4 drops from the dose bottle,
4 times a day; all essences can be used more frequently in crisis situations.

Arnica. To release lingering effects of emotional trauma (fear of the trailer after an accident, touchy about a healed injury area, ear shy, head shy, and so forth).

Aspen. For donkeys who are afraid to enter trailers, stalls, stocks; for the donkey who senses impending danger (use before or during intense storms or earthquakes, for instance).

Borage. Instills strength, resilience, and courage; for the grieving jenny after the loss of her foal.

Cayenne. When training, to break old, ingrained habits; encourages rapid change.

Centaury. For timid donkeys who are bullied by herd mates; for shy, previously abused donkeys.

Cerato. For inattentive, slow learners who are easily distracted; during competitions so that your donkey will be attentive to rider/handler cues.

Chamomile. For moody donkeys; improves disposition.

Chestnut Bud. To help break long-standing bad habits; to improve concentration, focus, and memory retention during training sessions.

Chicory. For demanding, pushy donkeys; overvigilant new mothers who "smother" their foals; jennies and foals at weaning time; donkeys recently separated from a favorite caretaker or stable mate; and donkeys who hate to be alone.

Cosmos. To establish a bond between donkey and rider or caretaker.

Dandelion. For stiff-muscled, tense, anxious donkeys.

Elm. For all donkeys while trailering; for resistance to farrier and vet work and grooming; for easily overwhelmed, nervous donkeys.

So, try Elm for your donkey when you trailer him to his first show; a Honeysuckle, Chicory, and Mustard blend for a homesick new donkey; or Cerato for one who is easily distracted. Honor your hunches when deciding which remedies to choose. Single essences work well, or combine two to five (but no more) essences (Rescue Remedy counts as a single remedy) into custom remedies.

Essences are sold in ⅛- or ¼-ounce stock bottles; the fluid inside is called the "mother essence." To mix a remedy in a dosage bottle for your donkey, you'll need:

- ✘ A stock bottle of each essence
- ✘ A 1-ounce brown, amber, or blue dropper bottle
- ✘ A small funnel
- ✘ Brandy or cider vinegar
- ✘ Water (spring, distilled, bottled, or rainwater, but don't use water directly from the tap)

Gorse. For broken spirit, depression; for donkeys who give up or won't eat during illness or after surgery, or after losing a home, caretaker, or stablemate.

Holly. For temperamental donkeys; biters, kickers, jealous donkeys; jennies who reject their foals.

Honeysuckle. For homesick donkeys when away from caretakers and stablemates.

Hornbeam. For depressed or barn-sour donkeys who are tired or lethargic during work times but have plenty of energy while free.

Larch. Increases presence, confidence, charisma; especially good for donkeys who are bullied in a herd.

Madia. To boost confidence during competitions or training.

Mallow. To ease depression associated with old age.

Mariposa Lily. To create mother-foal bond (including fostering); for orphan foals raised by humans.

Mimulus. For known fears (crossing water, trailering, clippers, loud noises, plastic) and timid donkeys.

Red Chestnut. For separation anxiety: weaning, a new home, away at competition.

Rescue Remedy. For trauma, upsets of every kind, sorrow, fear, competition nerves, confusion, terror, and hysteria.

Scleranthus. For moodiness; for unpredictable donkeys who behave erratically or vacillate between extremes.

Tiger Lily. For aggressive, hostile donkeys or cranky jennies in heat.

Vine. For herd aggressors and other domineering donkeys.

Walnut. For drastic change; helps equines to accept new caretakers, new home, extensive traveling, motherhood, and so forth.

Wild Rose. Helps old, listless, or cranky donkeys who seem angry or dejected; for broken-spirit and hard-to-catch donkeys.

Here's what you do:

1. Screw off the dropper cap and boil it with the bottle and funnel in a glass or enamel-coated pan for 10 minutes. Drain everything on a towel.

2. Drop two to four drops of each essence into the clean, dry dosage bottle, then cap it.

3. Hammer ("succuss") the bottle against the palm of your hand about 100 times.

4. Fill the bottle nearly to the brim with water. Pound again.

5. Add 1 tablespoon of brandy or cider vinegar as a preservative (brandy works best).

6. Cap the bottle and succuss it another 100 times. (That might seem like a lot of effort, but this is the proper method for preparing vibrational therapies.) Now the essence is ready to use.

A standard dose is four drops from the dosage bottle, given four times a day. But essences can be administered much more frequently, even minutes apart, during emergencies. Giving more drops at a time doesn't increase the medicine's potency; just administer the standard essence more frequently. Some folks administer mother essence directly from their stock bottles, usually in two to four drop shots. This works; it's just not the traditional way to do it.

Homeopathy

One hundred years ago, 15 percent of all American doctors were homeopathic physicians. They studied their craft at one of 22 homeopathic medical schools and practiced in more than 100 homeopathic hospitals. One thousand homeopathic pharmacies dispensed homeopathic medicines, as did conventional pharmacies and general mercantile stores throughout the land.

Because of its successful use in the treatment of the killer infectious diseases of the day, such as typhoid, cholera, and scarlet fever, homeopathy skyrocketed in popularity during the late nineteenth century. Records show that the death rates in homeopathic hospitals were from one-half to one-eighth that of conventional institutions of the day. Presidents James Garfield and William McKinley, Mark Twain, Daniel Webster, John D. Rockefeller and William Wrigley were all avid advocates.

However, the American Medical Association considered homeopathy a brand of quackery, and it actively discouraged its physicians from practicing it. In 1883 it expelled the entire New York State Medical Society for including homeopaths on their roles. As the century progressed, homeopathy gradually fell from favor — on our continent, that is, but not abroad.

Dosing with Flower Essences

- Dribble four drops of essence inside your donkey's lower lip or under his tongue. If you can avoid it, don't let the dropper touch skin, either his or yours. If it does, sterilize the dropper again when you're finished.

- Drip the dose into his drinking pail. Mix it with a dab of another liquid and squirt it into his mouth with a disposable syringe. Dilute it with a cup of plain water and sponge it onto his skin, apply it as a compress to a painful injury, or spritz him with a spray bottle.

- Gouge a hole in a morsel of apple or carrot, stuff the cavity with bread, and drip the essence onto the bread. Or drizzle a few drops on a sugar cube or frosted wheat square and feed it from your hand.

- Massage drops on his gums or on acupressure points any place on his body. Add essences to unscented cream, oil, or lotion. Rub it in.

- Even hard-to-approach donkeys can be dosed with flower essences: squeeze four drops into a water-filled household sprayer, shake, and squirt from a distance.

- Homeopathic physicians have attended the British royal family for more than 100 years; there are currently five homeopathic hospitals working within the British National Health Service (some have two-year waiting lists for nonessential procedures), and homeopathic animal care has a large following throughout Great Britain today.
- According to a survey taken in 1992, 39 percent of all French citizens purchase homeopathic remedies, and 70 percent of France's physicians approve of or prescribe them. Homeopathic medicine is taught at seven French medical schools, 21 schools of pharmacy, two dental schools, and two veterinary colleges.
- According to the same survey, 10 percent of all German physicians are practicing homeopaths, and 10 percent more routinely prescribe remedies; German medical and veterinary colleges require course work in homeopathy; 3,000 of Germany's 11,000 natural health practitioners specialize in homeopathy, and 98 percent of all German pharmacies handle homeopathic remedies.
- Homeopathy is popular in Asia, too, especially in India, where 120 four- and five-year homeopathic medical schools train physicians. At least 100,000 homeopathic doctors and veterinarians currently practice medicine in modern India.

In the Beginning

Homeopathy (pronounced home-ē-OP-a-thē) was born in 1870 when German physician Samuel Hahnemann translated a medical treatise written by a Scottish doctor who claimed *Cichona officialis,* a Peruvian bark that we now know contains quinine, cured malaria. Ever the curious researcher, Dr. Hahnemann boiled the bark into a hearty brew and downed it for several days. As he expected, his healthy body developed acute malarian symptoms, which, when he stopped drinking the stuff, promptly went away.

Hahnemann and his followers began experimenting with other plants, minerals, and animal substances, conducting "provings" (clinical trials), and entering their findings into Hahnemann's *Materia Medica,* a master reference still used to determine which remedy to prescribe.

His conclusion: Any substance that causes a symptom in a healthy person or animal can, in extremely teensy doses, cure the same symptom in a healthy one. And that is how homeopathy works today.

How Remedies Are Made

Skeptics sometimes sneer at homeopathy because remedies are concocted using strange-seeming substances (dog's milk, toad venom, cuttlefish ink, mare's placenta) and they're so extremely diluted. To prepare a remedy, manufacturers dilute one part mother tincture (the concentrated plant, mineral, or animal extract that gives it its name) with either 10 parts distilled water (1:10) or 100 parts water (1:100). This mixture is vigorously shaken, diluted, shaken again, and so on until the desired potency is achieved.

A 1:10 mother tincture further diluted six times is considered 6x potency (x is the Roman numeral for 10); when two hundred times it's called 200c (c is the Roman numeral for one hundred), and so it goes. This procedure is called "potentizing"; the more potentized the remedy, the longer and more deeply it acts and the fewer doses needed for treatment.

However, because higher potencies sometimes trigger "aggravations" (temporary but nonetheless unpleasant worsening of symptoms), lower potencies (6x–30x or 6c–30c) are usually recommended, especially for use by beginning practitioners. Lower potencies work just as well; they simply have to be administered more frequently.

Suggested uses and dosages are printed on remedy labels, but those are only guidelines. Because the same remedies in the same doses apply to the same maladies in donkeys, humans, and all other animals of all ages and sizes, beginning practitioners can diagnose and dose their donkeys using information intended to diagnose and dose people or pets. It all works the same.

Why try homeopathy? For starters, it's admirably safe medicine. Homeopathic remedies are, excepting in isolated cases when incorrectly given in massive doses and for far too long, utterly nontoxic. Dosed in low potencies, they have no side effects whatsoever. If you choose the wrong one for your donkey, it simply has no effect. Remedies are inexpensive and readily available; most American chain drugstores carry them, food

WHEN TO USE HOMEOPATHIC REMEDIES

Suggested standard dose for an equine of any age or size: 1–2 pills, 5–6 pellets, or 10–15 drops. If desired, c-potency remedies may be substituted for x-potencies. *(For more information on potencies, see page 179.)*

Aconite (monkshood). Allays fear, stress, and shock associated with injury. Give Aconite at the onset of laminitis, to jennies and foals at weaning time, and to calm nerves prior to and during trailering or competition.
- Prior to or during trailering, competition, weaning: one 6, 12, or 30x potency dose once an hour for 3 hours prior to need, then at 15-minute to 1-hour intervals for as long as needed.
- For fear or mild shock associated with injury: one 6, 12, or 30x potency dose immediately, then hourly (or more often if necessary) for 4 doses.

Argent Nit (silver nitrate). Calms anxious, nervous donkeys prior to training, transporting, or relocating them, or before a competition, especially those who suffer from nervous diarrhea.
- One 6, 12, 30x potency dose the night before, another the morning before, and 1 dose each just prior to and following training, shipping, or competition.

Arnica (leopard's bane). Arnica is the premier remedy for physical pain and trauma. Give Arnica for bruising, soft tissue swelling accompanied by lameness or strains and to alleviate inflammation

and stiffness, to help prevent infection following injury, before a workout to prevent stiffness after, before and after foaling or surgery, and to ease shock associated with injury or trauma.
- One 6, 12, or 30x potency dose once an hour for 4 doses, then 4 times daily. Topical Arnica cream can be applied 3 times daily (not to broken skin).
- Before and after surgery: one 6, 12, or 30x potency dose 3 times daily for 3 days prior to surgery; 6 times daily after surgery.

Apis (honeybee). Use for bee and insect stings and other hot, red, swollen conditions such as hives.
- One 6, 12, or 30x potency dose every 15 to 30 minutes for 4 doses, then twice daily for 2 or 3 days.
- Externally: Rub directly on the gums every 10 minutes for an hour, then once an hour for acute, severe allergic reactions.

Bellis Perennis (daisy). Eases the discomfort of and heals hoof bruises caused by stones or concussion. Given postpartum, it heals bruised tissue and aids in pelvic recovery.
- One 6, 12, or 30x potency dose 4 times daily for 3 days.

co-ops and vitamin outlets stock them, and equine-specific products can be mail-ordered or purchased at saddlery shops. They can be given along with conventional and other alternative medicines. And they work. Beautifully.

Throughout Europe and Asia, the homeopathic system has heroically withstood the test of time. Dr. Hahnemann conducted extensive provings on each

BRAY SAY: *Homeopathy is a soup made from the shadow of a pigeon that starved to death.*

— Abraham Lincoln
(himself a staunch supporter of homeopathy)

Calcarea Flourica (fluoride of lime). Promotes the growth of hoof horn.
- One 6, 12, or 30x potency dose twice weekly for 8 weeks.

Calendula (marigold). Helps all injuries, internal or external, to heal faster. Give for thrush, slow-healing ulcers, and to heal open wounds. Especially useful for wounds containing pus. Apply Calendula cream to cuts, scrapes, burns, and bruises.
- One 6, 12, or 30x potency dose 3 times daily for 3 days.
- Calendula cream: twice daily for 1 week, or longer if needed.

Caulophyllum (squaw root). Used during the final weeks of pregnancy, it assists in the slackening of a jenny's pelvic ligaments and strengthens uterine contractions.
- One 6, 12, or 30x potency dose twice weekly during the final 4 weeks of pregnancy.

Echinacea (purple coneflower). Enhances the immune response, allays infections, and hastens the healing of all wounds and injuries. Good for insect bites and snakebites.
- One 6, 12, or 30x potency dose twice daily for 4 days.

Hypericum (St. John's wort). Hypericum is the best remedy for nerve-end damage and acute pain. Give to jennies who have delivered large foals and to newly gelded geldings. Good for painful puncture wounds, postoperative pain, wire cuts, splinters, and for bee and insect stings.
- One 6, 12, or 30x potency dose 4 times daily for 4 doses, then once daily for 7 days.

Ignatia (St. Ignatius bean). For upset donkeys experiencing sudden grief. Dose jennies and foals at weaning time and nervous, anxious, new donkeys separated from and missing old caretakers and friends.
- One 6, 12, or 30x potency dose 1 to 6 times over a 2-day period.

Ledum (marsh tea). For deep puncture wounds and adverse reactions to vaccinations or other injections and to alleviate the swelling and pain of horsefly bites, bee stings, and snake and other insect and animal bites. Can be given with Hypericum.
- One 6, 12, or 30x potency dose hourly for 4 doses (then 3 times daily for severe pain).

Natrum Mur (salt). Eases the distress of a donkey who is grieving or depressed.
- One 6, 12, or 30x potency dose 1 to 6 times over a 2-day period.

Sepia (cuttlefish ink). Helps to regulate jennies' estrus cycles and encourages maternal instinct in jennies who are indifferent or hostile toward their offspring.
- One 6, 12, or 30x potency dose 4 times daily for 3 days.

remedy. More recently, remedies have been subjected to extensive clinical research in Great Britain, the Netherlands, Italy, France, and Germany, and they're currently being tested in the United States under the direction of the National Institutes of Health (NIH).

Treating Donkeys with Remedies

To choose the correct remedy for whatever ails your donkey, carefully assess his symptoms, then consult a diagnosing reference (a book of symptoms cross-referenced with appropriate remedies; you can buy them wherever homeopathic remedies are sold). Carefully compare the symptoms with those listed under each remedy description until you isolate one or two or three that jibe pretty closely.

Begin with the remedy that matches best. Give your donkey one dose; wait the suggested amount of time before administering another. If his symptoms improve, even just a bit, after the first dose, stop. Don't give another unless his improvement plateaus or his symptoms return. If they do, administer the same remedy again, but only if it still matches his symptom profile. If it's changed and a different remedy more closely matches the new set of symptoms, dose with that remedy instead. It's rare to need the same remedy for more than three days. If you don't see results in two days, you're probably using the wrong remedy.

Problems that come on quickly will usually resolve themselves quickly, and the results can be astoundingly dramatic. Chronic illnesses, old injuries, and long-standing behavioral problems usually take longer to mend. And often, the older your donkey, the slower he'll heal.

Homeopathic remedies are sold as single and as combination remedies. Combination remedies are fine if you can't decide what remedy to use (your donkey gets the one he needs and the others have no effect), but many users prefer single remedies as Dr. Hahnemann thought that they worked best.

Remedies are marketed as FDA-approved milk-sugar-based (i.e., lactose-based) pellets, pills, and

This postcard reads, "The donkey in the wheel, Carisbrooke Castle, Isle of Wight."

Don't touch homeopathic remedies with your fingers. Instead, tip them into your donkey's mouth using folded paper.

BRAY SAY: *[Referring to a donkey that he once owned] The case of that jackass was most singular. When everything was pleasant all around he would kick the worst. When his rack was fullest and his stall fixed with new straw and everything real comfortable, that jackass would start in on the almightiest spell of kicking that was ever seen. All the veterinary surgeons in the neighborhood came and tried to find out just what made the critter kick so. They never could agree about it. One thing we all noticed was that he always brayed and kicked at the same time. Sometimes he would bray first and then kick, but other times he would kick first and then bray, so that confused us, and nobody in that whole country was ever able to find out whether that jackass was braying at his own kick or kicking at his own brays.*

— Abraham Lincoln

DONKEY TAILS

The Twelfth Donkey

When Johha grew old enough to work for his living he became a donkey driver. One day, being in charge of twelve donkeys employed to carry earth to the city, it occurred to him, before starting with the laden animals, to count them. Finding the tally complete, he took them to their destination and unloaded them.

He then mounted one of them, and was going to return when he found one donkey missing. At once dismounting, he put them all in a row, and was astonished and greatly relieved to find the twelve there. He thereupon remounted and set off again, wondering as he rode along how it was that he had missed one donkey. Suddenly the suspicion flashed upon him that possibly the second count had been faulty, so he counted again, to find once more that only eleven were racing along in front of him. Terribly disconcerted, he again got down off the creature he was riding and, stopping the others, once more counted them. He was puzzled to find that there were again twelve.

So absorbed was he by this mystery, that he went on counting and recounting the donkeys till his master, surprised at his long absence, came and solved his difficulty by obliging him to follow his asses on foot.

— As told by J. E. Hanauer in his 1907 book,
Folk-lore of the Holy Land,
Moslem, Christian and Jewish

granules, as creams and gels, and in liquid suspension form. Try doses of 5 or 6 BB-size round pellets, 1 or 2 thin aspirin-size pills, or 15 to 20 granules. Creams and gels should be lightly glossed on the afflicted area.

You'll sometimes see "nosodes," or remedies prepared from diseased tissue, listed in homeopathic references. Nosodes are essentially homeopathic vaccinations. Use them only as directed by a homeopathic vet.

with Flower Essences on page 178.

To coax the remedies into your donkey, pellets, pills, and granules can be wrapped in folded paper until you reach the barn, then tipped from the paper to the inside of his lower lip. Milk-sugar-based remedies are designed to dissolve quickly, but if he spits them out before they do, administer them as suggested in Dosing with Flower Essences on page 178.

Stored correctly, homeopathic remedies last for decades. Here are some pointers:

- ✘ Try very hard not to touch the remedy itself (gels and creams excepted).
- ✘ Dropped pills, pellets, or granules should be discarded; don't dump them back into the bottle.
- ✘ Store remedies away from strong-scented oils or camphorated and mentholated liniments and rubs.
- ✘ Always store remedies in a cool, dry, reasonably dark place.
- ✘ Don't transfer remedies from one container to another.
- ✘ Uncap and recap remedies quickly.
- ✘ Stash all remedies out of the reach of children; they're nontoxic, so they won't harm your kids, but the milk-sugar pills are so tasty that your supply might vanish in one fell swoop.

Winchy My Donkey

Winchy My Donkey was a popular game among pre-Victorian-era boys living on the Isle of Man, says J. J. Joughin in a 1916 issue of *Mannin,* a folklore magazine. The rules were simple:

"One boy, called the *standard,* stood with his back against a wall, a second placed his head against the *standard's* middle, a third stood bending back to back with the second, a fourth like the second, and so on, all except the *standard* being called *donkeys.* The last tried, straddle-wise, to reach the *standard* over the *donkeys'* backs, the *donkeys* trying to 'winch' him off. If he succeeded he had another turn, if not he took his place at the end of the line, and the *standard* in his turn tried to advance, the second boy becoming *standard.*"

PART 3

Donkey Fun

CHAPTER 12

CLICKER TRAINING

A wink is as good as a nod to a blind donkey.

— Irish proverb

There are many gentle, effective ways to train a donkey, but none is easier or more fun than clicker training. If you try the bare-bones procedures in this chapter and like what you see, visit some clicker training Web sites, buy a book, take a class, or join a listserv. The sky is the limit — there is *nothing* you can't teach your donkey with a clicker and a bag of tasty treats!

This postcard from 1902 reads, "A hungry donkey ('Jack') is eating a crying little girl's dress!"

What Is Clicker Training?

Clicker training is a form of operant conditioning used to train animals, especially dogs, and animals at marine parks and zoos around the world. In clicker training, a *conditioned reinforcer* (a click that tells an animal "well done!") is delivered at the precise moment that he performs a desired action. Followed by a *primary reinforcer* (a tidbit of a favorite food), this conditioning helps an animal to understand precisely what action his trainer expects.

Using clicker training, donkeys can learn anything under the sun, from picking up their feet to tricks and intricate routines under saddle or in harness. It's ideal for helping rescues and wild burros work past their fears, and it's the perfect, force-free way to train young donkeys. What follows is a stripped-down, basic introduction to clicker training, including a few things you can teach your donkeys right now. *(See* Resources *for book titles on clicker training.)*

Who Started It?

Clicker training's roots trace back to the late 1920s, when Harvard psychologist B. F. Skinner began devising a means to change everyday behavior, something he called *operant conditioning,* which the *American Heritage Dictionary* defines as "a process of behavior modification in which the likelihood of a specific behavior is increased or decreased through positive or negative reinforcement each time the behavior is exhibited, so that the subject comes to associate the pleasure or displeasure of the reinforcement with the behavior."

However, not much progress was made with theory until 1943, when two of Skinner's students, Keller Breland and Marian Breland, shaped his teachings into practical training methods that they ultimately taught to trainers at Marineland, Busch Gardens, and Sea World. Thousands of students came to study at their Animal Behavior Enterprises headquarters in Hot Springs, Arkansas.

Then, in the 1960s, Karen Pryor, a scientist with an international reputation in the fields of marine mammal biology and behavioral psychology, coined the term "clicker training." She began giving dog-training seminars in 1987, and twelve years later published her book, *Don't Shoot the Dog! The New Art of Teaching and Training.* Clicker training officially came of age.

Clicker Training Online

The best possible way to learn about clicker training, ask questions, and share your triumphs and tribulations with kindred souls is to participate at clicker training listservs. The best ones for donkey lovers are Donkey Click and Click Ryder *(see* Resources *for the Web sites).*

Gaining Trust

Before you can clicker-train a rescued donkey, you must gain his trust. Each animal reacts to new life changes in his own manner, yet there are basic steps rescuers take to bond with these special equines. We use the following procedure, adapted from British horseman Henry Blake's excellent book, *Talking with Horses.*

First, we isolate new rescues for both quarantine and training purposes in a safe, substantially fenced small pen shelter.

For thirty days I provide the new donkey's every need. He's given free access to grass hay and water and fed tiny amounts of grain or treats from a pan at least twice a day. At first, I drop the treats into the pan and walk away, but as soon as he'll allow it, I hold the pan.

I (and no one else) spend at least thirty minutes each day quietly speaking or singing to the newcomer, first from a distance and eventually while scratching or running my hands over his body. If I miss a day, the thirty-day bonding period begins anew.

Slow and easy movements are the rule. I approach his shoulder — not his head — gazing at the ground, speaking or singing, hands at sides. If he's frightened, I step back and wait until he's comfortable again.

The first touch is a gentle scratch on the withers, shoulder, or chest. As he allows it, I'll stroke my hands across his body, incorporating Tellington-Jones TTouches (*see* Resources) until a bond is established and he fully relaxes. Then clicker training begins.

Always remember:

- Frightened or angry donkeys may react with tooth and hoof. Don't let the donkey get between you and your planned escape route.
- Use a breakaway halter on an aggressive donkey, and leave it on until a bond has formed.
- Never move quickly. Don't rush, don't grab, and don't raise your voice. Be willing to back up and give your donkey space to feel safe.

Patience Pays Off. Don't expect overnight miracles when you're working to gain a donkey's trust. Some animals respond in a few days, others require a month or more of patient handling. If you persevere, in time it will happen and you and your donkey will forge a team.

Why Click?

Alexandra Kurland brought clicker training to the horse world with her first edition of *Clicker Training for Your Horse*. Since then, clicker training has taken the horse world by storm, and for a number of very good reasons.

Clicker training is appropriate for both on-the-ground and mounted training. Virtually anyone can learn it quickly and easily from a book or video. Clicking can be combined with any other sort of positive training method; I combine it with Linda Tellington-Jones's TTEAM techniques, and many natural horsemanship practitioners swear by clicker training.

CLICKER LINGO

You'll encounter the following basic lexicon of words at clicker-training listservs and in clicker-training books and articles online.

Bridging. Teaching your donkey to recognize a sound that will tell him that he's chosen the correct response; in clicker training, a well-timed click is a *bridging signal.*

Clicker. A small, handheld plastic box with a metal strip inset; when the strip is pressed it emits a distinct, two-note click.

Conditioned reinforcer. A *neutral stimulus* (a click) paired with a *primary enforcer* (food) until the neutral stimulus takes on the reinforcing properties of the primary enforcer.

Continuous reinforcement. Every correct response is reinforced (rewarded).

Event marker. A signal used to mark correct behavior at the moment it occurs; a click is an event marker.

Extinction. The weakening of behavior through nonreinforcement; for example, ignoring an incorrect behavior.

Free shaping. To train using *free shaping,* wait until your donkey does a desired behavior on his own, and the instant he does, click and reward him. For example, to teach him to bray, wait until he starts to bray of his own volition, then click and reward.

Jackpot. A mega-reward given for exceptional effort.

Molding. To train by *molding,* place your donkey or one of his body parts in the desired position and then click and reward. An example: In teaching him to load into a trailer, you could place one forefoot on the trailer ramp, then click and reward.

Negative reinforcement. Taking something away to discourage an unwanted behavior. For example, if your donkey mugs your pockets for unearned treats, you could turn your back or walk away.

The same techniques can be used for training all sorts of animals; I use many of the same techniques I learned from my all-time favorite clicker-training book, Peggy Tillman's *Clicking with Your Dog: Step-by-Step in Pictures* (*see* Resources), to train (in addition to dogs) donkeys, horses, sheep, pet pigs, and goats.

Though donkey owners who have never tried clicker training often think that feeding food rewards will make mouthy pests of their donkeys, the exact opposite is true. If the only time a donkey is given a food reward is when he *earns* it, he knows that randomly mugging for feed simply won't work.

Clicking works for saddle training too. This young woman clicks her donkey for crossing some scary water.

Positive reinforcement. Adding something that your donkey will work for; for example, rewarding your donkey with a tidbit of food for picking up his foot increases the probability that he'll pick up his foot again next time.

Primary reinforcer. A reinforcer that your donkey was born needing, such as food.

Punishment. In clicker training, a consequence to a behavior in which something added to or taken away from the situation makes the behavior less likely to occur; for example, leaving the area if your donkey walks away during a training session makes it less likely that he'll do that again.

Rate of Reinforcement. The number of reinforcers doled out in a given amount of time.

Ratio. A schedule of reinforcement in which you reinforce your donkey based on a predetermined number of responses; in other words, you'd reward the first correct response after a specific number of correct responses.

Reinforcement. The act of adding or removing something from a situation that makes a given behavior more likely to occur in the future.

Reinforcer. Anything your donkey will work to obtain (generally, food).

Secondary reinforcer. A *conditioned reinforcer* using something he wasn't born needing; a click that tells your donkey that he's done something correctly and has earned a reward.

Target. An object that you teach your donkey to touch with his nose.

Three-fer. In a *three-fer,* your donkey must perform three correct behaviors to earn one click and one reward.

Timing. Ideally, the click should occur the exact moment a correct behavior occurs.

Two-fer. When your donkey must perform two correct behaviors to earn one click and one reward.

There are countless types of handy clickers to choose from. My favorite is the bungee-style clicker in the front.

Getting Started

You'll need a few basic items to try your hand at clicker training, though it's easy to improvise if you can't find them close at hand.

A CLICKER. Clickers are readily available at most pet stores. Another way to get one is to buy Alexandra Kurland's introductory clicker kit, *Getting Started: Clicker Training for Horses,* on the shelf at most large bookstores. If neither option is available, you can click your tongue, but the sound must be loud, crisp, and not too similar to the "clucks" some owners make to encourage their donkeys (which might confuse your donkey). Store-bought clickers are better. If you have a choice of clickers, choose a bungee-style model that you can snap to a belt loop or wear around your wrist; that way you can drop the clicker when you need to and not lose it.

> **BRAY SAY:** *One has to fear in front of a goat, in back of a mule, and on every side of a fool.*
>
> — Yiddish proverb

A TARGET. I like the marine float targets Shawna Karrasch sells on her On-Target Training Web site (*see* Resources), although I make my own. Making one is the essence of simplicity. I have two: a short one for close-up work (mine has a 10-inch handle, made from a 15-inch dowel rod) and a longer one for leading (with an 18-inch handle, made from a 22-inch dowel rod). If you're not sure of the length, make it long; it can be shortened later if it's too unwieldy. Here's what you do:

1. Buy marine floats at a department store (they come two to a package) and a ½-inch dowel rod at a hardware or lumber store.
2. Cut the dowel rod to the desired length.
3. Squirt a small amount of strong glue just inside both ends of the hole in a marine float.
4. Add the handle, and there you are!

However, other targets work well, too, especially once your donkey knows what a target is all about. Our Curly horse stallion preferred a dog-toy target shaped like a jack from an old-time ball-and-jacks game. He could easily pick it up with his teeth when we played

Bluestone Mamagwezy, a yearling Miniature jenny, plays peek-a-boo.

target-and-fetch. When no other target is close at hand, I frequently use a halter squashed flat in my hand. A brush, the lid from a small bucket, or a soda bottle would work just as well.

REWARDS. It's important to have rewards handy, so have something to carry them in. Don't plan to carry rewards in your pants pockets. It's difficult to remove a food reward from a tight pocket in a timely manner and depending on what you use for rewards, it can be messy! I carry food rewards in a two-pocket carpenter's apron. Others favor hunter's or photographer's vests with loads of easy-access pockets. Fanny packs and ready-made reward buckets and bags also work; inexpensive clip-to-the-belt bags designed for dog training are especially good buys.

HOW-TO
Targeting

By touching his nose to a target, your donkey learns to perform a task to earn a reward. Simple targeting leads to more complex maneuvers like standing still to be haltered, leading at your side, and fuss-free trailer loading. To me, the best part of clicker training is seeing the lights go on as your donkey realizes what you want him to do. It's magic — and you make it happen!

First, teach your donkey that a click means food. I do this several times prior to the first formal training session by simply clicking, then immediately handing my donkey a tidbit, and I do this about ten times per session. Donkeys are smart; they quickly

connect the click with the treat, so don't carry this to extremes. To teach actual targeting:

1. Stand at your donkey's head with the target and clicker in your left hand (this is easier with a marine float target than with something bulky) and your right hand free to quickly dish out goodies.
2. Hold the target close to your donkey's nose. Chances are, he'll be curious and reach out to see what it is. The moment his nose brushes the target, click, reward, and lavishly praise your donkey.
3. Keep the target close to his nose; if he doesn't immediately touch it again, position it where he's apt to bump it if he moves his head. When he does, click, reward, and praise.
4. Keep it up until that magical moment when your donkey pauses, thinks, and then reaches out to touch the target on his own. Jackpot! Haul out the grandest goodies in your reward bag and celebrate.

The *eureka!* moment may happen during your first session or it might take several sessions before it occurs, but once your donkey makes that vital connection, the rest of clicker training is a breeze.

This mare is learning to lower her head to touch the target with her nose.

Using Food Rewards

Clicker-training rewards should be tiny tidbits of something your donkey really likes. The object isn't to feed your donkey. Big lumps of stuff he has to stop to chew won't do. You'll need two kinds: standard rewards and something extra special for jackpots. Dry things are more pleasant to carry and handle, but if your donkey is ga-ga over cantaloupe and you don't mind digging through a squishy treat bag, go for what your donkey loves best. Some suggestions:

- ✘ Biscuit cereals (examples: Kashi or Chex, Spoon-size Shredded Wheat, Frosted Mini Wheats) broken into two or three pieces
- ✘ Cheerios
- ✘ Pretzel pieces
- ✘ Animal Crackers (buy them in the huge economy bags)
- ✘ Commercial or homemade dried bread cubes (flavored ones are especially tasty)
- ✘ Dry (uncooked) pasta
- ✘ Sunflower seeds (in the shell or not)
- ✘ Shelled peanuts
- ✘ Sugar cubes (best saved for jackpots), small jelly beans, gummi bears, peppermints smashed into manageable pieces, after-dinner mints, teensy valentine hearts, candy corn, and other small, hard or semihard candies (*Note:* don't feed chocolate to your donkey; it's toxic to equines of all kinds.)
- ✘ Alfalfa pellets (*small* ones so that your donkey doesn't choke), pelleted rabbit chow, nonmedicated sheep or goat pellets (make certain that they aren't treated with Rumensin), and pelleted horse feed
- ✘ Carrot coins (carrots sliced into thin slices) or baby carrots cut into three or four pieces

- ✘ Raisins or dried fruit (apples, apricots, pineapple) cut into pieces
- ✘ Commercial horse treats whacked with a hammer and reduced to small pieces
- ✘ Minute amounts of your donkey's favorite sweet feed

Important Note: Once you begin clicker training, *never* indiscriminately feed treats from your hands! You can still treat your donkey, but place the treats in his feed pan and walk away; once you begin clicker training, anything eaten directly from your hand must be earned.

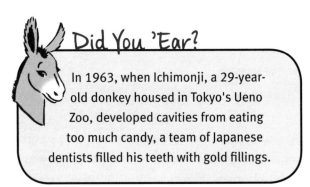

Did You 'Ear?

In 1963, when Ichimonji, a 29-year-old donkey housed in Tokyo's Ueno Zoo, developed cavities from eating too much candy, a team of Japanese dentists filled his teeth with gold fillings.

Following a Target

Once the donkey understands that touching the target elicits a treat, teach him to follow it with his nose. Start by moving it around within arm's length so that your donkey can touch it without moving his feet. Then hold it up high so that he has to reach up for the target and then low near the ground so that he lowers his head. Finally, take a step back so that he has to step forward to reach it. Keep moving around a small area until he understands that following the target is a very good thing.

THE TEN COMMANDMENTS OF CLICKING

1. Keep sessions short and make them fun. Three five-minute sessions per day is ideal (and infinitely more productive than a single one-hour session). Donkeys hate boring repetition, and bored donkeys simply tune out.

2. Correct timing of the click is crucial. Picture this: you're cleaning your donkey's foot and he holds it up like a prince; you put the foot down, then click and reward. You think you're clicking for him holding his foot up; he thinks you're clicking because he put it down. Which is he likely to do the next time? Click *during* the desired behavior, not after.

3. Click once (click-click, once in and out) per correct behavior. If your donkey does something especially well, give him a jackpot reward or extra food, but don't reel off a series of clicks.

4. While training, click for both voluntary and accidental movements toward your goal, and set your donkey up for success. Hold the target where his nose will bump it, coax or lure him into a position you want, or briefly place him there, but don't pull, push, or hold him in place; he needs to think that he earned the reward of his own accord.

5. At first, don't hold out for perfect behavior. Once he understands what he needs to do to earn a reward, then you can be picky, but in the early stages click and reward for effort.

6. Correct bad behavior by clicking good behavior. Mugging for food is a classic example: If your donkey nudges and nibbles at your pockets looking for food, ignore him. Look away or turn your back and count to five. Then give him a chance to earn his treat by targeting or performing some other simple task. He'll quickly learn that mugging is counterproductive.

7. By the same token, don't reinforce undesirable behavior. If your donkey does something wrong or engages in spontaneous behavior during training (such as targeting when you didn't ask him to), stop what you're doing and count to five to let him know that you're not going to click and reward that action.

8. Once you're positive that your donkey understands a behavior, don't click for every correct response. Start clicking every second or third correct response, then stretch it out until you're clicking correct responses at irregular intervals. While it seems as though the donkey would find this discouraging, quite the opposite is true — he'll wonder which response will be the magic one that earns the reward, and he'll try all the harder to get it.

9. Take your donkey's prior history into consideration when formulating training goals. A donkey who has never been loaded into a trailer will be easier to teach to load than one who was once beaten while loading or who fell and injured himself while on the road. A naive donkey often follows a target right into the trailer; the frightened one may need the process broken down into many small, rewardable steps before he willingly walks up the ramp.

10. If you get mad, stop the training. Walk away and cool down. There is always another time, another day. Don't link lead-jerking and shouting with clicker training in your donkey's mind. Clicker training is meant to be fun.

Clicker training works for all sorts of animals — including goats!

When you're positive that he understands the concept of targeting and you're sure that he'll do it 100 percent of the time, add a spoken cue as you offer the target ("touch" and "target" are logical choices). Once he understands, reward *only* when you ask him via the cue word to perform; ignore spontaneous touches. Now you're ready to start shaping behaviors.

> **BRAY SAY:** *An ass does not stumble twice over the same stone.*
>
> — Turkish proverb

Haltering

If you've ever dealt with a hard-to-halter equine, you know how frustrating it can be. By using clicker training you can change that annoying habit in a few short sessions. Here's how:

1. Once your donkey is targeting reliably, scrunch up his halter (with the crownpiece buckled) and hold it together with the clicker and the target handle in your left hand. Gradually move the halter forward so that eventually the donkey's nose touches it along with the target. Continue doing this until your donkey accepts touching the halter.

2. Put down the real target and hold the scrunched-up halter toward the donkey. Give your verbal cue. Most donkeys touch the halter right away. Gradually unscrunch the halter until you're holding just the crownpiece. At this point it looks like a halter, but the donkey is touching it anyway.

3. Unbuckle the crownpiece, let it dangle, and begin touching the donkey's face with the halter, clicking the moment it makes contact and he doesn't pull away. Then hold it so that you can slip the noseband over his nose. Next, start putting the halter farther up on the donkey's head each time you click (it takes longer to deliver

Offer the scrunched-up halter to the donkey (step 2) and give your verbal cue.

rewards when both hands are involved in haltering, but clicks tell the donkey that food is on its way).

4. Fasten the halter and offer a giant jackpot! Take off the halter and put it on a few more times.

This really works. I trained an untouchable, formerly abused rescue horse to be haltered in three five-minute sessions. In another seven, I taught her to safely lead.

HOW-TO
Leading

Once your donkey reliably follows a target, it's a snap to teach him to lead freely and willingly at your side.

1. Use your longer-handled marine float target and wear your reward pouch on the right side of your body.

My husband, John, prepares to teach Ishtar to lead willingly at his right side.

2. Stand on the left side of your haltered donkey facing forward with your shoulder even with the middle of the donkey's neck.

3. Keep the lead rope safely gathered into a figure eight in your right hand (not coiled around your hand in case the donkey pulls back) with the float part of the target facing away from the donkey.

4. When you're ready to begin, bring the float around in front of the donkey's nose about two feet ahead of his muzzle and give your verbal cue ("Touch!").

5. When the donkey steps forward and touches the target, click, reward, and lavish the donkey with praise. Repeat again and again until the donkey understands what you want.

6. Then hold the target in front of the donkey and click as he takes a step — but *before* he reaches the target. Then click for every two steps, then three.

7. Before long, the donkey will be following the target and you can increase the distance he travels before clicking and rewarding.

8. Introduce a new cue (such as "Walk!") as he starts to walk forward.

When he understands, put the target away. Now you have a happy donkey walking calmly at your shoulder. In this manner you can teach your donkey to load in a trailer, to trot in-hand for halter classes, or perform tricks like standing with his front feet on a pedestal. When you want him to go someplace new, just bring out the target again.

Thanks to clicker training, Ishtar doesn't mind having her ears handled one bit.

Handling Ears

With some donkeys, touching their ears is next to impossible; not so with the clicker-trained ass. To train him to accept ear handling:

1. Halter your donkey and hold the clicker and lead in your left hand.
2. Touch his face with your right hand. Click, reward, and gradually move your hand closer to his ears.
3. Continue doing this until you reach the point where he starts to object.
4. Go back to a place on his face where he doesn't object and start advancing very slowly again.
5. Click and reward for each improvement as you gradually inch your way to his ears; ignore (but don't reinforce by click-rewarding) any objections.

It may take several sessions to overcome this common fear, so be patient and take things slowly.

Picking Up Feet

Approach handling feet in the same way that you approached touching ears. Start at your donkey's shoulder or hip and run your hand down his leg, clicking and rewarding until you reach a spot where he starts to object. Proceed more slowly, always clicking and rewarding positive responses, until he's okay with your touching his hoof.

Next, lean against your donkey so that he shifts his weight to his opposite leg; when he does this, click-reward. Then, run your hand down his leg, grasp his hoof, and pick it up just a bit. *Click while his foot is coming up, not as he slams it back onto the ground.* If his hoof starts down before you have time to click, count to five without reinforcing the behavior and try again.

When he lets you pick up his foot and hold it up for a few seconds, click-reward (still holding his hoof off the ground) and set it back down again. Work toward three or four minutes of patient standing without leaning on you for support, then slowly increase the length of time before he earns each click.

John clicks and Ishtar picks up her hoof.

Balaam and His Ass

There are scores of references to asses, both domestic and wild, in the Jewish Torah and the Talmud as well as in other Hebrew texts. Some texts put forth that donkeys are bad on the outside and good on the inside, and material but loving and loyal. One of the most interesting stories is that of Balaam's ass.

This story takes place near the end of the Israelites' 40 years of wandering, shortly before Moses's death. The Israelites have already defeated two kings on the east of the Jordan River, so Balak, king of Moab, is getting jumpy. He sends elders to Balaam, a non-Israelite prophet, to convince him to come to Moab and curse Israel. Here is the rest of the story as told in Numbers 22:21–33. Note: Balaam's ass and the serpent in the Garden of Eden are the only talking animals in the Hebrew Bible.

Found at an auction, this antique 5 × 7" engraving was produced in the mid-1800s and depicts Balaam and his ass.

[21] And Balaam rose up in the morning and saddled his ass, and went with the princes of Moab.

[22] And God's anger was kindled because he went: and the angel of the Lord stood in the way for an adversary against him. Now he was riding upon his ass, and his two servants were with him.

[23] And the ass saw the angel of the Lord standing in the way, and his sword drawn in his hand; and the ass turned aside out of the way, and went into the field; and Balaam smote the ass to turn her into the way.

[24] But the angel of the Lord stood in a path of the vineyards, a wall being on this side, and a wall on that side.

[25] And when the ass saw the angel of the Lord, she thrust herself unto the wall, and crushed Balaam's foot against the wall: and he smote her again.

[26] And the angel of the lord went further, and stood in a narrow place, where there was no way to turn either to the right hand or the left.

[27] And when the ass saw the angel of the lord, she fell down under Balaam: and Balaam's anger was kindled, and he smote the ass with a staff.

[28] And the Lord opened the mouth of the ass, and she said unto Balaam, What have I done unto thee, that thou hast smitten me these three times?

[29] And Balaam said unto the ass, Because thou hast mocked me: I would there were a sword in mine hand, for now I would kill thee.

[30] And the ass said unto Balaam, Am I not thine ass, upon which thou hast ridden ever since I was thine unto this day? Was I ever wont to do so unto thee? And he said, Nay.

[31] Then the lord opened the eyes of Balaam, and he saw the angel of the Lord standing in the way, and his sword drawn in his hand: and he bowed down his head, and fell flat on his face.

[32] And the angel of the Lord said unto him, Wherefore hast thou smitten thine ass these three times? Behold, I went out to withstand thee, because thy way is perverse before me:

[33] And the ass saw me, and turned from me these three times: unless she had turned from me, surely now also I had slain thee, and saved her alive.

CHAPTER 13

DONKEY RIDING

If one man calls you a donkey, ignore him.
If two men call you a donkey, think about it.
If three men call you a donkey, buy a saddle.

— Yiddish proverb

There are lots of ways to saddle train a donkey; this chapter reviews the approach that works best for me. It sounds easy, and it is. Virtually anyone can train an everyday trail- or pleasure-riding donkey using these low-key, tried-and-true techniques.

Donkey Dos and Don'ts

The way to a donkey's heart is through his stomach; that's why clicker training works so well with our long-eared friends. If you choose not to clicker train per se, still consider using food treats to reward a job well done. However, don't distribute treats without a reason — make your donkey earn them. Use them judiciously or not at all. The donkey mugging your pockets for treats is not a donkey focused on training protocol.

Donkeys are intelligent and learn quickly, so they don't do what they perceive as mind-numbing repetition. Forget the longe line and round pen, and don't drill your donkey until he's bored to tears. To keep him fresh and attentive, limit most sessions to 30 minutes or less, and strive to make sessions interesting and fun.

Donkeys don't respond to stimuli the way that horses

do. If they're frightened or confused they don't go forward; they freeze. It can be tremendously frustrating when he freezes, but you can't afford to lose your temper while working with your donkey. If persuasion isn't working and you feel you're ready to explode, stop what

These three ladies are setting out on a picnic at Cheyenne Canon, Colorado, in 1908. Although there were posed photo concessions in operation at that time, all of these women are obviously real riders. Note the seat, hand, and foot positions, and the tack has been adjusted to fit.

you're doing and walk away. A single spate of explosive temper can undermine your relationship with your donkey for life. It's better to yield a battle than forfeit your donkey's trust.

Punish your donkey only if he seriously misbehaves and he knows what he's done. Dangerous behaviors such as biting and kicking are never all right. However, punishment must be applied within seconds of the unacceptable behavior. And don't overdo it; donkeys remember perceived abuse for a very long time. Sweet as they are, most donkeys occasionally test their trainers to see what they'll do. When it happens, take a deep breath and stay calm. Repeat after me: patience, patience, patience.

Why Ride a Donkey?

If you enjoy horseback riding, you'll love riding donkeys instead. Horses are good but donkeys are better! Here are just a few good reasons to ride a donkey:

- ✖ Because they're intelligent and affectionate, and they aim to please, everyday pleasure-riding donkeys are incredibly easy to train. They need not be sent to a trainer; owners can do the work themselves.
- ✖ They're reliable; donkeys don't put themselves or their riders in harm's way. They rarely spook, they don't bolt, they're surefooted, and they won't venture into danger of any kind. These traits make them superlative mounts for children, the physically challenged, and nervous adult riders of all kinds.
- ✖ Donkeys are superbly smooth gaited. Some pace, some single-foot, but even trotting donkeys' gaits are smooth as silk. And a surprising number of donkeys are extremely fast walkers; your friends on horses may have to hustle to catch up.
- ✖ Donkeys stand out in a crowd. A handsome, well-trained riding donkey turns heads wherever he goes.

Donkeys vs. Horses

How does riding a donkey differ from riding a horse? The first thing you'll notice is that you feel as though you're riding downhill. Two donkey- and mule-specific anatomical features contribute to this sensation.

1. Because of the way a donkey's neck is constructed, he's unable to break over at the poll with his nose tucked back like a dressage horse. When in motion, the crest of a donkey's neck is correctly carried level with or slightly above parallel to the ground, with his head in a relaxed position.

2. Donkeys have very low withers; furthermore, most stand taller at the top of the hip than they are when measured at their withers. This makes saddle fitting problematic (we'll talk about that in a moment) and can add to that riding downhill sensation.

Donkeys' shoulders are more upright, and their withers higher, than horses' are.

Also, donkeys are more phlegmatic than the average horse; this is both a blessing and a curse. Because fast, jerky movements aren't part of a donkey's repertoire, he's an ideal mount for riders with fear issues and anyone else who treasures relaxed, easy time in the saddle free of unwanted surprises. However, any donkey can be trained to respond to cues in a timely if not flashy manner. The downside: riders who prefer highly responsive, spirited mounts will need to pick and choose to find a donkey with these characteristics.

How Much Weight Can a Donkey Carry?

The average donkey can comfortably carry 25 percent of his own body weight; that includes his rider and all his tack. However, a number of variables should be factored into the equation, including:

- ✗ *The donkey's age.* Donkeys mature slowly and shouldn't be ridden, even by children, until they're at least four years old. Overtaxing a young donkey can lead to problems later, so err on the side of caution.
- ✗ *The donkey's conformation.* Wide-set, heavy-boned donkeys with stout, strong legs can safely carry more weight than narrow, spindly-legged individuals.
- ✗ *Where and how the donkey will be ridden.* A donkey sedately ridden in an arena situation can handle more weight than one ridden on steep, hill-country trails or in barrel races at a youth rodeo.

Don't overtax your donkey. If he's not up to the task, choose lighter-weight equipment or a larger, sturdier donkey and use the donkey you already have in another capacity, such as driving or packing.

BRAY SAY: *If you pray for a Cadillac and God sends a jackass, ride it.*

— Alcoholics Anonymous

Donkey Tack 101

Buying tack isn't as difficult or as expensive as you might think, but you need to know what types of riding gear fit donkeys. Here are some things to consider.

Saddles

Fifty years ago, horse folk thought that a saddle was a saddle — with the right amount of padding, one saddle fit every size and style of horse. Nowadays we know better. For our animals to perform well and stay sound in body and mind, the equipment we use must comfortably conform to their bodies. Finding a ready-made saddle to fit a horse is difficult; finding one to fit a donkey is harder still.

Donkeys, especially larger ones, can wear any type of saddle that's used on a horse: Western, dressage, English general-purpose, Australian, endurance, McClellan military, sidesaddles, and so forth. The trick is choosing one that fits your specific donkey's back. Pre-fifties-era Western saddles (designed at a time when narrower horses were the norm), McClellan saddles, and modern English saddles built on narrow trees are often excellent choices for slender donkeys, while today's Western and English saddles built on Quarter Horse or Arabian saddle trees tend to fit donkeys with flat, broad backs.

Careful shoppers can outfit their donkeys in duds designed for horses, but not just any piece of equipment will do.

WESTERN

ENGLISH

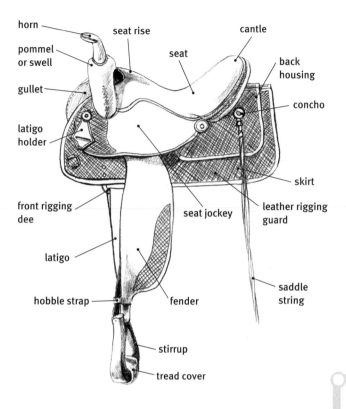

horn

pommel or swell

gullet

latigo holder

front rigging dee

latigo

hobble strap

seat rise

seat

cantle

back housing

concho

skirt

seat jockey

leather rigging guard

fender

saddle string

stirrup

tread cover

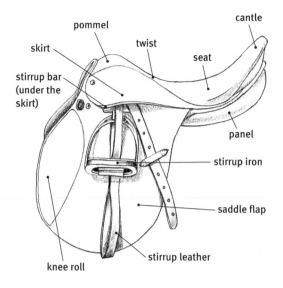

pommel

skirt

stirrup bar (under the skirt)

twist

cantle

seat

panel

stirrup iron

saddle flap

knee roll

stirrup leather

However, type is less important than fit. I suggest you buy and read a book about saddle fitting, because the better your donkey's saddle fits him, the happier he'll be. Keep in mind that most saddle resources address fitting a horse's back — but most of the same points apply to donkey saddle fitting.

Signs of an ill-fitting saddle include:

✗ Pressure-generated permanent white hairs on the withers area or rib cage

✗ Sores or raised, hot, swollen, or sensitive spots anywhere that the saddle makes contact

✗ Resistance to being saddled or ridden indicated by a swishing tail, pinned ears, attempts to bite or kick when the saddle is placed on the back or the cinch is tightened, mincing gaits, and body tension

Bareback Pads & Stirrups Don't Mix

Donkey riders are naturally drawn to bareback saddles — inexpensive English-saddle-shaped pads with girths, stirrup leathers, and stirrups — because they fit a donkey's back just right. However, if you use one, have a saddler attach sturdy dees for your breast collar and crupper *(see pages 205 and 206)* and always use them, or remove the leathers and stirrups and throw them away. Without a breast collar and crupper to stabilize it, the moment that you put more weight on one stirrup or the other, the pad will slide to the side; if your foot gets hung in the stirrup as the pad turns, you'll fall and can be dragged. Riders are injured or killed in this manner each year. Saddle pads with stirrups are unsafe at any speed.

Checking for Fit

To check a saddle for basic fit, do the following:

1. Stand the donkey on level ground with his head in a natural position.

2. Place the saddle on his clean, dry back but don't use a pad or cinch. If you're trying on a new saddle that you may want to return, place a thin piece of bed sheet over the entire area.

3. Set the saddle slightly ahead of where you think it should go and slide it back until it settles into a natural pocket on the donkey's back. Do this several times, making certain that it's settling into the same place every time.

4. Step back and look at the saddle. The flat area of its seat should be parallel with the ground. If the saddle tips forward, it's probably too wide; if it tips back, the tree is probably too narrow.

5. In either case, examine the gullet area (the cutout, or tunnel, that sits over the withers) of the saddle. You should be able to insert three stacked fingers between the gullet and your donkey's withers (this amounts to roughly two inches when you're seated in the saddle). If the saddle is too wide, there won't be enough clearance. If it's too narrow, the saddle will sit too high in front and dig into the donkey's back.

ENGLISH SADDLES

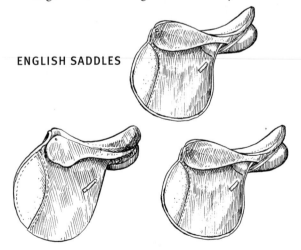

Here are some other things to consider:

✗ Check the sweat pattern on your donkey's back after a long, hot ride. His back should be evenly sweaty. Dry spots indicate areas where too much pressure inhibited sweating. Palpate these spots for tenderness; they're sure to be ouchy.

✗ Length is important, too, especially when fitting standard-size donkeys. A long saddle exerts undue pressure over the loin and kidneys of a short-backed donkey, causing pain and possibly permanent injury. A saddle's bearing surface should never extend beyond the vertebra corresponding with your donkey's last rib. Be aware that the large square skirts of today's Western saddles can poke and irritate a donkey's tender flanks; round-skirted saddles and barrel-racing saddles with small, square skirts are a better option.

✗ If a saddle doesn't fit, don't automatically add another blanket or pad. Judicious padding often eases minor problems, but if a saddle is already too tight, additional padding simply creates more compression and pain.

Should you avoid all this and buy a treeless saddle or ride bareback? Probably not. A well-fitted saddle distributes weight over the long muscles along your donkey's spine — weight doesn't bear on the spinal process itself, and this is good. A treeless saddle (or no saddle at all) puts all of your weight directly on your donkey's spine. This is fine for short riding sessions but can create painful pressure spots on longer rides.

Saddle fitting initially seems to be a daunting task, but in practice it's not. If you evaluate ready-made saddles designed for your donkey's type, you'll eventually find one that fits. However, if the saddle you own or want to buy doesn't fit your donkey's back, please choose a different saddle or buy another donkey. A seriously ill-fitting saddle is a bona fide torture device.

CHOOSING A SADDLE

Whether buying a brand-new saddle or a used one, choose a good one. These tips will help you narrow your search (points one through five apply when purchasing strap goods such as leather bridles and halters, too).

1. Don't buy a saddle constructed of poor-quality leather. Good leather lasts and lasts; inferior leathers stretches, tears, and never looks as good as the real McCoy. Quality leather feels smooth, supple, and substantial, even if it's 30 years old; avoid spongy, wrinkled, fibrous leather no matter what its age.

2. Make sure that all edges are burnished and sealed. Quality leather has nicely finished edges rather than rough, fibrous ones.

3. Layered leather should be sewn with small, neat stitches. When buying used tack, make certain that all stitching is sound.

4. Better equipment is laced or stitched, not riveted.

5. Choose tack with solid brass or stainless-steel fittings. Avoid "never-rust" (nickel silver) fittings, and refuse anything made with cheap, chrome-plated, pot-metal hardware.

6. Find out what sort of tree a Western saddle is built on (you can usually tell by lifting a side jockey and taking a peek). Well-made older saddles as well as top-of-the-line new ones incorporate sturdy, long-wearing rawhide- or bullhide-covered wooden trees. Mid-priced older saddles were often built on heavy-duty injection-molded Ralade plastic trees; they are lighter in weight but a bit more prone to breakage with age. Today's lower-priced Western saddles are usually built on fiberglass-covered wooden trees. It's wise to hold out for a saddle with a rawhide- or bullhide-covered tree; they're heavier than comparable synthetic and synthetic-covered trees but so much more durable.

7. Make certain that the tree isn't broken. Pick up the saddle with the pommel against your hip bone, grasping the cantle in both hands. Pull: there should be no obvious movement. Then place the saddle on a saddle rack or similar surface and press down on the pommel and cantle at the same time. Spring-tree English saddles will flex a wee bit, but if you feel much give, don't buy that saddle.

8. If you'd like to recoup your investment one day, be conservative. The rhinestone-spangled, pink-seated saddle you buy today will be hopelessly gauche just a few years from now. Even "conventional" styles change with the passing of time. A case in point: we own a wonderfully well-made, nice-riding custom Billy Royal Arabian saddle rendered hopelessly passé by the buckstitching that made it a head turner in its day.

9. When money is an object, consider purchasing a quality used saddle for the same price that you'd pay for a shoddily built new one. A tasteful, well-made saddle constructed on a rawhide tree and crafted using high-quality leather and fittings will always increase in value, and you'll be riding it many years after the el cheapos made of cardboard leather have self-destructed.

10. Alternately, invest your dollars in a good synthetic saddle manufactured by a major saddler; it'll last as long as a cheap leather saddle and be much easier to clean in the bargain.

ENGLISH BRIDLE

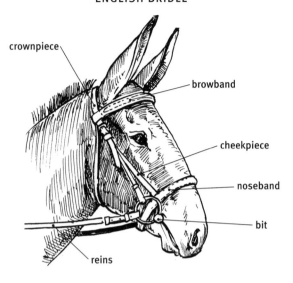

crownpiece

browband

cheekpiece

noseband

bit

reins

WESTERN BRIDLE

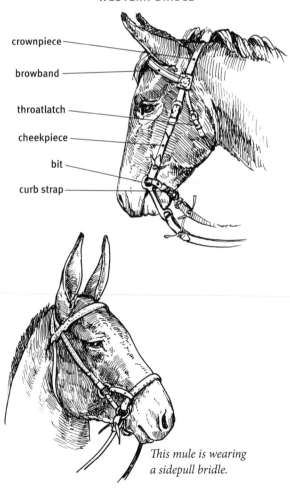

crownpiece

browband

throatlatch

cheekpiece

bit

curb strap

This mule is wearing a sidepull bridle.

Headstalls

Every donkey needs a bridle that fits. Because donkeys' foreheads are considerably wider than those of similar-size horses, off-the-rack headstalls require modification. This can be done in two ways: replace the browband with a longer one (this is generally easier when modifying English-style bridles because components can be purchased piecemeal), or purchase an oversize headstall and shorten the cheekpieces to fit your donkey's head.

Many donkeys object to having their sensitive ears handled as part of bridling. One solution: clicker train your donkey to accept ear handling as described in chapter 12. Another solution: buy a snap-crown bridle that can be fastened without handling his ears at all (*see* Resources).

Bits and Sidepulls

I prefer to ride donkeys, even fully trained ones, with snaffle bits. In reasonably sensitive hands most snaffles are mild bits and comfortable for a donkey to wear, yet in an emergency the donkey can be "doubled" (turned using a single rein) without damaging his sensitive mouth.

A snaffle bit works on a donkey's tongue, his lips, and the bars and corners of his mouth. A snaffle never has shanks. So-called Tom Thumb and Argentine snaffles are curb bits, not snaffles. Eschew snaffles with twisted wire or small-diameter mouthpieces; they're unnecessarily severe, and a well-trained donkey never requires such a bit.

Your donkey's first snaffle should be a full- or half-cheek model, or an O-ring or D-ring snaffle fitted with a chinstrap. Cheekpieces or a chinstrap prevent your donkey's bit from pulling through his mouth until he gains experience and responds to subtle pressure on the reins.

If you'd rather ride without a bit, try a sidepull, a bitless bridle that allows reins to be attached to your donkey's headstall at the side of his face. There are

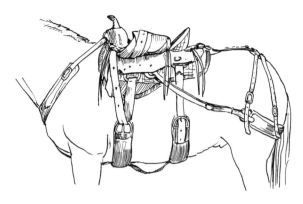

'Ear's a Tip

Before the Bit, the Dentist. Before introducing your donkey to a bit, schedule an appointment with an equine dentist or veterinarian. Wolf teeth (extras growing where teeth shouldn't be), sharp edges on molars, and the retained caps of baby teeth can interfere with the bit, causing considerable pain and resistance.

The saddle is unlikely to slip when stabilized by a breast collar and britchen.

Western sidepulls with nosebands made of leather, braided rawhide, or stiff lariat rope, and fancy general-purpose sidepulls fashioned of leather or Biothane. The Diana Thompson sidepull bridle (*see* Resources) is custom-made to measure, making it ideal headgear for saddle donkeys.

Breast Collars, Cruppers, and Britchen

Because donkeys are "mutton withered" (meaning their withers aren't prominent enough to properly stabilize a saddle), some trainers recommend using two extremely snug string cinches to hold a donkey's saddle in place. I disagree. I like to breathe freely and believe that don-

keys do, too, so I use a breast collar and a crupper or britchen when I saddle up my long-eared friends.

A breast collar prevents the saddle from sliding back while riding uphill, and it also helps to stabilize the saddle during mounting. There are many styles on the market, and most work well when they're properly adjusted. I prefer Y-style breast collars as they allow more adjustment than other types, and I recommend 100 percent mohair string models because they absorb sweat and conform to a donkey's shoulders.

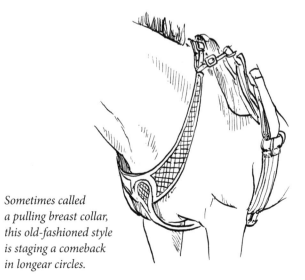

Sometimes called a pulling breast collar, this old-fashioned style is staging a comeback in longear circles.

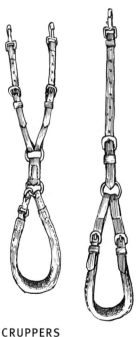

CRUPPERS

A crupper is a strap designed to keep the saddle of a donkey, mule, or mutton-withered horse or pony from sliding forward; it's also used with driving harnesses. Most riding cruppers have a snap (or snaps) used to connect the crupper to D-rings on the saddle, a buckle to adjust the crupper's length, and a padded, rounded, buckle-in strap that passes under the donkey's tail. It should be snug enough to keep the saddle in place but not so tight that it annoys the donkey or that it abrades the skin under his tail.

There are two types of riding cruppers: a straight strap that attaches to a single saddle dee and a Y-shaped strap that attaches to an equidistant pair of dees. They work equally well as long as the padded tail loop is soft, smooth, and well made.

Cruppers work best for donkeys ridden on flat or gently rolling terrain. In hilly or mountainous conditions, britchen is a better bet.

Britchen (pronounced BRITCH-en) is slang for "breeching," the part of a driving harness that drapes across an animal's hindquarters and acts somewhat similarly to a vehicle's brakes. "Britchen" is the term most donkey and mule owners use to describe a harness-like series of straps used like the crupper to prevent a donkey's riding saddle or pack saddle from sliding forward in steep hill country. Most ready-made britchen is sized for mules and Mammoth Jackstock, but custom saddlers will craft it to order in smaller sizes.

HOW-TO
Saddle Up!

Assuming that you've purchased a trained riding donkey and use a Western saddle, here's how to tack up to ride.

1. Halter your donkey and tie him to a stout, stationary object, then collect his riding gear. Hang up his bridle and set the saddle on its gullet, draping your saddle blanket or pad over it so that neither it nor the saddle lining picks up dirt or burrs.

2. Groom the donkey, paying special attention to areas where tack will lie. Take time to pick his hooves carefully.

3. Standing on your donkey's near (left) side, place the pad on his back well ahead of where the saddle will rest and then slide the pad back into place. This smooths the hair on his back and helps to prevent rubbing and chafing. If you use additional padding such as a Navajo blanket, place it atop the pad.

4. Drape the cinches and any additional harnessing (breast collar, crupper, or britchen) over the seat of the saddle, then hook the off side (your donkey's right side) stirrup over the saddle horn. Place the saddle on the donkey and gently shake it so it settles into place. Go to the off side and make certain that the saddle sits squarely on the pad. If not, don't scoot the pad forward — it will bring the hair with it. Either lift everything clear of the donkey's back and then move it forward and slide it into place, or remove the gear entirely and start over.

Britchen should ride low on the hip. This is a perfect fit.

5. Take down the off-side stirrup and cinches, then go back to the near side, leaving the additional harness strapping draped across the saddle. Hook the near side stirrup over the horn or lay the fender across the seat of the saddle. Always fasten the front cinch first. If you're using a breast collar and a crupper or britchen, cinch up snugly but not so much that you cut the donkey in two. If your saddle has a rear string cinch, it should be snugged up using roughly the same amount of tension as the front one; if

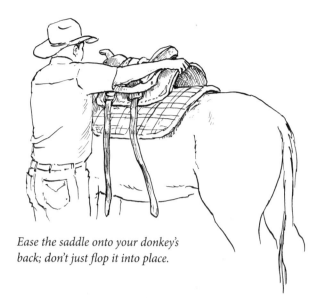

Ease the saddle onto your donkey's back; don't just flop it into place.

SYNTHETIC SADDLERY 101

Leather saddlery is wonderful, durable stuff, but it has to be cleaned frequently. It's also more expensive than the synthetic saddles, bridles, britchen, and cruppers that many riders favor today. If tradition beckons, leather will always be a logical first choice, but when price and practicality are more important, choose synthetics instead. Donkey strap goods such as bridles, cruppers, and britchen are crafted in leather, nylon web, Biothane, and Biothane Beta models.

- **Nylon web tack** has been around for a long time. Nylon tack is strong, inexpensive, and easy to clean but not often especially comfortable to wear, so Biothane and Biothane's Beta series are much better choices.

- **Biothane** is polyurethane-coated nylon. It's strong and durable, requiring little or no care. Biothane doesn't crack or peel, it won't absorb sweat, and it's easy to clean by just hosing it off. It keeps its flexibility even in subzero temperatures and stays flexible for its lifetime. Biothane won't chafe your donkey, and it conforms to his shape with use. It should outlast and outperform both leather and nylon, and it comes in a dazzling array of colors (one company offers Biothane tack in 12 opaque and 9 translucent colors).

- Made by the same company that developed Biothane, **Beta** is covered with a soft PVC vinyl

coating instead of urethane, it comes in two thicknesses ($1/16$" and $1/8$"), and it looks and feels like leather. It's easy to clean (just dip it into warm water and wipe it dry) and a bit more flexible than regular Biothane. However, it has a softer coating, so it isn't as scratch and wear resistant as regular Biothane.

Synthetic saddles are another logical option. Some synthetic English, Australian, and endurance saddles have adjustable trees, making them ideal for hard-to-fit donkeys. For instance, owners use a special Allen wrench to adjust the trees of Thorowgood saddles, and Wintec saddles (*see* Resources) are built on the company's patented Easy-Change Gullet System that allows an owner to easily remove the old pommel arch from his saddle and replace it with a new one in narrow, medium-narrow, medium, medium-wide, wide, or extra-wide width.

it has a leather back cinch, snug it up, but not too tightly — you should be able to force the flat of your hand between the donkey's belly and the fastened cinch. Do not let it hang loosely in the way that some horse riders fasten rear cinches. It can't do its job unless it's snug, and if your donkey should catch a hind foot in that dangling cinch, someone is bound to get hurt.

The rear cinch should be snug but not too tight.

6. Fasten the breast collar, making certain that it's adjusted properly and that you can insert no more than two stacked fingers between the front of it and your donkey's chest.

7. If you're using a crupper, unsnap it from your saddle. Feed your donkey's tail through the padded tail loop, gently pull the crupper forward, and reattach it to the saddle dee. Some donkeys prefer that the crupper be kept snapped in place and the padded tail loop unbuckled on one side, passed under the tail, then rebuckled. Follow whichever protocol you and your donkey prefer.

8. If you're using britchen, drop it into place, then buckle it to whichever saddle and/or cinch dees your particular britchen attaches to (this varies by make and model).

9. When you're finished, take down the near side stirrup and walk to your donkey's front end. Facing him, lift each foreleg up and straight forward to stretch the skin beneath his front cinch; this helps prevent pinching and cinch rubs.

10. Next, pick up your bridle. Unfasten the crownpiece of your donkey's halter and rebuckle it around his neck so that he's still safely tied.

No, this woman isn't shaking hands with her donkey; she's unwrinkling the skin in his "armpits" so that the cinch won't rub painful sores.

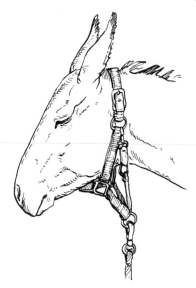

Refasten the halter around your donkey's neck so that he doesn't stage an impromptu escape.

11. If he doesn't mind having his ears handled (if he strongly objects, you'll have to teach him not to mind or switch to a snap-crown head-stall), face in the same direction that he's facing and, holding the bridle's crownpiece in your right hand, pass your hand between his ears so that your wrist or forearm rests between them. Use your left hand to gently guide the bit into his mouth. If he resists taking the bit, insert your left thumb and first and middle finger into opposite sides of his mouth and press gently on his tongue. When the bit is in, still grasping the crownpiece with your right hand, use your left hand to carefully draw his ears through the headstall, then slide it up into place and fasten the throatlatch.

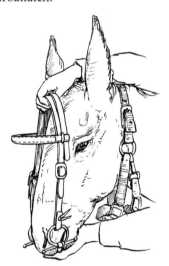

Gently ease the bit into your donkey's mouth, and then ease the bridle into place with your right hand.

12. Before mounting, walk your donkey for a few minutes, then check to make certain that both cinches are still snug and nothing is amiss. And that's it — you're ready to ride!

When unsaddling, repeat this process in reverse. Fasten the donkey's halter around his neck, remove the bridle, and slip the halter onto his head. Unbuckle any harness strapping and drape it over the seat of the saddle. Unfasten the back cinch first, then the front.

Go to the off side and drape both cinches across the saddle seat, hook the stirrup over the saddle horn, then return to the near side and remove the saddle by lifting it and gently sliding it toward you, off your donkey's back. Place it on its gullet and drape the pad over it, sweaty side up. If your donkey is hot, walk him dry and then groom him, pick his feet, and put him away.

Training Your Donkey to Ride

Sensible, sweet donkeys take to riding like goslings to water — it doesn't take much to teach a tame, well-mannered donkey to accept a saddle and bridle, stand to be mounted, rein reasonably well, and reliably walk and trot along the trail.

Before You Begin

Before training your donkey to ride, he should lead at your side, back up when you ask him to, and stand patiently wherever he's tied. You'll need an assistant for certain phases of his training and a safe enclosure where you can work. You will also need:

- ✗ A training surcingle
- ✗ Side reins
- ✗ Riding equipment, including a snaffle bit and (at the minimum) a crupper
- ✗ Clickers and your donkey's favorite food rewards (optional)

Getting Started

First, accustom your donkey to training gear. Tie your donkey in the training enclosure using a comfortable halter and a strong lead rope, and allow him to examine his surcingle. When he touches it, praise him or click and reward. When he's comfortable with it, place it across his back, then gradually tighten it until it's snug but not too constrictive. Continue praising or click/rewarding each step. If at any point he becomes

Riding with a Switch

Have you noticed that many photographs of donkey riders in lesser-developed countries show them carrying sticks? No, they don't use them to beat their long-eared friends! The sticks are used to guide them precisely as stated in the quote below.

[In Mexico] the donkey is guided by the voice, a stick, or a rope-halter. The halter rope lies on the left side and is pulled to turn him to the left, or borne across the neck to turn him to the right. The stick is used on either side if you desire to turn him to the other. Or the least rising of the stick suffices; while, if you are walking behind him, a mere touch on either flank will turn him quickly and surely.

— Theodore Ayrault Dodge,
Riders of Many Lands (1894)

This early postcard shows a Haitian woman guiding her donkey with a stick.

frightened or resistant, ignore misbehavior and back up a step or two so that you can click/reward for positive behavior, then proceed again, a bit more slowly.

When the surcingle is old hat, introduce the saddle and his other gear, starting with his saddle pad or blanket, following the same procedure. Reward him with praise or a click/reward at each step, and be certain that he's comfortable before taking another.

Many donkeys learn this much in a single session. If yours does, quit for the day to allow him to "chunk down" what he's learned. If it takes more than one session, that's okay, too, but don't move on until he's comfortable wearing tack.

At the next session, saddle him, then lead him around wearing his tack. It will feel strange, so reassure him and praise or click/reward when he relaxes. That's enough for this lesson; it's an important one, so stop when he gets it right.

Wearing a training surcingle while tied introduces your donkey to pressure around his heart girth.

Bitting

His first few bitting sessions are very simple: without fanfare, gently put the snaffle bit and headstall (no reins) on him and turn him loose to walk around and think about this thing in his mouth. Stay nearby in case he gets into trouble, but give him space to figure out the bit on his own. As part of the next few sessions, slip the headstall off and back on several times, always praising or click/rewarding him for taking the bit.

When he's comfortable carrying a bit, add a surcingle and side reins. Adjust the side reins so that there is no pressure on the donkey's mouth unless he tries to drop his muzzle to the ground, then turn him loose to digest this new turn of events. Allow several sessions for this important step, shortening the reins slightly each time until your donkey is holding his head in a natural, relaxed riding position as shown in How to Make Side Reins.

Now shorten one rein — not a lot, but enough so that there is slight pressure on that rein. After 15 minutes, switch reins. Allow several sessions for your donkey to learn to respond to bit pressure before moving on.

First Rides

If you've taken time to accustom your donkey to his tack, the first ride should be a nonevent. Tie him and saddle him as usual, but leave him unbridled (but haltered) for the first few rides. Ask your assistant to stand at his head, holding him, and praise or click/reward him while you sit in the saddle the first few times. You can approach this in two ways.

1. Mount from a mounting block. This is the best way, because it's easier on his back, the tack stays put, and it's much less likely to frighten him. Stand on the block next to him while the assistant rewards him for being good, then gradually lean out over the saddle so that he sees you looming above him. When he's relaxed,

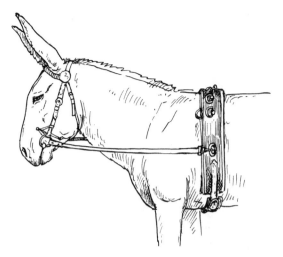

Side reins teach a donkey to yield to mild pressure on the bit.

How to Make Side Reins

Make your own inexpensive, stretchy side reins out of rubber bungee cords:

- With your donkey's head held in a natural position, measure from his bit to the uppermost rings on his surcingle and add three inches.
- Purchase two rubber bungee cords of this length.
- Fasten snaps to the S-hooks on the ends of both bungees and hammer the S-hook closed.
- Adjust their length by tying knots in the bungee cords.

These plain Jane side reins are stretchier, more adjustable, and work just as well as expensive ready-mades!

slide smoothly into the saddle, then make a huge fuss telling him what a fine donkey he is.

2. Mount from the ground. This is scarier for your donkey, so be lavish with praise or click/rewards when he's good. To mount from the ground, place your foot in the stirrup and put some weight on it. If he isn't alarmed, repeat this step several times. If he's frightened, let him settle down, then use your hand to put weight on the stirrup until he's relaxed and then work up from there. Next, step up and drape your body across the saddle so that you can slide off or step down if you need to. When he accepts that, swing your leg up over his back and settle into the saddle. Be sure to praise him to the stars.

Studies indicate that using a mounting block is much easier on your donkey's back than mounting him from the ground.

When he's comfortable with you on his back, ask your assistant to lead him while you ride. Use a halter so that the assistant is in control. Stop, start, and turn both ways. When she stops him, both you and your assistant should tell him to *whoooooa*. Repeat this for two or three sessions.

The next step is to add the bridle and ride by yourself, but ask your assistant to walk by your donkey's head in case he becomes frightened or confused. Concentrate on smooth, easy stops and turns.

To stop, squeeze the donkey lightly with both legs and alternate gently pulling back on the reins and then releasing them. The moment he starts stopping, praise and click/reward; never simply haul back on the reins.

To turn right, lightly squeeze the donkey with your right leg, while moving your left leg forward a bit and pressing lightly on the donkey's left side (imagine that your right leg is a post and that you're using your left leg to help him to turn around the post). At the same time pull the right rein out to the side at withers height, while keeping the left rein still. Lightly pull and release the rein, pull and release, until your donkey takes a step (or even just leans) in that direction. Reverse the process to turn to the left. Praise your donkey or have your assistant click/reward his first successful efforts and build on that solid foundation.

Finally, ride him totally on your own. Continue praising him lavishly for good behavior (you can click/reward from the saddle, if you like). Congratulations — you've saddle-trained your donkey!

Additional Basic Training

When your donkey is at ease with being ridden in the training enclosure, you can start riding him out in the big, wide world!

One of the best ways to teach a donkey about many things is to ride him in the company of well-trained donkeys, horses, or mules (trail rides are the perfect

Donkeys are rarely swift enough to compete against horses in speed events, but they shine in these events at donkey and mule shows.

training venue). Their presence will calm him, and he'll want to do what they do, like walking past scary things, crossing water, and standing quietly when tied to a picket line.

On the Trail

For one reason or another, many of us usually ride alone. When it's just you and your donkey on the trail, there are lots of things that you can do to make riding a safer venture. An important one: Always tell someone where you're going and when you think you'll be back. If no one is at home when you leave, call someone or leave a note where it's certain to be seen. Then go where you said you would, and try to be back on time. If you're late, that's a signal that something may be wrong, and if anything should happen to you or the donkey you're riding, searchers will know where to find you.

Resist the urge to explore new territory while riding alone. Follow trails familiar to both you and your donkey. And don't ride by yourself unless your don-

key is reasonably well trained or you're an extremely accomplished rider. School a green donkey at home or when riding with others, not when you're out on the trail all alone.

Even if you're a strong, experienced rider, saddle up when you ride by yourself. Riding bareback is fun and you may prefer it, but even trailwise equines sometimes act up. If your donkey and you come to a parting of the ways, you'll have to walk (and he'll be in danger if he heads on home without you), so stack the odds against being tossed, every way that you can.

Be at home before nightfall. Trail hazards — holes, old wire, fallen tree branches, and boggy mud — are much harder to see in the dark, and if your path follows a roadway, drivers probably can't easily spot you and your donkey.

Check the weather forecast before hitting the trail. Getting caught in a downpour isn't much fun, and you might not find shelter from the storm. Finally, when riding out alone, always wear a helmet. The skull you save will be your own.

HEY HO AND AWAY WE GO

The rousing campfire tune we know as "Donkey Riding" has nothing at all to do with riding real, live donkeys. However, the lyrics imply actual donkey riding, and because my donkeys love to hear it sung (and yours will, too), here's the story behind the lyrics.

"Donkey Riding" is a traditional work chantey sung by sailors of British descent as they loaded goods on the decks of sailing ships. When loaded, the ships sailed away to England, Fortune Bay in Newfoundland, or around Cape Horn in South America. "Riding the donkey" refers to loading cargo with the assistance of a steam-powered donkey engine. The donkey engine was a single-cylinder model invented in 1881. It quickly revolutionized logging and the loading of ships.

"Donkey Riding" should be sung with a hearty, swinging beat. The same tune but sung with different lyrics was a favorite of well-known folk groups like the original Kingston Trio, Bonnie Rideout, and the Tannahill Weavers, who recorded it as "Hieland Laddie, Highland Laddie," and "High Caul Cap," respectively.

Donkey Riding

Chorus (sung after each verse):
Hey ho and away we go,
Donkey riding, donkey riding,
Hey ho and away we go,
Riding on a donkey.

Was you ever in Quebec,
Donkey riding, donkey riding?
Stowing timber on the deck,
Riding on a donkey!

Was you ever 'round Cape Horn,
Donkey riding, donkey riding?
Seen the Lion and the Unicorn,
Riding on a donkey.

Was you ever in Mobile Bay
Donkey riding, donkey riding?
Loading cotton all the day,
Riding on a donkey.

Was you ever in London-town,
Donkey riding, donkey riding?
See the King in his golden crown,
Riding on a donkey?

Was you ever in Broomilaw,
Donkey riding, donkey riding?
Where the boys dance heel and toe,
Riding on a donkey.

Gearing Up for the Trail

Invest in a horn or cantle bag or a saddle pad with pockets to carry your gear. An alternative: Add saddle strings to your Western saddle if it doesn't already have them; you can use them to carry a jacket, a halter, or a bag containing your lunch.

Pack along a halter and a sturdy lead rope. You'll need them in an emergency or even if you just dismount to eat or rest. Never, for any reason, tie any equine by his reins! If you do and he yanks back, he'll hurt his mouth and the reins will probably snap — and he might decide not to be caught. Knotted rope halters with attached leads are especially handy to pack along; they can be tightly wadded and stuffed in small saddle pockets.

This postcard is labeled "Scenes et types — Jeune Arabe" (an Arab girl). It is addressed to "Francisco," in 1911, in Italian.

No pockets? No rings? You still need a halter and lead, so if there's no other way to carry it, leave your donkey's halter on under his bridle when you tack him up. Then double an unattached lead rope so that its ends are even and, leaving no looped ends to snag on branches, knot it snugly around your waist like a belt.

One important piece of equipment should not go into your saddle pouch, and that's your cell phone. If you have one, take it with you, but carry it in one of your pockets. Then in an emergency you can call for help, even if you've fallen or been thrown.

Unless you're riding on your own property, don't take your dog along. Loose dogs get injured in traffic and attacked by other dogs, and they can also chase livestock or get lost. You can't control your dog from the saddle, so it's in everyone's best interest if he stays at home.

On the Road

If your trek includes roadway riding, walk (don't trot or canter) on the roadside, not on the pavement itself. Hardtop is slippery, so avoid it if you can.

In most states it's correct to ride on the right, moving with the flow of traffic. Stay alert and stay as far to the right as you can. If you hear a noisy vehicle approaching from either direction, keep your donkey walking but turn his head away from the racket (shorten your right rein so that he's looking a bit to the right) and take a good, deep seat. If you think he might spook, dismount and hold him until the vehicle has passed.

As you ride along the grassy berm, watch for broken glass, wire, rotted metal, and other junk. Holes — sometimes deep ones left when signs were removed — pose serious hazards, too.

If you pass playful pastured equines, scary farm animals (like pigs or llamas), or anything else that you think might upset your donkey, dismount and lead him past. The exception: unfriendly dogs. Stay in the saddle. In your most authoritative voice, order them to "Down!

Stay!" or point and tell them to "Go home!" If that fails and they attack or worry your donkey, swallow your pride and yell for help.

Across Country

It's easy to get into trouble while riding across country, especially when you're riding alone. Don't cross deep water, even if you're familiar with that particular crossing. Avoid bogs and mire where your donkey might become stuck. Walk slowly up and down steep hills; don't let your donkey rush. Watch for rocks, slippery mud, low-hanging branches, old wire, holes, snakes, and wasps.

Don't trot or canter along any trail or across any meadow unless you know that the way is safe. If you'd like to jump something, examine the footing on both sides before you do it. And don't jump unless you know that your donkey is capable and reasonably willing. If you fall, either of you might be seriously injured.

If, despite your planning, you get caught in a storm, try to find shelter at once. If the wind is strong and gusting, shelter in brush but away from taller trees. If you suspect that a tornado is approaching, quickly untack

'Ear's a Tip

Riding Etiquette. When attending functions where horses are present, remember that many horses have never seen a donkey firsthand, much less heard one bray. Don't press close to a horse on the edge: Move off at a distance and ask the horse's rider how she'd like to proceed. Most riders will want to introduce their horse to your donkey — but slowly.

and release your donkey so that he can run away, lie down in a shallow gully as far from tall trees as possible, and stay there until all danger is past. If you're caught in a thunderstorm with lightning splitting the air close by, stay away from lone trees and open spaces, water, and hilltops. If you must seek shelter in woods, avoid the tallest trees. Weather emergencies are scary, especially when you're on your own, but if you stay calm and know what to do, you and your mount should be okay.

Riding in General

By walking the last mile home, your donkey will arrive nicely cooled down and ready for turnout. Unsaddle and treat him to a nice, quick grooming, and then give him his freedom. He's earned it!

Once your donkey is comfortable walking with you on his back, add trotting to his riding repertoire. Some donkeys trot readily; others need considerable encouragement. It's always best to reward good behavior until he understands what you want; however, if he's simply lazy you may have to use negative reinforcement to create behavior that you can reward. If you must, use a riding crop or a switch, but only enough to encourage forward movement. When he responds, praise him. He'll soon learn exactly what you want.

A well-trained donkey can do anything that a horse can do (and sometimes better) from winning Western pleasure classes to jumping to dressage. While these highly trained donkeys readily lope or canter, your everyday pleasure-riding donkey probably won't unless you insist quite vigorously. It can be done, but why? Trail riding a donkey is such a sweet, laid-back pleasure that I'm simply not in that big of a hurry. If I want to lope, I ride a horse. Your donkey will be glad if you do, too.

Muhammad's Talking Donkey

And be moderate in thy pace and lower thy
voice; for the harshest of sounds without doubt
is the braying of the donkey.

— Qur'an (31:19)

The prophet Muhammad took pity on beasts of burden. He forbade the beating of animals as well as branding or striking them on the face. When he encountered a donkey that had been branded on the face, he exclaimed, "May Allah condemn the one who branded it." He also once said, " . . . there exist among the ridden ones some who are indeed better than their riders, and who praise their Lord more worthily."

At various times Muhammad rode a white mule (Duldul) and a talking donkey (Ya'fūr). The fourteenth-century Islamic scholar Ismail Ibn Kathir wrote of the sayings of Muhammad about Ya'fūr:

> When Allah opened Khaybar to his prophet Muhammad . . . he [Muhammad] received as his share of the spoils: four sheep, four goats, ten pots of gold and silver and a black, haggard donkey. The prophet . . . addressed the donkey asking, 'What is your name?'
>
> The donkey answered, 'Yazid Ibn Shihab. Allah brought forth from my ancestry 60 donkeys, none of which were ridden except by prophets. None of the descendants of my grandfather remain but me, and none of the prophets but you, and I expected you to ride me.'
>
> The prophet . . . said to him, 'I will call you Ya'fūr, oh Ya'fūr.'
>
> Then the donkey replied, 'I obey.'
>
> The prophet then asked, 'Do you desire females?'
>
> The donkey replied, 'No!'
>
> So the prophet used to ride the donkey to complete his business and if the prophet dismounted from him he would send the donkey to the house of the person he wanted to visit and Ya'fūr would knock at the door with his head. When the owner would answer the door, the donkey would signal to that person to go see the prophet.
>
> When the prophet died, the donkey went to a well belonging to Abu Al-Haytham Ibn Al-Tahyan and threw himself in the well out of sadness for the prophet's death, making it his grave.

CHAPTER 14

DONKEY DRIVING

*It is not the ass that draws
the cart, but the oats.*

— Russian proverb

There are few finer ways to enjoy the company of donkeys than through driving, but only if it's done right. Done haphazardly or incorrectly, driving is fraught with danger. Beginners should take lessons or ask an experienced driver to teach them to harness, hitch, and handle the whip and reins.

However, any reasonably experienced driver can choose appropriate equipment and teach most donkeys to drive, because donkeys take to harness like mice to cheese.

Getting Started

Before choosing a driving donkey and his gear, decide what sort of driving goals you have in mind. Will you be leisurely walking and trotting down dirt roads by yourself or with another person by your side? Giving rides to a group of passengers? Showing? Showing requires fancier equipment and a flashier donkey who responds to cues in an instant. Trail driving in groups of horse-drawn vehicles? Trail driving means keeping up with horses, so driving a particularly phlegmatic Miniature probably won't do. More passengers? You'll need a buggy or wagon instead of a cart.

Because this is an introduction to donkey driving, let's keep it simple. Let's say that you want to train and drive an easygoing donkey to a basic two-wheeled cart. From there you can move on to showing, trail driving, multiple hitches and wagons (if you like) or simply enjoy your everyday, pleasure-driving donkey as is.

A (Very) Basic Driving Lexicon

Cart. A two-wheeled drawn vehicle.

Blinkers. Also called blinders and winkers. Flaps on a driving bridle that are positioned beside the wearer's eyes so he can see straight ahead but not to the sides and rear.

Buggy. A basic four-wheeled vehicle.

Carriage. A fancier four-wheeled vehicle.

Dash. The section of floor that curves up in front of the driver.

Groom. A second person in a vehicle whose job it is to assist the driver.

Header. Someone who jumps out of the vehicle, goes to the donkey's head, and steadies the animal in an emergency.

Putting-to. Also called hitching. The act of attaching a harnessed animal to a vehicle.

Shafts. Two metal or wooden poles attached to the front of a vehicle and that parallel the donkey's sides when he's driven.

Singletree. Also called a whiffletree or swingletree. A wooden piece, with a swivel point in the center, to which the harness traces are attached.

Wagon. A heavy-duty, four-wheeled vehicle (the pickup truck of the driving world).

Whip. Not only a piece of gear that a driver carries, but also a recognized term for the driver.

Choosing a Driving Donkey

The best pleasure-driving donkeys aren't sluggards, nor are they excitable donkeys who spook at inconsequential things or overeager animals who constantly worry the bit. Fortunately, most donkeys neatly fill the bill.

Choose a donkey big enough to do the job that you have in mind. This brings us to the oft-asked question: How much weight can a donkey pull? The answer is: it depends. Donkeys can pull 15 to 20 percent of their own weight, but this isn't based on the weight of a vehicle and its passengers; it's based on the amount of weight on the donkey's back.

Try this. Ask two people (a typical load) to sit in a typical cart while you pick up the shafts. If the cart is well balanced, there should be very little weight on the shafts when they're held at working height, parallel to the ground. Keep in mind that balance changes as additional passengers are added to the mix; good carts have a means of adjusting balance so that even when loaded there is still little weight put on the harnessed donkey's back. A Miniature Donkey hitched to a finely balanced cart can pull more weight than a Standard donkey hooked to a poorly balanced cart that exerts a great deal of pressure on the animal's back — as is often the case with homemade carts.

Other factors enter into the amount a donkey can comfortably pull. It's easier to pull a cart on level terrain than up and down steep hills; easier on smooth surfaces than across rough, rocky pastures; and easier when you drive 3 miles at a walk than trot full-out for 10. Even so, there is a limit to how much weight a donkey can manage, even hitched to the best of carts. Four-wheeled vehicles remove most of the weight from the hitched donkey's back, and they're easier to pull; thus a better choice for increased carrying capacity. However, buggies and wagons jackknife easily, so four-wheeled vehicles and inexperienced donkeys (especially with novice drivers at the reins) don't mix.

DETERMINING SHAFT SIZE

WITHERS HEIGHT	LENGTH (A)	MINIMUM WIDTH (B)	MAXIMUM WIDTH (C)	BREECHING STRAP IRON (D)
56" (14 hands)	72"	22½"	26"	28"
50" (12.2 hands)	72"	22"	26"	26"
48" (12 hands)	65"	18"	24"	26"
44" (11 hands)	63"	18"	24"	24"
40" (10 hands)	63"	16"	24"	24"

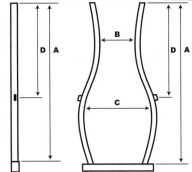

Adapted from "Harnessing Guidelines for Single Donkey Carts," Henk J. Dibbits (an article written in *The Netherlands*; look for it online).

Choosing a Vehicle

There are scores of wonderful donkey-size carts on the market, including old-fashioned pony-cart styles with balloon tires, fancy big-wheeled Meadowbrooks, cute restored governess carts, and various elegant gigs. Though many seasoned drivers disagree, I prefer a simple easy-entry metal cart with motorcycle tires. Because I clicker train, I need to get in and out of the cart I drive with relative ease. However, ease of entry is a primary consideration even when clicker training isn't part of the game. Most of the serious driving accidents I know of occurred when drivers' hands, arms, or legs got caught between wheel spokes, often while dismounting to address emergency situations.

Easy-entry carts have their drawbacks, too. They don't pull as smoothly as larger-tired vehicles, and their balloon tires are prone to punctures. Also, their metal shafts will bend rather than break, which is a bad thing in a serious wreck. Some are poorly constructed using shoddy materials. Still, a high-quality, well-balanced, easy-entry cart is a safe, easily handled vehicle for beginners (between the shafts, at the reins, or both). It's what I drive and what I recommend. However, no matter what type of cart you choose, it's wise to keep the following points in mind.

Start with a vehicle of recent make. Avoid the antique jog cart stored in Grandpa's barn or the cute but unrestored governess cart for sale at the Amish buggy maker's shop until your donkey is fully trained and you know how to evaluate and properly restore vintage vehicles. Most 50-year-old and older vehicles require complete restoration, and even newer ones usually need at least new (expensive!) wheels and often shafts.

Make absolutely certain that the cart you buy can be adjusted so that it's perfectly balanced when passengers are aboard. Some carts have movable seats; others move the axle via levers or winding mechanisms. If you don't

Did You 'Ear?

In 2002, a Poitou jennet foal was born in Australia, the result of artificial insemination of her biological dam and subsequent transplant of the resulting embryo into a Standardbred (horse) mare. In 2004, Quasar du Vermont, a purebred Poitou jack foal was born at the Hamilton Rare Breeds Foundation in Hartland, Vermont, the first Poitou ass conceived using frozen semen.

choose an easy-entry cart, make certain that you and any passengers you plan to carry can enter and exit the vehicle with relative ease and safety.

Choose a cart with a singletree *(see page 222)*. Cheap carts don't have them, and they are essential.

If the cart that you choose is fitted with foot brakes, you can slow a runaway or hold the cart back when traveling downhill. They aren't essential, but they're very, very nice to have.

A good suspension system equates with a comfortable ride. Easy-entry carts have coil springs; leaf suspension is usually a feature of better, big-wheeled wooden carts. Given a choice, remember that leaf springs absorb more shock. It also helps to have a cart with a padded seat and backrest. These are essential on long drives or rough, rutted roadway and pasture driving. By the same token, make certain that the seat's height is in proportion to the length of your legs; your balance will be off if you can't brace your feet behind the dash.

Some carts are designed with easily replaceable shafts and tires so that you can use them with animals of more than one size. With a change of tires and shafts and a few other minor adjustments, one of my carts fits my Boer goats and my Miniature Horse; if I had a Miniature Donkey, it would fit him, too!

Choose a cart with sturdy, well-made shafts. Pony carts and easy-entry carts with metal shafts should be fitted with welded end caps, not rubber stoppers that pop out and expose ragged ends. To determine whether a cart's shafts will fit your donkey, measure his length and his width at shoulder and hip. Proper shaft height can be figured at roughly 3 inches per hand (e.g., the top edge of the shafts, held horizontally, should be 33 inches from the ground when buying for an 11-hand donkey). Even better: Measure and match tug-height on the donkey to the corresponding shaft-height with a driver in position and the floor of the vehicle parallel to the ground.

Once you have a cart, take care of it so that it lasts. Check the air in the tires before and after you drive, airing up as needed. It's best to store your cart out of the weather, covered with a fabric throw or canvas tarp. At least twice a year (more often is better), tend to the maintenance suggestions on the next page.

This postcard is labeled, "An Irish Donkey Cart."

- ✗ Grease or oil the wheel bearings (some wheels have enclosed bearings; if yours does, you can skip this step).
- ✗ Wash it (including the wheels), dry thoroughly, then wax metal and wood parts with appropriate products.
- ✗ Treat the seat with vinyl or leather conditioner.

Driving Gear

Driving gear initially seems complex, but on closer examination it isn't. It consists mainly of a few essential parts and the straps and hardware needed to keep them in place.

BRIDLE. While well-trained donkeys should respond to oral commands, the bridle is worn for added control. It should be adjusted so that:
- ✗ The bit fits against the corners of the donkey's mouth
- ✗ The throatlatch is snug enough to prevent the bridle from slipping off but loose enough for the donkey to flex his neck
- ✗ The noseband is tight enough to keep the bridle close to his face but loose enough to easily insert at least two fingers laid flat between the lower jaw and the noseband

BLINKERS are used to help the donkey focus on the road ahead. The donkey's eye should be at the center of the blinker; the blinkers shouldn't touch the donkey's eyelashes.

REINS. Reins connect the donkey's mouth to the driver's hands. They fasten to the donkey's bit and pass through terrets (rings) on the neck strap and/or saddle.

BREAST STRAP. The breast strap passes around the donkey's chest just above the point of his shoulder. Adjusted too high, it would press against the windpipe and choke the donkey; too low, and much of the draft power is lost. The neck strap keeps the breast strap in the correct position, especially important in donkey harness. Without it, a working donkey's slanted chest causes the breast strap to slip down while he's pulling.

TRACES (AND THE SINGLETREE). A trace on both sides of the donkey transmits the force from the breast strap to the cart. They're fixed to a ring or buckle on the breast strap and connect to the singletree. The singletree should be long enough to prevent the traces from coming in contact with the donkey's body and the traces long enough to prevent his hind legs from touching the singletree.

When in motion, a donkey's legs and shoulders move backward and forward and so do the ends of the breast strap; therefore, the traces must be free to move as well. The singletree swings with the donkey's shoulder movements, thus preventing the breast strap from rubbing sores on his chest and shoulders; this is why cheap carts without singletrees are a very bad investment. Traces must never be fixed directly to the cart.

SADDLE AND GIRTH. The primary function of the saddle is to take the vertical force of the shafts over the donkey's back and spread the weight of the load over the donkey's rib cage. The girth prevents the cart from tipping backward. The saddle and girth work together to balance the weight of the cart.

BREECHING. Breeching acts as the cart's braking device. It prevents the cart from running forward into the donkey's rear end, and it enables the donkey to push the cart backward while backing up. Breeching straps attach to the shafts of the cart via a loop-type iron or to a hook. Some pointers:
- ✗ The breeching band should lie halfway between the donkey's tail head and his hocks; if adjusted too low, it interferes with the movement of his hind legs.

Miniature Donkeys are favorite driving animals and strong enough to pull a large adult in a well-balanced cart.

✖ A hip strap secures the breeching in place; the top should lie just behind the point of the hip.

✖ When the donkey is pulling, there should be about three to five inches (depending on the donkey's size) between the breeching band and the donkey's rump.

WHIP. The driver carries a whip to help cue the donkey's movements. It should be long enough to touch the donkey's shoulders and is never properly used behind the saddle area.

Choosing Harness

There are two types of pleasure-driving harness: breast-collar- and collar-style harness. Donkeys can pull more weight with the collar-style, but fitting a collar is tricky business, even for experienced drivers (collars are neither flexible nor adjustable, so they either fit precisely or not at all), and collar-style harnesses cost a good bit more than breast-collar models. Breast-collar-style harnesses offer more adjustability if the donkey who wears it gains or loses weight or if

At the other end of the spectrum, Mammoths make fine cart asses too.

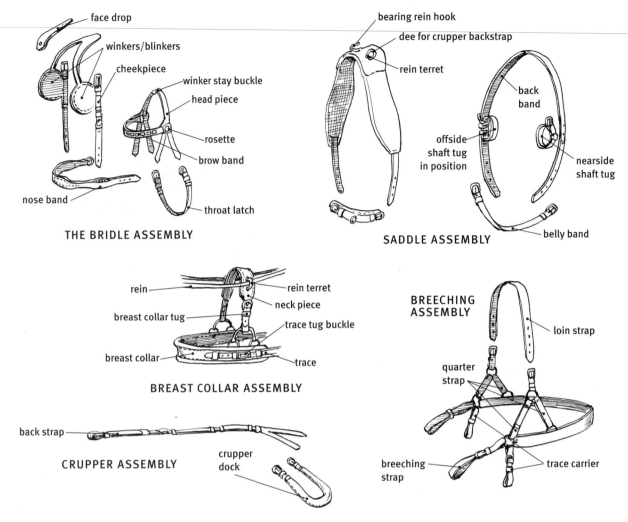

THE BRIDLE ASSEMBLY

face drop
winkers/blinkers
cheekpiece
winker stay buckle
head piece
rosette
brow band
nose band
throat latch

SADDLE ASSEMBLY

bearing rein hook
dee for crupper backstrap
rein terret
back band
offside shaft tug in position
nearside shaft tug
belly band

BREAST COLLAR ASSEMBLY

rein
rein terret
breast collar tug
neck piece
trace tug buckle
breast collar
trace

CRUPPER ASSEMBLY

back strap
crupper dock

BREECHING ASSEMBLY

loin strap
quarter strap
breeching strap
trace carrier

more than one donkey uses the same harness; some prefer it to collar harness (I do), and it's a better buy for beginning and occasional drivers, so that's what I recommend.

A good harness costs as much as a good saddle; plan to spend between $300 and $1,200 for a quality pleasure-driving harness, depending on size, craftsmanship, and what it's made of (leather or synthetics). *Don't* buy a harness manufactured of low-quality leather and shoddy hardware! Harness or hardware failure can and often does cause serious accidents resulting in injury to both donkey and driver. Eschew $120 eBay harnesses built in India of porous, weak leather. Better a second-hand, quality harness than a new one that fails under pressure.

Quality harness comes in leather and synthetics including plain nylon strapping, Biothane, and Biothane Beta *(see* Synthetic Saddlery 101 *on page 207)*. Choose between leather and synthetics, and when harness shopping, keep the following tips in mind:

✖ Reread tips one through five under Choosing a Saddle in chapter 13; all of those points apply when buying a harness, too.

✖ It's best to buy a harness designed for donkeys. A driving bridle designed for a pony or horse won't fit your donkey, and because driving bridles are more complex than most riding bridles, simply replacing a short browband won't do the trick.

✖ Alternately, order a custom harness constructed to your donkey's measurements.

- ✘ When properly adjusted, a harness should have extra holes for increasing and reducing size in the event that your donkey gains or loses weight or you'd like to use it on another donkey.
- ✘ If you're a beginner, take along an experienced driver when buying a used harness to make certain that all necessary pieces are there and in working order.

Teaching a Donkey to Drive

This is how I start my donkeys in harness. Some will say that it's simplistic, but this is what works for me. The process may take four or five short sessions or up to several months. The important thing is to make certain that the donkey thoroughly understands each

Begin your donkey's harness training by ground driving him in an open bridle (one without blinkers) and a training surcingle.

step before taking another. Donkeys are clever; given a chance to understand what you're asking, they're very, very easy to teach to drive.

I train my donkeys to ride before I drive them; that way they're familiar with the bit and giving to pressure before we start. Also, I don't drive before my donkeys are four years old; thus, they come to this training with mature minds, ready to absorb what I teach them.

Desensitizing Your Donkey

It's important to introduce your donkey to scary situations *before* he encounters them while hitched to a cart. Using your clicker, plenty of rewards (don't forget jackpots), lots of patience and praise, and the assistance of one or more helpers, introduce your donkey to umbrellas, barking dogs, children on noisy bikes, people behaving in unaccustomed ways (limping, leaping, lying on the ground), and any other such things that you can think of. If he hasn't already learned to walk through puddles, navigate bridges, cross painted road lines, and pass scary items — like pigs, road signs, bicycles, joggers, skateboarders, and manhole covers — take him to places where he can see and experience these things.

Asses on the Standard of Ur

The earliest depiction of vehicles in the context of warfare is found on the Standard of Ur (2600 BC), discovered by archaeologist Sir Leonard Woolly in one of the oldest of the royal tombs at Ur.

The vehicles depicted are skin-covered, double-axled battle wagons pulled by asses or onager mules. Although sometimes mounted by a spearsman as well as driver, these heavy protochariots may have been used for show rather than as actual vehicles of battle. The Sumerians also used a lighter, two-wheeled type of chariot. Both types had solid wooden wheels; the spoked wheel didn't appear in Mesopotamia until much later.

Groundwork First

Start by teaching the necessary commands on foot. You'll need verbal cues to move out ("get up," "hup," "step out"), to stop ("whoa"), stand still ("stand" or "stand up"), reverse ("back" or "back up"), trot from a walk ("trot," "trot on"), walk from a trot ("walk"), and verbal cues for right and left (traditionally "gee" and "haw," respectively).

Get out your clicker and bag of rewards and work until the donkey shows that he knows these cues forward and backward. "Whoa" is the most important command. Draw it out — "whoooooa," so that it sounds like no other word, and say it *only* when you expect him to stop. Don't use "whoa" once he's stopped and you want him to stand still. Whoa should mean one thing and one thing only: stop right now! Don't proceed until he has an absolutely foolproof whoa.

Next, teach him to ground drive. If you possibly can, recruit a helper to walk beside your donkey during the first few sessions to help him to understand what you want. Fit him with a standard riding bridle,

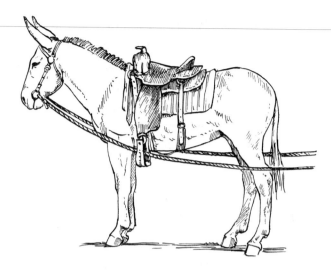

No surcingle? Fasten your stirrups together under your donkey's belly, and then run the long lines through the stabilized stirrups.

a cheeked snaffle or cheeked mullen mouth (straight bar) bit, and a training surcingle (or use your saddle with the driving reins run through the stirrups). Start in a round pen or similar small enclosure, then move to a larger enclosure, and finally out into the big, wide world. Walk, trot, turn, drive around pylons; click and reward until he's stopping, starting, and turning with precision.

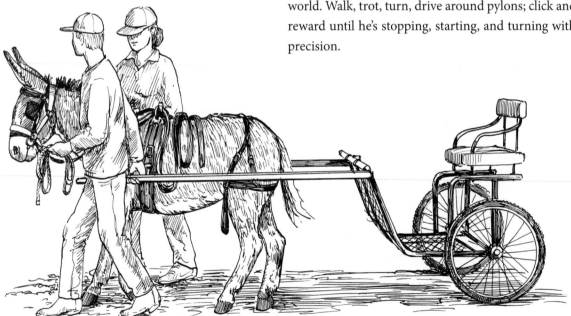

Lead a newly harnessed and hitched donkey until he's comfortable and then get into the cart.

Bring on the Harness

Harnessing most donkeys is no big thing. However, allow the donkey plenty of time to examine each item as you harness him up. If you ride with a britchen, the crupper will be new, so give him plenty of time to figure that out.

Most harnesses come with a blinkered bridle. This is something the donkey needs to get used to. Lead, then ground drive the donkey in harness until he's comfortable with the equipment that he's wearing and he willingly and consistently responds to verbal cues.

Next, the Cart

Ask your assistant to help again as you do the following:

- ✘ Take your harnessed donkey, wearing a halter with an attached lead over his driving bridle, to his usual enclosed training area and allow him to examine the waiting cart.
- ✘ Ask your helper to pull it around while the donkey watches.
- ✘ When he doesn't mind that, hold the donkey still and have your assistant bring the cart up behind him and lower the shafts.
- ✘ With you on one side and your helper on the other, walk the donkey forward, *unhitched,* while the two of you pull the cart along in place.

If your donkey is typical, he'll take these steps in stride. If he doesn't, *don't go any further* until he does.

Putting-To

Now it's time to hitch the donkey. Do so with a minimum of fanfare, then start by leading the donkey around. When he's comfortable, ask your assistant to walk near the donkey's head; you drive the donkey while walking beside or behind the cart. When all is well, either you or your helper can board the cart and drive, with the other walking along beside the donkey's head. Repeat this step for several sessions, making certain that he reliably stops, stands, and turns with relative ease, before you drive without the person walking by his side. Finally, drive outside of the enclosure.

It's always wise to take an able-bodied passenger along when driving, but it's *essential* when driving a novice donkey. If something frightens him, it's important to have both a header to assist and someone to steady him with the reins.

Harnessing Safety

Keep these tips in mind when harnessing and working around harnessed animals:

- If there is any doubt in your mind regarding how to harness and hitch correctly, ask an experienced person to show you how it's done. I can't stress this too strongly.
- Keep talking or keep one hand on a donkey wearing blinkers so that he knows where you are at all times.
- Never tie a harnessed animal by the bit.
- Don't leave a donkey hooked to a vehicle unattended, not even for a minute.
- Never hook an animal unless he is wearing a sturdy halter over his bridle and is safely and securely tied to a sturdy object (not a tree or fence post) by a strong lead rope.
- Don't stand directly in front of an animal who is hitched to a vehicle or in front of the vehicle's wheels.

Drive Safely

These tips and tidbits will give you and your harness donkey a safe and happy start.

- ✘ Double-check your equipment before setting out; make certain that everything fits correctly and is in good working order.
- ✘ Quietly enter and exit the vehicle; don't dawdle.
- ✘ The driver (with reins in hand) should be the first one in the cart and the last one out.
- ✘ Wear a riding helmet approved by the ASTM (American Society for Testing and Materials), particularly when driving a novice donkey; insist that passengers, especially children, wear them, too.
- ✘ Never tie yourself or any passenger into the cart.
- ✘ Keep your hands off the wheels; make certain that passengers do, too.
- ✘ Learn how to drive correctly, with both reins in one hand and the whip at the ready in the other. If you drive with a rein in each hand, you have no means of quickly shortening the reins in an emergency.
- ✘ Stay awake; don't let your mind wander. Watch your donkey's ears and his body language. Anticipate his actions and act accordingly.

Taking It to the Next Level

When you're ready to move on to more exacting driving, these helpful organizations will help to pave the way (*see* Resources *for their Web sites*).

- ✘ The American Driving Society is the primary driving organization serving the United States.
- ✘ Carriage Driving Community rightfully calls itself *the* online resource for carriage driving.
- ✘ Recreational Equine Driving, the largest, most active driving list at YahooGroups, actively encourages donkey drivers to join their friendly ranks.

BRAY SAY: *Being President is like being a jackass in a hailstorm. There's nothing to do but stand there and take it.*
— President Lyndon B. Johnson

Patron Saints

The patron saint of asses is Anthony of Padua. Born into a wealthy Lisbon (Portugal) family, Anthony gave up the privileged life to become a traveling evangelist. In addition to asses, St. Anthony is patron saint of (among many other things) seekers of lost articles, travelers, starving people, pregnant women, fishermen, sailors, swineherds, and amputees.

Two stories link St. Anthony with asses. The first is more commonly known. A heretic named Bonillo once challenged St. Anthony to prove the "fable" of the Holy Eucharist. Bonillo devised a contest whereby he would starve a donkey for three days, withholding all food and water from the poor beast, which he did. St. Anthony spent those three days on retreat in the forest in prayer. At the end of three days, St. Anthony went to the church to receive the Holy Sacrament, and then he met Bonillo and the donkey in the town square.

Bonillo placed a stack of feed 20 feet from the starving donkey, then he released the poor animal and it rushed toward the feast. At that moment St. Anthony exposed the Holy Sacrament and called to the ass, "Donkey, in the name of God I command you to come here and adore your Creator!"

The donkey came to an abrupt halt, turned, and walked to St. Anthony, where he dropped to his knees with his head toward the ground. In astonishment, Bonillo ceded the contest and converted to Christianity on the spot.

The other tale is more pleasing to donkey lovers: One day when St. Anthony went out to tend his garden, he found a group of asses dining on his vegetables. He gently took hold of one of the beautiful creatures and said, "Why do you eat what you have not sown, and why do you injure one who has never done you any harm? Go in the name of God and return no more." The donkeys never raided his garden again.

While St. Anthony of Padua is the official patron saint of asses, another saint, St. Germanus of Auxere, is nearly always depicted with an ass lying at his feet. According to the 1929 issue of *Blackfriers,* a publication put out by the Dominican order, "St. Germanus of Auxere deserves to rank as the patron saint of donkeys, seeing that he so preferred his own humble beast to the splendid horse offered him by the Empress Placida, that though it (the donkey) was in a dying condition, he raised it again to perfect health."

CHAPTER 15

FUN WITH DONKEYS

*It's only the Lord can make a racehorse
out of a jackass.*

— Irish proverb

Apart from riding and driving, are there other great things to do with donkeys? Oh, you bet! Here are a few more activities to consider.

Showing

Showing donkeys is like showing horses but a lot more fun. Donkey and mule shows are more laid back than most horse shows, and they're specifically planned so that competitors have a good time.

Donkey Treats

Pack along some of these for your next show:

- Apples, pears, pitted peaches or plums, peeled pineapple, rind-on cantaloupe and watermelon (to avoid the chance of choking, always cut fruit into bite-size chunks before serving)
- Carrots, turnips, kohlrabies (ditto)
- Unsweetened or lightly sweetened Chex and Kashi cereals, doled out one square at a time
- Commercial horse treats (for Miniatures, broken into donkey-size portions)

Make Mine Miniatures

Miniature Donkey owners can show at all-Miniature events or open shows for donkeys and mules of all sizes and kinds. *(See* Resources *for contact information for Miniature Donkey associations.)* Miniatures generally show in one of the classes listed below; the first two are about appearance, with the remainder judging performance — although a good many of these are primarily about having fun! (Showmanship is judged on the handler's performance along with the manner in which she has fitted her donkey for the class.)

- ✗ **Halter.** Judged on the donkey's conformation and movement.
- ✗ **Color.** Judged 50 percent on color and 50 percent on conformation.
- ✗ **Showmanship.** The handler's ability to prepare and show a halter donkey, rather than the donkey itself, is judged.
- ✗ **Driving.** Pleasure driving, obstacle driving, team driving, reinsmanship, turnout, and harness races.
- ✗ **Costume.** Riders and handlers deck their donkeys and themselves in costumes ranging from the sublime (Mary and the baby Jesus)

to the ridiculous (Bill Clinton perched atop the Democratic donkey). Costume divisions may be further divided into youth and adult versions of historic costume and just-for-fun events.

- ✘ **Trail and jumping.** Handlers lead the donkeys in these events.
- ✘ **Games.** Snigging, coon jumping, and musical tires (played like children's musical chairs).
- ✘ **Catch your ass.** A timed event in which donkeys are released at the opposite end of the arena and competitors run, catch their own donkeys, and race them back to the starting line.
- ✘ **Diaper race.** Contestants vie to be the first to lead their donkeys to the other end of the arena, diaper themselves, and race the donkey back to the starting line without the diaper falling off.

To learn more about the ins and outs of showing Miniature Donkeys, visit the Missouri Miniature Donkey Breeders Association Web site (*see* Resources) and download their show manual — it's free!

Coon Jumping

Coon-jumping donkeys and mules jump from a standstill from a box on the ground drawn with chalk; at North American Saddle Mule Association shows, coon jumping is a timed event. These are the Missouri Miniature Donkey Breeders Association's rules.

REQUIREMENTS. Standing, the donkey must stop after entering the 10' by 10' marked space in front of the jump. He/she must clear the jump starting at 12" and increasing 2" with each round. Clearing the jump means not knocking the bar down. This is done on an elimination basis: When a donkey fails to clear a height with two attempts he/she is disqualified. The handler may use his own blanket to place over the jump.

CRITERIA. Height of jump, with two tries.

DISQUALIFICATION

- ✘ Not clearing a given height within two attempts
- ✘ Stepping on the line or outside the 10' box before jumping
- ✘ Use of edible treats to encourage the donkey
- ✘ Inappropriate attire or equipment
- ✘ The donkey jumping before all four feet have stopped moving

Le Roman de Fauvel

The *Roman de Fauvel* is a long, satirical poem accredited to the fourteenth-century French royal clerk Gervais du Bus. Copies began circulating on the Paris literary underground in 1310, and it was published in 1314.

The poem, ostensibly about a donkey named Fauvel, is a biting commentary on the corruption of the royal court of France and the papal court at Avignon. The allegorical plot revolves around Fauvel who, unhappy with life in the stable, moves into his master's home, where he sets up shop and has a custom hayrack installed. Dame Fortune, the goddess of Fate, smiles on Fauvel and appoints him the leader of the house. Fauvel is subsequently visited by a parade of leaders (symbolizing church and state rulers of the day) who fawn over and groom Fauvel, thus coining the term "to curry Fauvel," which later became "to curry favor."

Fauvel's name is an acronym composed of the first letter of each of several sins: *Flaterie* (Flattery), *Avarice* (Greed), *Vilanie* (Guile), *Variété* (Inconstancy), *Envie* (Envy), and *Lacheté* (Cowardice); broken down to *fau-vel,* it also means "veiled lie." The poem was banned for being antiestablishment and heretical, yet it was so popular that a second book of Fauvel soon followed. Twelve copies still survive, including a lavish presentation volume prepared in 1316. This masterpiece features scores of illuminated miniatures and 167 musical inserts.

✖ The distraction of other handlers or donkeys by you or your donkey

✖ Mistreatment of the donkey

Snigging

Snigging (log dragging) is a popular timed event at donkey and mule shows. These are the Missouri Miniature Donkey Breeders Association's rules. The International Mule and Donkey Association adds that for Miniature Donkeys, singletrees are to be 18 inches in length and the log not over 8 feet long and 3 to 6 inches in diameter. North American Saddle Mule Association rules for donkey snigging are far more involved; consult the NASMA rulebook for further information.

REQUIREMENTS. The donkey, in harness, is ground driven and moves/drags a log through a serpentine course of cones. The log will be attached to a singletree and will be provided by show management. This is a timed event, so any gait is allowed.

CRITERIA. How quickly the donkey and handler can move a log through the course without knockdowns. Knockdowns will have a 5-second penalty and displacing cones will have a 2-second penalty.

✖ Equipment failure

✖ The handler moving the log

✖ The distraction of other handlers or donkeys by you or your donkey

✖ Mistreatment of the donkey

Open Shows

Open shows are generally not tailored for a specific type, height, or breed. Most every state and regional donkey and mule club sponsors one or more shows each year, ranging from purely-for-fun events up to huge shows with lots of great prizes.

To locate these shows, visit the American Donkey and Mule Association Web site (*see* Resources) for further information.

North American Saddle Mule Association Shows

The North American Saddle Mule Association is a large, active organization devoted to showing and promoting the mules and Standard-size or larger donkeys recorded in the NASMA studbook. If you're serious about show-

This undated vintage card is titled, "Learning to Ride."

ing, this is your group! Donkeys are allowed to show in many North American Saddle Mule Association classes, and they have their own show division as well. Visit the NASMA Web site (*see* Resources) to download their free, comprehensive book of showing regulations; it's an education in itself.

Of special interest are the Donkey Training Level classes written for novice donkeys. Sanctioned shows are encouraged to host at least three training level classes: Pleasure, Donkeymanship, and Trail. Donkeys are shown at a walk and trot (no cantering), and neck reining is not required.

Packing

When I was young I loved to backpack. Now, at 60 and counting, bulging packs of equipment have lost their allure. Fellow baby boomers, handicapped campers, and families with young children who have a lot of gear to backpack into the woods face similar problems: how to transport their equipment but save their backs. Enter the recreational pack donkey: a companionable, ecology-friendly answer to "Who's going to carry all of this stuff?"

Choosing a Pack Donkey

The donkey you pack should be stout (but not obese), healthy, have sound legs and good feet, and enjoy being around people. He should lead readily at your side, stand tied without fussing, and pick up his feet.

Miniature Donkeys are especially well suited for recreational backpacking. They're biddable, easy to handle, and close to the ground. There is no need to hoist packs as high as a horse's back, making it a pleasure for anyone (even children) to load or lead them, and they're small (and tidy) enough to transport to the trailhead in a van or an SUV.

Standard donkeys are traditional pack animals. They can pack more goods, yet they're still small enough to make loading and leading a breeze.

Few people recreational pack with Large Standards and Mammoths because they're traditionally ridden, not led. However, if your donkey is a big guy and you want to lead him, he can pack everything that you need to camp like a king! Make sure that your chosen donkey sees a veterinarian and farrier before undertaking a packing adventure. All shots must be up to date, and a Coggins test is necessary if you'll cross state lines.

BRAY SAY: *The donkey is for his size an excellent pack animal and is so employed in many countries including India, Egypt, Somaliland, Persia, and China. His pace is slow, compared with the mule, and his load is only 100 lb. against 160 lb., but they demand comparatively little attention and small rations; they will do well on poor classes of forage and are valuable transport on lines of communication. In India they are, like mules, selected by measurement at not more than eight or less than three years of age.*

They are particularly hardy and useful pack animals, if they are not overloaded or overdriven. Their pace is two and a half miles an hour, and they can cover 15 miles in a day (the ordinary distance of which mule transport is capable is 20 to 25 miles a day, carrying 160 lb. and the saddle; and this is only in good condition, if the rate is to be maintained for a considerable period).

— *Animal Management* (Her Majesty's Stationery Office, 1915)

Pack Donkeys in Ancient Assyria

Assyrian merchants in the ancient city-state of Aššur inscribed thousands upon thousands of cuneiform tablets between 2000 and 600 BC, many of which are still in existence. These records frequently refer to asses. From them we know that tin and textiles were transported from Aššur to central Anatolia on pack asses purchased in the Aššur area for about 20 shekels (170 grams) of silver each. Traders preferred large, black donkeys because these were considered hardier than the rest. Each ass packed about 200 pounds of goods in two sacks tied to the sides of a packsaddle with a third bag secured across the top. Caravans made up of 300 donkeys were common; the six-week, 650-mile trip was a hard one; some donkeys always died along the way. All but a few of the remaining donkeys were sold in Anatolia, usually for about 30 shekels each. The remaining few donkeys packed gold and silver back to Aššur.

Choosing a Pack

If when thinking of a donkey pack you visualize intricate backcountry packing gear, think again. Today's recreational packing equipment is streamlined and simple to choose, to adjust, and to use.

Recreational packing gear falls into one of four classes: training packs, companion packs, packsaddle and pannier combinations, and saddle panniers. Good packing gear of any type provides adequate padding, alleviates pressure on the spine, and is very stable, allowing for slightly different weights in each saddle bag without shifting to the side or moving while the donkey walks.

TRAINING PACKS (ALSO CALLED DAY PACKS). A training pack is a simple, soft-sided, saddlebag-like affair with a single girth and a built-in breast collar and britchen to keep it in place. The fabric connecting the two bags lies directly on the donkey's back; this is not a good thing — too much weight pressing directly on any pack ani-

Sagebrush Shorty poses with his hardworking team of donkeys.

mal's spine over a prolonged period of time can inflict permanent damage. Training packs are best used for carrying light loads on day hikes, and nothing more.

COMPANION PACKS. The companion pack is a nice choice for longer day trips or overnight camping. A good companion pack features a thick, divided pad that keeps the weight off its bearer's spine, detachable panniers (pack bags) like those used in full-scale packing gear, a single girth, and a built-in britchen and breast collar to keep it in place. Companion packs for packgoats fit most Miniature Donkeys, and llama-size versions adapt to fit Standard and Large Standard donkeys.

PACKSADDLE AND PANNIER COMBINATIONS. Packsaddle outfits include two large, detachable panniers and a sawbuck-style packsaddle secured with one or two wide girths and a britchen and breast collar. The sawbuck provides complete spinal relief, making it possible for the donkey to carry considerably heavier loads. Packsaddle and pannier sets designed for goats fit most Miniatures and smaller Standard donkeys, although minor adjustments are sometimes necessary. Llama packsaddles don't fit donkeys well; Standard and larger donkeys should wear packsaddles especially designed to fit them.

SADDLE PANNIERS (ALSO CALLED SADDLE PACKS). Saddle panniers are designed to drape over a full-size Western riding saddle. They're less expensive than packsaddle outfits, they're easy to use, and they come in a wide variety of fabrics and sizes. They're ideal for recreational packing with Large Standard donkeys and Mammoths.

DONKEY FAIRIES AND SCARIES

- The Portobello Brag, a *bogey beast* (a scary animal spirit) or *boggle* (a term applied to any sort of ghostly apparition) frequently spotted near Birtley, County Durham, in the North of England, frequently appeared in the form of an ass. Whoever mounted it was carried off at high speed, and then he or she was tossed into a bog or blackberry thicket. The creature would then run off "nickerin' or laughin' mischievously."

- The Phouka of Kildare was an Irish *pooka* (pronounced POO-ka and sometimes spelled *phouka*, *pooka*, or *pwca*; a type of mischievous bogey beast who was generally — but not always — helpful and well-disposed to the human race) who had the body of an ass but described itself as the ghost of an idle kitchen boy.

- The Hedley Kow was a bogey beast, mischievous rather than malignant, that haunted the village of Hedley, near Ebchester. He often took shape as a pony or donkey. He'd end his pranks with a horselaugh at the expense of his terrified victims.

- Other bogey beasts that sometimes appeared in donkey form were the *shock* and the *shag* (the latter favored the shape of a coal black donkey with fiery red eyes).

- Roman mothers warned their children of nasty beings that would come and eat them if they weren't good. One of these creatures was Mormo, a scary old woman with donkey's legs.

- In Greek myth the *Empousai* (or *Empusae*) were fearsome underworld creatures that, in the guise of beautiful young women, seduced gullible young men. The Empousai had flaming hair and mismatched legs: one was made of bronze and the other was the leg of an ass.

Unfortunately, packsaddles and panniers for Miniature Donkeys are hard to find (donkey entrepreneurs, take note), so until someone hops aboard the bandwagon and builds gear for Lilliputian asses, Miniature Donkey packers must adapt goat-packing gear (which fits most Miniatures surprisingly well) or custom-build their own (See Resources).

Several of the packing suppliers listed in the back of this book build "burro size" gear for Standard donkeys weighing 600 to 800 pounds; Mammoths generally wear equipment designed for mules.

Training and Conditioning a Pack Donkey

Training pack donkeys is the essence of simplicity. If your donkey is already trained to ride or drive, simply tack him up with packing gear (letting him examine each piece before putting it on him) and then head out!

Did You 'Ear?

The asses pictured in prehistoric rock carvings and cave paintings in Spain and France are probably onagers or Kiangs.

Pliny the Elder, Roman naturalist and author of *Natural History* (AD 23), advised donkey owners to stand their jennies facing north during breeding to assure the birth of jenny foals.

Thomas Pennant, writing in *History of Quadrupeds* (1793), cites a reference to donkeys in a document dated to the time of Ethelred II (AD 968–1016), making it the earliest reference to donkeys in England.

If he isn't already trained to ride or drive, turn back to chapter 13 and teach him to wear packing gear the same way that you'd teach him to wear a saddle. Once he's leading well with his gear in place, you're ready to hit the trail.

However, keep in mind that donkeys, like all backcountry hikers, should be conditioned before taking to the trail. Start with short jaunts, allowing your donkey to become accustomed to new and sometimes scary things like river crossings and camping overnight, and begin with relatively light loads. A well-conditioned donkey can comfortably pack 15 to 20 percent of his own weight (that includes the weight of his gear), but he has to build up to it slowly.

Packing Pointers

When loading training packs or panniers, place the heaviest items on the bottom of the pack. The idea is to maintain a low center of gravity and avoid a top-heavy load. To avoid pressure points and sore backs, each pannier should weigh the same, within a pound or two. For multiday trips, buy a portable scale that you can take along to weigh and balance your load every morning. Always enclose breakable items in plastic containers, don't let pointy items jab the donkey's sides, and place often-used items like maps and reading glasses in easily accessed sections of your pack.

On the Trail

Recreational packing with donkeys can be like backpacking by yourself. Leave camping areas in the same condition or better than you found them. Know and observe the rules of the trail. Be courteous; you're representing donkey owners wherever you go. On shared trails, be especially cautious when encountering horseback riders. Remember, some horses have never seen a donkey before. And have fun. Adding a donkey to the wilderness experience is just about as good as it gets.

Pack Burro Racing

Imagine you're running in a 29-mile marathon. The course, an unpaved four-wheel-drive trail, fords streams, twists around boulders, traverses snow fields, and climbs from 10,000 feet above sea level to Mosquito Pass (elevation 13,186 feet) on the Continental Divide. After you struggle to the top, you turn around and retrace the trail back to the beginning — while pulling or being pulled by an energetic donkey on a 15-foot rope. Congratulations; you've just completed the World Champion Pack Burro Race in Fairplay, Colorado, the crown jewel of pack burro racing.

Pack burro racing is a sport indigenous to the Rocky Mountains of Colorado. The first such race was run in 1949, when 21 entrants and their stalwart long-eared partners ran from Leadville to Fairplay for a $500 prize. Now there are several pack burro races run in Colorado each year, usually held in conjunction with town festivals such as those in Georgetown (an 8-mile race held during Railroad and Mining Days), Cripple Creek (a 3-mile amateur race and a 10-mile pro race run as part of Donkey Derby Days), Fairplay (15-mile and 29-mile races held during Burro Days and the Park County Fair), Leadville (women's 15-mile and men's 22-mile

Bill Lee poses with his teammate Bullwinkle at the Fairplay, Colorado, World Champion Pack Burro Race. This event is known as the crown jewel of pack burro racing, a sport that began in Colorado in 1949.

BRAY SAY: *One of the Eight Immortals was Chang Kuo, who lived the life of a hermit on Mount Chung-t'iao in Hêng Chou. Of Chang Kao and his donkey, it was written: "[An] attribute, distinctive of this hsien, is the white donkey upon whose back he rides. The association existing between the two is so close that frequently when Chang Kuo is represented unmounted (his ass presumably being tucked away in his cap-box), a miniature image of the animal may be seen amid a curling wreath of vapour emitted from the open end of his drum. . . . Whenever he stopped to rest, he folded his donkey up, when it was no thicker than paper, and slipped it into his cap-box. Then as soon as he wished to ride again he squirted water from his mouth over it, and transformed it back into a donkey."*

— W. Perceval Yetts, *The Eight Immortals*, 1916

races run during Boom Days), and Buena Vista (a 12-mile race held during Gold Rush Days).

The Fairplay, Leadville, and Buena Vista events constitute the Triple Crown of pack burro racing, but the granddaddy of them all (and the most prestigious) remains the World Champion Pack Burro Race at Fairplay.

Running with the Burros

While rules vary a bit from race to race, the rules governing the World Champion Pack Burro Race at Fairplay are typical. If you think that you'd like to run this grueling race, here is what you need to know.

EQUIPMENT. Your burro must be outfitted with a regulation packsaddle and outfitted with prospectors' gear, including mandatory equipment such as a pick, shovel, and gold pan. Altogether the load must tip the scale at 33 pounds or better. This weight will be checked at the beginning and end of the race. The loss of all or part of the pack and its mandatory trappings is cause for disqualification.

The race committee strongly recommends that you also pack along at least one quart of water, food such as energy bars, and a windbreaker or the like. These items are in addition to the required 33 pounds of packsaddle and paraphernalia.

Your burro must wear a regulation halter fitted with a lead rope no more than 15 feet in length and a recommended 1 inch or greater diameter. The lead rope may be knotted or looped. A jack chain (a lead with a chain or strap over your burro's nose) may be used, but participants injuring their burros' noses with a jack chain are swiftly disqualified.

BURRO. "Burros" come in all sexes and sizes; your burro will qualify if he's iron tough and fit for the job. He will be carefully checked by a veterinarian before and after the race. You must keep him under control at all times,

and any burro or handler that interferes with another burro or its handler may be disqualified.

COURSE ROUTES. The race routes are well marked and must be followed exactly. If your burro leaves the course at any time, he must be returned to the place where he left it before he may resume the race. The long course is approximately 30 miles in duration. The short course is 15 miles.

ADDITIONAL RULES

- ✘ You and your burro must remain a team throughout the race.
- ✘ No one may accompany or assist you or your burro during the race.
- ✘ You can push, pull, or carry your burro but under no circumstance may he carry you.
- ✘ You are not allowed to startle, scare, or deliberately slow the pace of another runner or his long-eared teammate.
- ✘ Mistreatment of your burro is cause for swift disqualification.

THE WINNERS. Teams must cross the finish line as a unit. The runner may be leading the burro or the burro may be dragging the runner, but the burro's nose crossing the finish line signals the win.

Prizes are awarded for up to sixth or eighth place and range (in 2008) from $50 to $500 for placing on the short course and $50 to $1,000 for the long course. In case of a tie, prize money is split between the tied teams.

If You Want to Participate

Fairplay, population 500, is located 85 miles southwest of Denver via U.S. 285; the race is held on the last weekend in July. You can take your own burro or rent a trained one locally. According to race tradition, winners split their earnings with a rented burro's owner.

"Burros" range in size from Standard donkeys to

Mammoth Jackstock. Bigger isn't necessarily better; it's the fast, cooperative donkey that helps his partner to win. According to veteran racer and donkey breeder Curtis Imre (who rents racing burros by the season or the race), trained feral burros adopted through the Bureau of Land Management Wild Horse and Burro Program are ideal choices for pack burro racing.

Get yourself in condition; this is a grueling sport. Imre, who runs in addition to raising racing burros, suggests that you should be able to run a decent 10K or finish a marathon. And you'd better build your upper body strength for holding on to the burro's lead rope (most pack burros lead, rather than follow).

After Fairplay

The second leg of the Triple Crown is held at Leadville the weekend after Fairplay and the third at Buena Vista the weekend after that.

Pack burro racing has its own governing body, the Western Pack Ass-ociation (*see* Resources). If you're tough and you're a runner who loves a challenge, check out pack burro racing; it might be the perfect sport for you.

Widdershins around Ireland

In 1979 Kevin O'Hara was at loose ends. A Vietnam vet, he was finding it hard to fit back into the peacetime world. While visiting his grandmother in Ireland he came up with the notion of walking with a donkey and cart around the perimeter of his ancestral country. Opinions poured forth in the local pub on how to do it, bets were placed on whether he could succeed, and soon Kevin (knowing nothing about equines) was on the road with a young jenny named Missie.

Beginning in County Galway, the pair traveled around Ireland "widdershins" — counterclockwise — as the seasons turned through spring, summer, fall, and winter. The two never paid for lodgings but enjoyed the hospitality of countless homes along the road in accommodations ranging from farm fields to the best bedrooms. They met farmers, politicians, Irish Traveller families in their caravans, and children, and slowly, mile by mile (1,700, in fact), Kevin recovered his faith in humanity and in himself.

Kevin wrote about his travels in his book, *Last of the Donkey Pilgrims* (*see* Resources). *(See page 247 for a photograph of Kevin and Missie.)*

Prunes, Shorty, and Bub

A rustic, five-foot, rock-and-mortar monument stands on Fairplay's Front Street next to the Hand Hotel overlooking the river. Miner Rupert Sherwood and his friends erected it in memory of their burro pal, Prunes.

When prospectors discovered gold along the South Platt River in 1867, Prunes went to work in the neighboring mines just a few years after the sleepy town of Tarryall (later renamed Fairplay) came to life. Miners, it is said, could send reliable Prunes down the mountain with a shopping list fastened to his pack gear. When shopkeepers finished loading him up, Prunes faithfully trudged back to the mining camp, thus saving a trip to town.

When Prunes grew too feeble to work, he moved into town and panhandled food from the townsfolk, who treated him well. Prunes became friends with Rupe Sherwood, and the two spent many days together scouring the old mining sites for gold. In 1930, a blizzard caught Prunes unaware and stranded him for days without food in a snowed-in shed. The old donkey never fully recovered and died the following spring. Prunes was believed to be 63 years old. The townsfolk buried Prunes under the monument erected in his honor. When Sherwood died a year later, friends honored his wishes and his ashes were interred with his old friend, Prunes.

But Prunes's isn't the only donkey monument in Fairplay. A smaller, less conspicuous grave on the courthouse lawn marks the final resting places of a burro named Shorty and his canine cohort, Bum.

Shorty, a donkey with unusually short legs, was born at a gold mine in Mosquito Gulch in 1906. Though small in stature, Shorty had a big heart and worked the mines as hard and faithfully as any other mining burro. When mining wound down around 1940, his last owner abandoned him to forage for himself. When his eyesight began failing, Shorty wandered into Fairplay to mooch.

Fortunately the old, nearly blind donkey soon met a kindred soul: a scrawny, part Dalmatian stray, known as Bub. Bub became Shorty's eyes. Twice a day he led the donkey from door to door to beg for handouts. Bub whined and scratched doors; Shorty brayed to announce their arrival. The townsfolk loved it, and the pair quickly became the town's pets.

In the spring of 1951, a hit-and-run driver left Shorty dying in a ditch. Bub was found mourning at his old friend's side. Fairplay's residents were so moved by the bond that the two animals shared that they took up a collection to build a monument to the pair. Shorty's remains were buried beside the monument. Just three weeks later, Bub, who had always judged traffic precisely, dashed out in front of a semi truck and was killed; he was buried at his old friend's side.

DONKEY ACTIVITIES

SHOWS. *Driving, coon jumping, riding, speed events, or halter: there are classes for everyone at donkey and mule shows.*

JUMPING. *Coon jumping simulates the type of jumping from a standstill that raccoon hunters' stalwart saddle mules are famous for. Even some of the tiniest Miniature Donkeys are prolific jumpers. Try it with your donkey — it's fun!*

COON MULE JUMPING. *(right) In this photo you can see a box marked on the ground with chalk. The mule must jump, flat-footed and from a standstill, without moving out of the box. Some mules jump fences well over 6 feet high.*

SNIGGING. *(below) In this timed event, donkey and handler weave their way through pylons while dragging a log.*

PACK BURRO RACING.
In this grueling but rewarding sport, sometimes it's just you and your donk enjoying spectacular Rocky Mountain scenery . . . and sometimes you have company as you race to the finish line! Photos taken at the World Championship Pack Burro Race, in Fairplay, Colorado.

PACKING AND RIDING. *There are lots of great things to do with donkeys and mules, from packing (right) to pleasure riding (below left), and even to fox hunting on mule back (below right).*

DRIVING/DRAWING. *Miniature Donkeys are strong enough to pull large adults in wagons or in well-balanced carts.*

ON THE ROAD. *Kevin O'Hara with his young jenny named Missie, with whom he walked around the perimeter of the Emerald Isle for almost a year. (See page 239 for their story.) In this photograph the pair amble toward Slea Head on the Dingle Peninsula. Missie is hidden by the cart with only her long ears visible.*

ON GUARD. *(below) Guardian donkeys like this handsome fellow excel at protecting sheep.*

PART 4

Donkey Business

CHAPTER 16

BREEDING DONKEYS

*It is a good omen to hear a bell ring, an ass bray,
or a kite cry, when starting out to see a married
woman whose husband is alive.*

— Indian proverb

Sooner or later, most donkey owners decide to breed a jenny or two, and some breed as a monetary venture. Before you join these ranks, however, there are some things to consider. I'm not trying to discourage you. As a former breeder and equine rescuer, I'm simply asking you to give the topic serious thought before putting your jenny in foal. Then, if you do the homework and you still want a sweet, fuzzy baby (or two or three or more) all your own, this chapter contains what you should know about breeding donkeys.

Breeding Q&A

Q *Why do you want to breed your jenny?*

A If you think that it's an inexpensive way to add more donkeys to your herd, think again. There are countless fine donkeys looking for homes, so rescuing a donkey or paying $125 for an adult wild burro (or even $500 for a grown Standard donkey) is far more cost-effective than paying for a stud fee and jenny care, feeding your jenny through gestation and lactation, and raising a foal to adulthood. Also, the adult you purchase can be trained right away, if he's not already.

Everyone loves a foal, but what happens when that charmer grows into a rowdy teenager just lounging around, eating, and taking up space? Will you sell him? Are you willing to teach him ground manners, geld him (this costs $150 in my area and more if the lad is 18 months of age or older), and screen buyers so that he finds a good home?

Q *Is your jenny breeding quality, and if so, are you willing to pay to breed her to a comparable jack?*

A Review Evaluating Donkeys in chapter 3 and be brutally honest — if your jenny isn't breeding quality, don't breed her! If she is, realize that it's costly to breed her to a quality jack. Prebreeding vet work, $200 to $800 stud fees, $5 to $8 per day jenny care, and post-breeding vet work add up quickly. If your jenny doesn't conceive, at worst you're out your money, and at best you'll have to rebreed again next year.

Q Do you think you'll get rich breeding donkeys?

A Savvy breeders make a fair amount of money breeding donkeys, although most of them would question the word "rich." However, to earn a profit raising don-

Is this boy's large, fuzzy foal a Mammoth? We think so.

keys, you should plan to breed the crème de la crème of registered Miniatures, top-quality Mammoth Jackstock, or an imported heritage breed. It's hard to break even raising Standard donkeys, no matter their color or registration status, so if you do breed them, it had better be for love. Keep in mind that, no matter what you breed, you must begin with top-of-the-line breeding stock, maintain your stock under optimal conditions, plan breedings with utmost care, and aggressively market your product. Startup and ongoing expenses will be substantial. We'll talk more about this in chapter 18.

A Jack for Your Jenny

The first step is finding the perfect jack for your jenny. If she has flaws — and all donkeys do — the jack should be strong where she is weak. Breeding two individuals with the same faults is courting disaster. Search for jacks using the same resources cited in chapter 4; people who have foals for sale may own just the jack you hope to breed your jenny to. Once you've located two or three potential boyfriends for your jenny, contact their owners and ask these questions:

✘ What sorts of fees are involved? Some jack owners include a certain number of days of board for your jenny in the stud fee. If so, how many and what will happen after that? If you'll be paying across the board for jenny care, how much will it cost and what does the fee include?

✘ Where and how will your jenny be kept? Will she have her own shelter and exercise area or be turned out in a strange herd to fend for herself? What happens if she's injured while she's away being bred?

✘ Does the jack owner require proof of Coggins testing, health papers, or uterine culture results before your jenny sets foot on his farm? What about specific vaccinations that your jenny might not already have?

✘ Does he hand breed or pasture breed visiting jennies? For your jenny's reproductive health and her safety, hand breeding is probably the better option, but this is something you should hammer out in advance.

If you like what you hear, ask each jack owner to send you a copy of his breeding contract and any additional information you'd like to know about his jacks. When they come, read the contracts carefully and compare your options, then arrange to see the jacks that you're still considering.

When you visit, ask to see where your jenny will be kept. Are you comfortable leaving her there? If not, is the jack owner willing to provide other accommodations, perhaps at a greater price?

Look the jack over carefully and examine his offspring, especially the ones whose dams resemble your jenny. In regard to both conformation and disposition, are they the sort of foals you want for your farm?

After examining all contenders, make your choice and get a signed breeding contract outlining all of the specifics of this breeding. Congratulations; you've just completed step one!

A JACK OF YOUR OWN

Unless you breed a number of jennies each and every year, you probably won't need your own jack. Quality jacks are expensive. If you buy one, he'll require separate housing and substantial fencing. You'll feed him year-round, even if you breed only one jenny or mare a year. By using outside jacks, you can choose the ideal mate for each jenny. If you must have a jack, remember:

- Don't take jacks for granted, ever. Even if they're sweet as pie, they are hormone-driven breeding animals and under the right set of circumstances, they will crack. Case in point: While a friend was hand breeding a mare, his gentle, loving Mammoth jack swung around and snapped his upper arm with one crushing crunch of the jack's massive jaws. Stay on your toes when working around jacks, and make no exceptions!

- Most jacks breed jennies or mares but not both. A young jack raised with donkeys prefers jennies. In most cases, mule-breeding jacks must be raised with mares, isolated from female donkeys.

- In most states jacks, stallions, bulls, rams, and the like are considered dangerous animals. You are liable for any damage your jack inflicts on any human or animal, even if your land is posted "no trespassing." This is true off your land as well. Make certain that you can control your jack under all circumstances if you haul him to shows, parades, or exhibitions.

- If you breed your jack to outside mares or jennies, plan to market him aggressively (learn all about it in chapter 19). Investigate the legal angles, and hire a lawyer to draw up a breeding contract tailored specifically to your needs. Discuss jack handling and the merits of pasture versus hand breeding with experienced breeders who stand asses at stud, and build facilities accordingly.

Getting Ready for Breeding

Even if the jack owner doesn't require it, have your veterinarian run a uterine culture on your jenny. A number of diseases affect the reproductive tract, and infection is quite common, especially with older jennies. A qualified veterinarian can detect potential problems through a thorough physical examination or by use of uterine cultures. Most infections can be eliminated by appropriate treatment before breeding time, so it's to your advantage to have a culture done at least six weeks prior to your jenny's scheduled first breeding. Make sure that your jenny is healthy and in good flesh, neither too fat nor too thin, when she leaves for her date with the jack.

When your jenny comes home, there are several things you can do to keep her healthy throughout her pregnancy. For one thing, don't let her get fat. Feed good grass hay during early gestation, making certain that she has a plentiful supply of clean water and a properly balanced mineral available at all times. Unless she's grossly overweight, begin feeding a judicious amount of grain about six weeks before her first due date, beginning with a handful — if she's a Standard donkey, work up to a pound to a pound and a half of good grain mix per day; if she's smaller or larger, adjust the amount accordingly.

Have an ultrasound done on your jenny as soon after breeding as possible (adept equine practitioners can confirm pregnancy 12 to 14 days after conception) to make certain that she's pregnant and to see how many foals she's carrying. That's right: foals. Jennies are far more likely to conceive twins than mares are. This is not a good thing. Carrying twins is very hard on a jenny's body, and the chances of her safely delivering two healthy, normal-size foals are fairly slim. If your vet detects twins, he can apprise you of your options. Most vets advise that twin pregnancies be terminated and the jenny rebred.

Vaccinate. Your jenny should be vaccinated with killed rhinopneumonitis vaccine (Pneumabort or Prodigy) during her 5th, 7th, 9th, and possibly her 11th months of pregnancy. At the beginning of her 10th month (an estimated four weeks prior to her original estimated delivery date) she should receive at minimum a tetanus booster, although most vets recommend a four-way tetanus, flu, and eastern/western encephalomyelitis booster. If she doesn't foal at 11 months, give her a second booster and continue doing so every four to six weeks until she foals. This way she's sure to pass antibodies against these serious diseases to her foal via her colostrum (first milk). Because a foal's own immune system doesn't start functioning until he's three or four months old, he needs that passive transfer of antibodies from his dam to stay healthy.

At the same time that we boost our mares or jennies, we inject them with a selenium and vitamin E supplement called Bo-Se. This is a common (and essential) practice in selenium-deficient parts of the country; check with your vet or County Extension agent to see if it's needed in your locale.

Be sure to keep her hooves in prime shape (she's carrying a lot of extra baggage), and have her hooves trimmed a few weeks before you think that she's due.

If she's a Miniature Donkey, make arrangements to protect her foal from predation. If you have a number of Miniatures but not a large herd (a large herd of Miniature Donkeys can protect themselves from all but large predators like bears and mountain lions), use guardian animals — a Standard donkey, llama, or a livestock guardian dog — in pastures where neonatal foals are kept, or keep them penned near the house until they're bigger.

Ride or drive your jenny as usual until she's roughly six months pregnant; continue lightly riding and driving through her ninth month if you take it easy and she's in good condition.

If you live where fescue toxicity is a problem, it's best to remove your jenny from fescue pasture and stop feeding fescue hay 60 days before anticipated foaling (30 days prefoaling is the absolute minimum). If she's ingested fescue right up to her foaling date, ask your vet for advice as there are drugs to help counter the effects of fescue toxicity.

Preparing for Foaling

At least a month before your jenny's due date, assemble a complete foaling kit or check the one that you already have, making certain that you have everything that you need. Set up one or more foaling areas (in the barn or in easily accessible paddocks or pens) depending on the number of jennies expecting foals; don't assume that they'll take turns, because they often don't. To be on the safe side, if stores and veterinary practices are closed on Sundays where you live, add a package of mare's milk replacer to the list.

Be There

If you put your jenny in peril by breeding her, *be there* when she foals. Countless jennies, mares, and foals die every year when the presence of a human attendant could have made the difference between life and death, if only to call a vet.

Some folks claim that you're lucky to witness the birth of a foal, but that's pure fiction. Any reasonably observant person can predict imminent foaling. Once that baby is ready to be delivered, it's your responsibility to be there if something goes wrong — no exceptions.

A Well-Stocked Foaling Kit

We pack our foaling supplies in a Rubbermaid toolbox-stool. It's roomy, it has a lift-out tray for small items, it's easy to tote to the barn, and it's so much nicer to sit on than an overturned five-gallon bucket. The box holds:

- A bottle of 7 percent iodine (used to dip new-borns' navels).
- A shot glass (to hold the navel-dipping iodine).
- A digital veterinary thermometer.
- A bulb syringe (the kind used to suck mucus out of human infants' nostrils).
- Shoulder-length, preferably sterile, obstetric gloves.
- At least two large squeeze bottles of lube. We like SuperLube from Premier1 Supplies (*see* Resources); it's formulated for assisted kidding and lambing, so it's concentrated and it really, *really* works.
- Betadine Scrub (for cleaning jennies prior to assisting).
- Sharp scissors; disinfect them and store them in a sturdy resealable plastic bag (you never know when you might need them).
- A sharp pocketknife (ditto).
- A hemostat (ditto).
- A halter and lead (you won't want to leave the jenny to scare one up in an emergency).
- Two flashlights (we like to have a backup in case the first flashlight fails).
- Two large, soft, terrycloth towels.
- A full roll of paper towels.
- A clock or watch.

You will also need a cell phone (or a ground line to the barn), two clean buckets, a ready supply of warm water, and access to a trailer in case you have to transport your jenny to the vet.

Are YOU Ready?

While most jennies foal without assistance, unless your vet is certain to be no more than 15 minutes away from your farm at foaling time, be ready to help out if the need arises.

- Keep your fingernails clipped short and filed (you won't have time for a manicure when an emergency arises).
- Know how to determine what configuration a mal-positioned foal is in and precisely how to correct it. If you think that you might forget, photocopy and laminate instructions to keep in your foaling kit.
- Practice. Teach your fingers to "see." Borrow a pile of plush toy animals. Place one in a fairly close-fitting paper bag. Without looking, stick your hand in the bag and figure out what you're feeling. Switch animals. You'd be surprised at how helpful this exercise can be.
- Program appropriate numbers into your cell phone so that you can call at least three good vets at the touch of a button.
- Install a baby monitor or video camera in your barn, or better yet, round up a comfy cot and a sleeping bag. We'll talk more about this later.

Stall or Pasture Foaling?

Most breeders who monitor their jennies at foaling time prefer that the jennies give birth indoors. It's easier to keep track of a stalled jenny, and if something goes wrong, there are lights and running water close at hand.

If you do foal your jenny outside in a paddock or corral (especially if you won't be there when she foals), make certain that it's under hygienic conditions and that the enclosure is such that she can't lie down close to the perimeter and deliver the foal under the fence.

A foaling stall is a better bet. It should be completely stripped as needed, all surfaces sprayed with dilute bleach solution, and deeply bedded with dust-free straw (shavings and sawdust cause respiratory problems in neonatal foals). It must be roomy enough for the jenny to walk around, roll, and stretch out in comfort. There must be no protrusions along the walls (nails, splinters, bucket hooks) for a newborn foal to be hurt on. It should provide privacy for the jenny but offer an unobtrusive spot from which an attendant can observe the foaling. Watering buckets, especially when foaling Miniature jennies, should be small enough or hung high enough that a jenny can't drop her foal headfirst into the bucket or a wobbly foal can't tumble headfirst into it and drown.

Give Me a Sign

A jenny's pregnancy can last between 11 and 13 months (and occasionally longer); 12 months is said to be the norm. Because donkeys' gestation periods are so inexact, you must watch for signs of impending foaling every day.

Milk Testing

Milk testing refers to the practice of measuring the mineral content of a jenny's udder secretions prior to foaling. Within 24 hours of foaling the calcium content increases markedly. Water hardness test strips or commercial mare test strips can be used to check calcium content, but there's an easier way. I learned this technique from Dr. Karen Hayes's book *The Complete Book of Foaling* (*see* Resources), and in 15 years of foaling mares and jennies, it hasn't failed me yet!

Foaling Facts

- Studies indicate foaling begins when the unborn foal releases hormones that indicate that he is "done." However, the dam can hold off the birthing process for a short span of time if things aren't right. This is why attendants can sleep in the barn all night and the jenny will foal during the 15 minutes when they go to the house to brew a cup of coffee. When birth is imminent, stay put!

- Behaviorists say that equines are drawn to water when giving birth, theoretically because water helps to disguise birthing scents. Foals are sometimes lost when their dams drop them into ponds and creeks; our neighbor lost two donkey foals from the same home-raised jenny in just this manner.

- Several university studies indicate that most mares give birth between 11 p.m. and 3 a.m. Although some of our foals, both horse and donkey, arrived at daybreak and a few at various other hours through the night and day, nearly all have indeed arrived during that time span.

Give your jenny a spotlessly tidy, spacious, nicely bedded stall to foal in. It's easier (and sometimes safer) than monitoring the blessed event outdoors.

The more calcium prefoaling udder secretion contains, the whiter it will be. Therefore, by expressing a few drops of milk into your hand and examining it, you can accurately predict when your jenny will foal. Udder secretions progress in color and texture from:

1. clear and watery, to
2. thin but cloudy, to
3. yellow-tinged and sticky (like floor wax), to
4. definitely amber colored and syrupy, to
5. skim milk, to
6. opaque white milk.

Start checking when the jenny's teats begin to inflate. If you can't express anything, it's probably too early and you can wait a day or two before trying again.

When you're able to express clear, thin fluid, begin checking every other day, then every day when the fluid becomes cloudy, and twice a day at the floor-wax stage. Things can happen quickly from this stage onward, so at that point be ready to foal. One of my mares went into active labor at the amber-colored-syrup stage and had opaque milk in her udder an hour later when her filly nursed.

Contrary to popular belief, removing wax plugs (they're actually blobs of dried colostrum) from a female equine's teats prior to foaling will *not* cause her to drip milk. Milk won't flow from the udder until triggered by the hormone oxytocin; nothing you do (or don't do) will alter that fact.

Things to Watch For

Keeping in mind that no one jenny is likely to show all of these signs, here are some things to watch for.

- ✘ Udder development. Most jennies begin "bagging up" four to six weeks prior to foaling. However, this is a very individual thing; some jennies show little development until days or even hours before foaling.

When a jenny gets "pointy belly," she's almost ready to pop!

Read It, You Need It!

Check the Resources section for books on foaling, and read them cover to cover before your jenny is due.

* A hormone (relaxin) causes the muscles and ligaments in the pelvis to begin relaxing about three or four weeks before foaling. This causes the rump to become increasingly steeper from hips to tail. As this happens, the area along the spine seems to sink and the tail head rises.

* The perineum, the hairless area around the vulva, sometimes bulges during the last month. About 24 hours before foaling occurs, the bulge diminishes and the vulva becomes longer, flatter, and increasingly flaccid.

* A few days before foaling, muscles in the floor and walls of the jenny's abdomen begin relaxing and the shape of the jenny's abdomen changes, making her belly seem to come to a point when viewed from the side and to look narrower viewed from the front or back.

* As the cervix begins to dilate, usually a few days to a few hours before foaling, the cervical seal (wax plug) liquefies. When this occurs, jennies often discharge strings of mucus from their vulvas ranging from clear, thin goo to a thicker, opaque white substance to a thick, amber-colored discharge.

First-Stage Labor

The following behaviors indicate that the jenny is in first-stage labor; this generally lasts for 12 to 36 hours prior to actual foaling.

* Her udder will engorge with milk so that the teats are filled almost to bursting. If her udder is pink, it will blush a rosier red; pink skin or dark, the udder takes on a shiny, moist look as foaling approaches. Wax plugs may form on the ends of her teats, and she might drip or stream milk.

* She may drift away from the herd to seek a nesting spot, sometimes in the company of her dam, a daughter or sister, or a best friend.

This jenny in first-stage labor has hunkered down to meditate on her situation.

- She'll yawn and stretch (stretching helps to put her foal into birthing position), urinate frequently and/or pass small amounts of manure, and may go off her feed.
- When she begins pacing, pawing, kicking at or watching her belly, circling, or lying down and getting up again, especially with her tail cocked out behind and kinked to the side, get ready — here he comes!

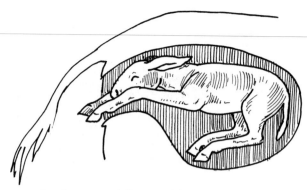

Front-feet-first, normal diving position.

Here Comes Baby!

Second-stage labor begins. The jenny lies down and rolls onto her side when a contraction hits, rides out the contraction, then rises and repeatedly repositions herself until she finds a spot that she likes. Once she does, she may roll up onto her sternum between contractions, but she'll usually remain lying down. However, some jennies deliver standing up, and that's normal, too.

The first thing to appear at the jenny's vulva is a fluid-filled, water-balloon-like sac called the chorion, one of two separate sacs that enclose a developing fetus within its mother's womb (the other is the amnion). Either or both sacs can burst within the jenny or externally as the foal is delivered.

In a normal front-feet-first, diving-position delivery, a hoof appears inside the chorion (or directly in the vulva if the chorion has already burst), followed by another hoof three to six inches behind the first one (depending on whether the foal is a Miniature, a Mammoth, or something in between), and then the foal's nose. Once the head is delivered, the rest of the foal follows quickly.

Be Prepared for Trouble

On pages 260 and 261 are details about some birth positions that are considered dystocia (difficult birth). If the worst happens, there are several points to keep in mind. First and foremost: **Call your vet!** Do it immediately. If he can't reach you promptly, he may ask you to get the jenny up and walk her until he arrives, or he may talk you through handling the dystocia yourself. If you can't reach your vet, try another. If no one can

The big moment is at hand — here comes baby!

reach you in time, take a deep breath, stay calm, and handle it the best you can. It's scary, but with care and determination most dystocias are correctable, and if you don't do it, your jenny and her foal will probably die.

Because of anatomical differences, you can't pull a foal the way you'd pull a calf. If you must pull, use *lots* of lube and pull only during contractions. Grasp the legs, preferably above the pasterns but below the knees, then pull. Don't pull straight back — pull out and down in a gentle curve toward the jenny's hocks.

Before entering a jenny, make absolutely certain that your fingernails are short and that you've removed your watch and rings. Wash her vulva using warm water and mild soap or a product like Betadine Scrub. Pull on an obstetric glove if you have one; if you don't, scrub your hand and forearm with whatever you used to clean the jenny and then liberally slather the glove or your hand and arm with lube. Now pinch your fingers together and gently work your hand into the vulva.

The jenny will be doing her best to push the foal out, and late labor contractions are *forceful*. Be prepared for that. When contractions hit, hold still and ride them out; when they let up, work quickly but carefully until the next one occurs.

Determine which parts of the foal are present in the birth canal. Closing your eyes and moving your awareness to your fingertips will help. If the foal's toe points upward and the big joint above it bends away from the direction the toe is pointing, it's a foreleg. If the toe points down and the major joint bends in the same direction, it's a hind leg. Follow each leg to the shoulder or groin if you can, making sure the parts you're feeling belong to the same foal (remember: jennies sometimes have twins). When you're certain that they do, if you can manipulate the foal into a normal birthing position, do so, and then help pull the foal out (by this time the jenny will probably be too exhausted to do it all by herself).

The inside of a jenny is extremely fragile and if you or the foal tears her, she'll almost certainly die.

When repositioning the foal, cup your hand over sharp extremities like hooves and work carefully and deliberately; her life depends on your gentle technique.

Foaling Tips and Timeline

Witnessing the birth of a foal is an exciting and moving event. However, for your jenny and her foal's sake, don't turn it into a three-ring circus. Don't invite the kids, the neighbors, and the boss's wife to come and watch your jenny foal. Avoid bright lights, noise, and dogs. Snap photos if you like but do it from a distance. Keep everything low key, and unless something goes wrong, don't interrupt her until the foal is born. Even then, work quickly and quietly, then get out of the way and let the little family bond.

The usual timeline for foaling is as follows:

1. In a normal foaling, the actual delivery (second-stage labor) takes approximately 20 to 40 minutes.

2. The jenny should deliver the placenta (third-stage labor) within an hour after foaling; if she hasn't passed it in two hours, call a vet.

3. The foal should stand within about an hour and nurse within a half hour after that.

Did You 'Ear?

British chimney sweeps carried their working tools on donkeyback or in donkey-drawn carts well into the early 1900s.

The Donkey Party in Paris put up a donkey for election in the 1890s; a similar party was established in Sulaimaniya, Kurdistan, just a few years ago.

Correcting Dystocia

To learn more about dystocia and how to correct it, consult *The Complete Book of Foaling* by Dr. Karen Hayes (*see* Resources). Meanwhile, here are some basics on problematic birth presentations.

BREECH. *Both hind legs, no head.* Two feet followed by hocks appear. Most jennies can handle this delivery by themselves, but because the umbilical cord is pressed against the rim of his dam's pelvis during this delivery, it's wise to *gently* pull the foal once his hips appear.

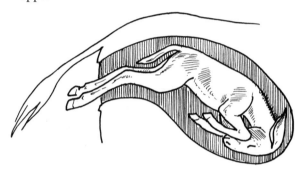

BREECH POSITION WITH LEGS BACK

FULL BREECH. *Butt first, hind legs tucked forward.* It's best not to try to do this yourself. However, if you must reposition this foal, try to elevate the jenny's hindquarters before you begin. Push the foal forward, work your hand past his body (it's a tight squeeze) and grasp one hock. Raise the hock up and rotate it out away from the body. While holding the leg in that position, use the little and ring fingers of the same hand to try to work

the foot back and into normal position; repeat on the opposite side. The umbilical cord will be pinched, so pull the foal as quickly as you safely can.

ONE LEG BACK. *Head, one leg.* Push the foal back just far enough to allow you to cup your hand around the offending hoof and gently pull it forward.

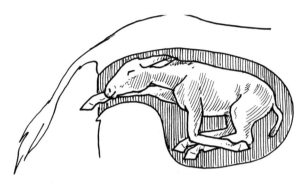

ONE LEG BACK

HEAD BACK. *Two legs, no head.* Push the foal back in the jenny as far as you can and then bring the head around into position. Alternate problem: Sometimes the front legs are presenting but the head is bent down. Attempt to correct in the same manner, but make sure that your vet is on his way.

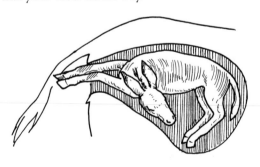

HEAD BACK

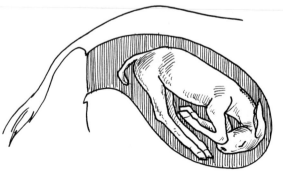

FULL BREECH

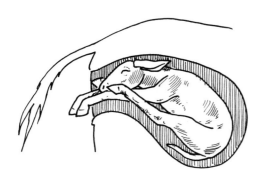

ALL FOUR LEGS COMING AT ONCE

ELBOW LOCK. *Head out, both forelegs partway out, one is trailing the other; the foal is stuck.* In normal deliveries, one leg is delivered so that the tip of one hoof is approximately at the level of the fetlock of the other leg. If the lead foreleg is too far ahead of the trailing foreleg, the trailing leg begins to flex and the elbow becomes lodged against the jenny's pelvis. This happens frequently and is easily corrected by pushing the foal back, grasping the hoof of the trailing leg, and pulling it forward into the proper position so that the delivery can proceed.

ALL FOUR LEGS AT ONCE. Attach a length of clean, soft, small-diameter rope to a set of legs, making certain that you have two of the same kind, and then push the foal back as far as you can. Reposition the foal for either a diving position or hind-feet-first delivery, depending on which set of legs you've captured.

HIP LOCK. *Foal is halfway out but stuck.* One edge of the foal's pelvis is hooked on the edge of his dam's pelvis; the foal's umbilical cord is pressed between his weight and the jenny's pelvic brim, and he's likely to suffocate. Pull the foal down and to the right, then down to the left, then to the right again and so on until you've "walked" the foal's pelvis through the jenny's pelvic structure.

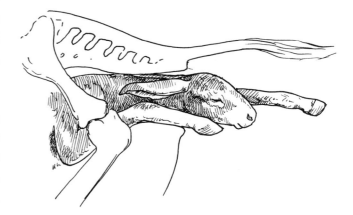

ELBOW LOCK

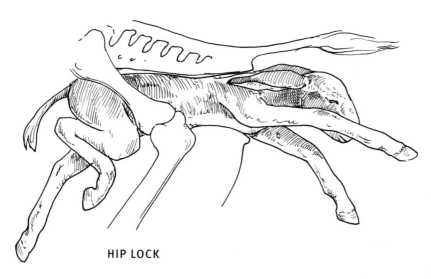

HIP LOCK

After the Birth

When the foal arrives, remove birthing fluids from his nose by stripping your fingers along the sides of his face. If he's struggling to breathe, use the bulb syringe from your foaling kit to suck fluids out of his nose and then tickle the inside of his nostrils with a piece of straw or hay. If he's really struggling to breathe or not breathing at all, and he's small enough, take a secure grip on his hind legs between the hocks and pasterns, place your hand behind his neck (near the withers) to support it, and swing him in a wide arc to clear his airway of fluids and jump-start his breathing. Grasp larger foals around their loin and groin and swing them in the same manner. Don't give up; chances are that he'll start to breathe and be perfectly okay.

Don't intentionally cut or break the umbilical cord; it will break on its own at a place about 2½ to 4 inches below the foal's belly (depending on his size) but not until the jenny stands up or the foal flops around enough to break it.

CPR for Foals

Realistically, if a newborn foal's heart isn't beating, you probably won't be able to save him. However, if his heart is beating and swinging him doesn't jump-start his lungs, this procedure might do the trick.

Check the foal's mouth and clear it of any foreign material, then lay him flat on his right side. Close his mouth and pinch the nostril closest to the floor shut. Now place your mouth over his upper nostril and blow a lungful of air into his lungs until you see his chest rise. Remove your mouth from his nostril and let the air escape. Repeat this again and again for about three minutes, giving the foal one breath roughly every three seconds. If it doesn't work in this amount of time, quit. The foal would likely be brain damaged if you persevere and resuscitate him beyond this point.

Once the cord breaks, think "dip, strip, and sip" — dip the cord in 7 percent iodine (fill a shot glass or empty film canister with iodine, hold the container to the foal's belly so the cord is completely submersed, then tip the foal back to effect full coverage; don't omit this important step); strip the jenny's teats to make certain that they aren't plugged and that she indeed has milk; then make sure the foal sips his first meal of vitally important colostrum within an hour or so after he's born.

Next: The Placenta

Within an hour after your foal arrives, your jenny will go into third-phase labor and deliver the placenta (or afterbirth), the tissues in which her foal developed inside her uterus. It is vitally important that these tissues come out promptly and intact. If she doesn't pass the placenta within two hours, call your vet without delay. Here are some steps for helping this process along:

1. Encourage the jenny to lie down and work at expelling the placenta by getting out of the stall and reducing distractions to a bare minimum.

2. **Don't** pull on exposed portions of the placenta in an effort to help! Pulling might cause the uterus to prolapse (turn inside out and hang outside the jenny's body), and then you'll be in big trouble.

3. Always wear protective gloves when handling the placenta. It's unusual, but if the jenny is infected, placental tissues can transmit diseases such as brucellosis (undulant fever) to humans who handle them.

4. After it's expelled, spread the placenta out and see if pieces are missing. It should be T-shaped, inside out (the purple part on the outside), with one torn spot from which the foal emerged. If you think something is missing, place it in a bucket of cool water, cover it, and call the vet.

5. If everything seems all right, dispose of the placenta by burning or burying it.

Nursing and Colostrum

Colostrum is a thick, yellowish milk produced by jennies beginning about two weeks prior to foaling and ending 12 to 36 hours after giving birth. It's jam-packed with important nutrients, but more important, colostrum contains vital immunoglobulins (antibodies) that foals need in order to survive. They must ingest an adequate amount of colostrum within 6 to 8 hours of birth as their gastrointestinal systems can absorb the immunoglobulins in colostrum only during a short window of time — usually not longer than 12 hours after birth.

Immunoglobulins are proteins used by the immune system to identify and neutralize foreign bodies such as bacteria and viruses. Foals that don't ingest colostrum lack immunity to diseases like tetanus and flu until their own immune systems kick in at three to four months of age. Foals also need the nourishment in colostrum to prevent hypoglycemia (low blood sugar). Foals have little or no fat stores and require frequent meals. Going several hours without colostrum or milk can leave a newborn foal very weak and unable to stand. Unless foals ingest colostrum or they're given a suitable substitute like a size-appropriate dose of oral IgG (immunoglobulin) supplement or a plasma transfer, they rarely survive; it's that important. Here are some ways to make sure that your new foal will thrive:

✕ If a foal can't nurse from his dam but she's still available, milk her and bottle-feed her colostrum to the foal in 2-ounce portions for Miniature foals and 3-ounce portions for larger foals until he's ingested 10 percent of his total body weight in fluid. Don't feed milk or milk replacers during the first 12 to 24 hours!

✕ Properly frozen colostrum stays good for up to one year. Quick-freeze it in 2- or 3-ounce feedings in small-portion human infant bottles or in double-layered resealable sandwich bags. Avoid storing it in self-defrosting freezers; constant thaw and refreeze cycles affect its integrity. *Never* microwave colostrum, which kills the protective

'Ear's a Tip

Colostrum Backup. If you can't use colostrum from a foal's dam, fresh or frozen colostrum from another jenny on your farm will do the trick.

antibodies; instead, immerse the container in another container of hot water until the colostrum registers 100 degrees (check it with a thermometer; don't wing it).

✕ Many equine hospitals, university vet schools, and large breeding farms stock horse colostrum and freeze it for emergencies. If you need colostrum, these are good places to look.

✕ If no colostrum is available, the foal will probably survive if promptly given a weight-appropriate portion of an oral immunoglobulin (IgG) supplement like Seramune, Foalimmune, or Lyphomune; some breeders give them to each foal as a matter of course (*see* Resources). These are *not* the same thing as the inexpensive, powdered supplements based on cow colostrum that you'll find on the shelf at your local feed store; if you have nothing else, try them, but most who do report very limited success.

✕ Colostrum-deprived foals (as well as foals from unvaccinated dams) should be given a tetanus antitoxin injection within a day after birth and at two-week intervals until they're four months old.

✕ If there is any doubt about whether your foal received adequate protection from his dam's colostrum, have your vet run a test to measure the amount of gamma globulins in his bloodstream (the target amount is a concentration of 800 milligrams or more per deciliter of blood; 400 milligrams is usually adequate) when he is 12 hours old; that way if his immunoglobulin level is low, there is still time to administer more colostrum. At 24 hours of age, the only alternative is (expensive) intravenous treatment with immunoglobulin-rich plasma.

Foals are programmed by nature to seek sustenance in dark places, but to a foal that could be any dark area in the stall. Experienced jennies circle and nudge their foals to put them into the correct position to nurse, though first-timers may have to be haltered and held until their foals make the right connection. Once you're sure the baby is nursing (watch his throat to see if he's swallowing), leave the jenny and foal alone to bond.

Neonatal donkey foals are delicate creatures; they must not be allowed to get wet or chilled. Keep your

IMPRINTING: YES OR NO?

Foal imprinting is the act of desensitizing foals to a host of stimuli immediately after birth. Pioneered by veterinarian Dr. Robert M. Miller, imprinting was a hot topic during the 1990s (*see* Resources). Since then, studies conducted at Pennsylvania State University, the University of Washington, the University of Rennes (France), and the Iceland University of Education all concluded that it really doesn't make that much of a difference. But does it?

The premise is simple: When a foal is born, he rests for about 30 to 45 minutes before struggling to his feet and drinking his mother's milk for the first time. Imprinting takes place during this brief window of opportunity. The following is Dr. Miller's basic routine (from "Intensive, Early Handling of Neonatal Foals: Mare-Foal Interactions" by Nancy Kate Diehl, Brian Egan, and Peter Tozer; University of Pennsylvania). Each process is repeated a minimum of 30 times but continued, if necessary, until the foal stops resisting.

- Rubbing the hand over the face, muzzle, upper gum, ear, poll, neck, pectoral region, thorax, flank, back, rump, ventrum, and down each leg
- Insertion of a finger into each nostril, each ear, the diastema of the jaw, and the rectum
- Slapping the bottom of each foot with the palm of the hand
- Rubbing a plastic bag over the face, muzzle, ears, poll, neck, pectoral region, thorax, flank, back, rump, ventrum, and down each leg
- Applying running clippers over the muzzle, the outside and inside of the ears, the poll, and the lower limbs

- Spraying water around, but not directly on, all parts of the body

We've followed the whole progression with most of the horse foals born on our farm for the past 14 years, and we like the results; each individual, without exception, matured into an unflappable, human-oriented riding horse. But would we do it with donkey foals? Probably not, for two good reasons:

1. Donkeys, being more intelligent and less reactive than horses, really don't need extensive early desensitization.
2. Donkey breeders who have used the full sequence claim that jennies are more likely to object to extensive early handling of their foals than mares do, and imprinting is more apt to interrupt the jenny-foal bonding process.

However, after dipping the foal's navel, we'd briefly imprint each foal by spending a quick five minutes running our hands over his head, ears, body, and legs. With donkeys, that's probably sufficient.

foal inside during inclement weather, and if he should get soaked, bring him in, dry him off, and warm him up as quickly as you possibly can.

The first manure a foal passes is a black, tarry substance called meconium (once it's ejected, he'll make soft, yellow-tan stools). Occasionally foals have a good deal of trouble passing meconium. If your foal seems to be straining, or if you don't see traces of black goo on his butt or meconium in the stall, give him a child-size Fleet enema. Lubricate the applicator tip well and be gentle; those are very tender tissues back there.

Bringing Up Baby

Unfortunately, the ideal scenario doesn't always play out to the end. The jenny dies, she rejects her foal, she has no milk, or a foal is too weak to nurse from his dam. What then? You raise the little guy yourself.

Your job is relatively easy if the dam is alive and is interested in the foal but has no milk. She'll socialize him and teach him to be a donkey; all you have to do is furnish his feed. When Mama is out of the picture, things are harder (but still doable). Here's what you need to know.

What to Feed

Orphans need colostrum even more than the average foal. If you can't milk the foal's dam and don't have the real thing in the freezer, discuss IgG supplementation or a plasma transfusion with your vet as soon as you possibly can.

After the first 24 hours have passed, you'll need milk to feed him. If you use powdered mare's-milk replacer, choose a product based on milk protein, *not* soy, and one designed specifically for feeding baby equines. There are many good ones on the market: PetAg Foal-Lac, Land O'Lakes Mare's Match, Merrick's Grow-n-Glow, and Buckeye's Mare's Milk Plus to name but a few. If mare's-milk replacer isn't available when you need it, Purina Kid Milk Replacer (for goats) can be fed to foals, but only when mixed according to the foal-feeding directions on the label; don't use the instructions for feeding kids.

Mix the product you choose according to directions on the label, in most cases by adding a measured amount of powder to a portion of 110-degree water, whisking thoroughly, then adding enough additional warm water to make a gallon of fluid. For best results, mix only enough for one or two meals.

Female goats sometimes adopt a foal so completely that they allow nursing on demand; no wonder some folks call them nannies!

A perfectly acceptable substitute for jenny's milk is fresh goat milk, particularly from a breed that gives relatively low-fat-content milk (such as Saanens, Sables, and LaManchas), but don't use devitalized canned goat's milk from the supermarket. In fact, if you're lucky enough to own a milk goat, place her on a milking stand with a handful of grain to keep her happy and allow the foal to nurse directly from the goat. After a while she might even adopt him and allow him to feed on demand.

However, don't feed a foal whole cow's milk; it has nearly twice as much fat content and only two-thirds of the sugar content of mare's milk, and it generally causes diarrhea in foals. In a pinch, low-fat cow's milk (2 percent fat) can be substituted for mare's milk by adding ¼ cup of white corn syrup per gallon of 2 percent milk.

Whichever option you choose, stick with it; constantly altering a neonatal foal's diet is flirting with disaster. If you must switch to a different type of milk or even a different brand, do it gradually over a course of six or seven days. Do your best to prevent serious scours that can lead to dehydration and possibly even death.

How Often to Feed

Some people say it's okay to feed neonatal foals two times a day. Don't believe it! Neonates require small amounts of nourishment, fed often and at regular intervals. Here's a guide for what works:

- ✘ Days 1 though 7, feed every two hours around the clock.
- ✘ Days 8 through 14, feed every four hours around the clock.
- ✘ Days 15 through 21, feed four times per day, omitting the nighttime feeding.
- ✘ Days 22 through weaning, feed three times per day.

Unlike lambs and kids, foals rarely tend to overeat. Many people give them all the milk that they'll drink, but remember, feed nothing but colostrum for the first 24 hours.

Digestive Aids

Foals, even dam-raised foals, are prone to developing ulcers. To help to head off problems, our equine practitioner recommends crumbling one Tagamet or Zantac tablet in an orphan's milk at any one feeding of the day. Some other strategies:

- ✗ To head off digestive difficulties in general, add one heaping tablespoon of live-culture plain yogurt to one bottle of milk per day.
- ✗ Alternately, stir a portion of equine-specific probiotic powder into one feeding, or dose your orphan with probiotic gels like Probios or Fastrack for horses.
- ✗ Foals are born with a sterile gut, which they populate by nibbling their dams' manure. If you see your foal doing this, don't stop him! It's perfectly normal (and important) neonate behavior.

Bottle Feeding

To start a foal on a bottle:

1. Stand him in front of you so that you're both facing the same direction.
2. Brace his butt against your legs and his forequarters with your elbows (you may have to sit on a bale of hay or straw to do this with Miniature foals).
3. Open his mouth with your left hand, insert the nipple, and with your palm under the foal's jaw, use your fingers to keep the nipple aligned with his mouth.
4. With your right hand, elevate the bottle just enough to keep milk in the nipple, adding more tilt as he empties the bottle.
5. If he doesn't suck, gently squeeze the bottle so a *tiny* amount of milk goes into his mouth. Don't drown him!

Some newborns take to the bottle right away. Others struggle and fuss. Keep trying. The little guy has to eat or he'll die.

When Baby Won't Eat

There are few things as frustrating as a foal that won't accept a bottle or drink from a pan. These are generally foals that nursed their dams and are pretty sure that the stuff you offer them is poison. Some ploys to consider include:

- ✗ Switching nipples until you find one that the foal likes.
- ✗ Lightly coating the nipple with fruit yogurt or sugar.
- ✗ Cupping your hand above the foal's eyes to simulate the darkness of his dam's groin.
- ✗ Offering him a bottle or pan of replacer or goat milk, and if he won't take it, putting him away. It may take two or three of these sessions until he's finally hungry enough to eat.
- ✗ Persevering. In the meanwhile, have your vet tube feed him if necessary. He *must* stay hydrated in order to survive.

Until he understands the process, it helps to stand astride the foal to keep his head in alignment with the bottle.

Pan Feeding

Some people prefer to pan- or bucket-feed orphan foals because it's faster and easier. We don't like it because pan-fed foals are wasteful and when left in the baby's stall, the milk collects flies and dirt. However, if you prefer that route, your foal will probably learn to drink quite readily if you place your finger in his mouth, then, while he's sucking, raise a small bowl containing milk or milk replacer up to his muzzle. Slowly remove your finger from his mouth while he's drinking. If he stops, repeat the above steps until he's drinking by himself. Always bring the milk up to the foal; **never force his head into a bucket.** Once he's mastered the skill, you can hang a shallow bucket in his stall and replenish the milk at every feeding. However, cleanliness counts, so buy two buckets so that you can thoroughly wash and disinfect one while the other is in use.

Feeding Gear

Every breeder has his favorite bottle-feeding gear. We feed our Boer goat kids with Pritchard teats attached to reused plastic water bottles; these are great for starting Miniature Donkey and Miniature Horse foals, too. The Pritchard teat is an oddly shaped red nipple attached to a yellow plastic cap ring sized to fit standard screw-top household containers (including recyclable 12-ounce and 1-liter soda and water bottles); its plastic base incorporates a valve that makes fluids flow smoothly through the nipple. Other folks prefer human baby bottles, especially brands with ergonomically designed nipples (NUK is a good one).

Soft-rubber lamb nipples from a feed store work well; hard-rubber calf nipples are too stiff. However, since lamb nipples pull rather than screw on, foals tend to yank them off the bottle at inopportune times.

Whatever type you use, enlarge the hole in the nipple so the milk flows as the baby sucks; otherwise, he might get discouraged and quit before he finishes his feed. However, don't make the hole so big that milk pours down his throat and he aspirates it into his lungs. The perfect solution: slice a ½-inch X in the nipple using a sharp knife.

When using plastic soda or water bottles, wash the nipple and bottle after each feeding in hot, soapy water, rinse thoroughly, and at the end of a 24-hour cycle, recycle the bottle. If you don't use a clean bottle every day, sanitize the one you use at least once a day using a solution of 1 part household bleach to 10 parts water and then very thoroughly wash and rinse it out. But don't bleach nipples; bleach tends to degrade them very quickly. Simply scrub them well with good old soapy water.

Dry Food and Water

Make certain that your foal has full-time access to clean, inviting water. He probably won't drink it until he's several weeks old, but a few foals drink water right from the start. By the same token, place small amounts of hay and milk pellets in your foal's stall beginning on day two. Remove it when it loses its freshness (another donkey or horse is sure to love it) and replace it with fresh, new feed. Place it at shoulder height so that your baby sees and nibbles at it until he begins eating in earnest. If the foal has an "auntie" (a companion to act as his surrogate mother) in his stall, set up a creep feed area that he can enter and auntie cannot.

Begin phasing out the bottle at about six weeks of age and continue feeding free-choice milk pellets and hay

'Ear's a Tip

Clicker Training Foals. Do foals respond to clicker training? Absolutely! As soon as your baby accepts food tidbits, he's ready to enter kindergarten.

until foals are five or six months old; some nutritionists recommend gradually replacing milk pellets with a concentrate mixture formulated for foals (e.g., Purina Junior, Buckeye Growth Formula, Nutrena LifeDesign Youth) until the foal is weaned from all milk products at three months of age, and the latter method is far less costly.

Socializing Baby

A bottle (or bucket) baby is in danger of growing up thinking that he's a human instead of a donkey unless you take steps to socialize him while he's young. This is important. He needs to know how donkeys act and think to fit into a herd later on.

The best thing to do is leave him with his dam if she's able and willing to keep him. He'll run to you for feeding, but she'll show him what he needs to know to fit in. Second best: Find him an auntie. Many female equines are perfectly willing to raise another's baby, even though they can't supply his grub. Try to set him up with a donkey auntie, but if none is available, try a friendly gelding (some geldings make outstanding aunties) or even a pony or horse.

Once you decide on a surrogate, *introduce them gradually,* especially if the little guy nursed his mom and knows where to find the lunch bar. Place them in adjoining stalls or pens where the auntie can see and smell her future charge. Encourage her interest in the little guy; then, after a day or two, if she shows no aggression toward the foal, halter her and snap on a lead, then introduce them in a fairly open area with a handler to control auntie and another to whisk the baby out of harm's way if things go wrong.

If no suitable equine auntie can be found, try a nonequine auntie. While nonequines can't teach the baby how to behave in donkey society, they provide comforting companionship, and this can help an orphaned baby thrive. A friendly, polled (naturally hornless) sheep makes a wonderful auntie, as do polled or disbudded (dehorned) goats, some llamas, and even calves. When choosing a ruminant auntie, make sure that the feed you provide for auntie doesn't contain Rumensin, urea, or any other additives toxic to equines because the baby will be sharing her fare.

If auntie is a nonequine species, be sure to get the baby out and around other donkeys as soon as you safely can. And nip unacceptable behaviors in the bud: kicking, biting, and crowding may be cute while he's little, but it's easier to prevent bad habits than it is to correct them later on.

Baby's Handy Handle

When you need to move an untrained foal from place to place or require a secure means of controlling one under trying circumstances, find a large halter and punch extra holes in the crownpiece at 2-inch intervals. To punch holes in nylon halters, burn them with a red-hot nail held in pliers, and be sure to use an oven mitt to hold the pliers.

Turn the halter upside down, and feed the baby's head through the noseband opening. Next, fasten the crownpiece under his barrel like a girth. Grasp the resulting handle, and there you are! The halter can be used as is, with the excess crownpiece dangling, but for best results, sacrifice the halter and trim the crownpiece to size.

Very Basic Training

Whether an orphan or dam-raised, your foal should learn to lead, stand while held, and pick up his feet at an early age. Don't tie him! Tying young foals and allowing them to struggle is a sure way to damage their fragile necks.

The accepted way to teach foals to lead is with a rump rope. This is simply a loop of soft rope draped across the foal's hindquarters that the person leading him can gently tug if the baby sets his brakes. If his dam or auntie is present, have an assistant lead his mom or auntie in front, so that the little guy is more inclined to follow.

To teach him to stand, stop and talk to him soothingly. Don't expect a lot; foals have short attention spans, so 5- or 10-minute leading/standing sessions are plenty.

When the foal stands well, run your hand down his shoulder to his pastern and give him a verbal cue to lift his leg. Pick it up and set it down, then praise him highly. Work up to holding each leg for 30 to 60 seconds; that's long enough for a suckling foal.

Other things to do while he's small include quietly loading with his mom or auntie into a trailer, "meeting" scary things like pigs or skateboarders, and generally introducing him to anything that might frighten an adult donkey. By showing him that scaries don't bite, he's far less likely to fear things later on.

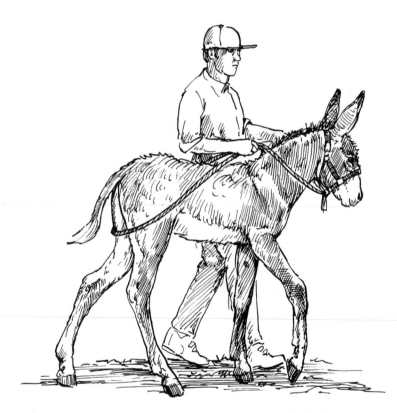

A rump rope is a handy tool for teaching foals to lead.

Out of the Mouths of Babes

hippomanes (hip-oh-MAH-neez; from Greek, meaning "horse, crazy"): circular, flat smooth body found in allantoic fluids, esp. in mares and cows.

— *Saunder's Comprehensive Veterinary Dictionary*

Hippomanes range in color from chalky gray to yellowish to darkest brown and are shaped like thick, oval pancakes that are soft, spongy, and rubbery to touch. They range in size from 2 to 6 inches long, 2 to 4 inches wide, and ¾ to 1¾ inches thick. No two are alike; even twins foals' hippomanes may be different sizes and different colors.

Prior to birth, a foal's hippomanes free-floats in a sea of allantoic fluid, the watery liquid that fills the outer sac of his placenta. The first trace of hippomanes forms in a mare or jenny's allantonic fluid around the 85th day of pregnancy; it will be ejected when the dam's water ruptures, or it might be found nestled in the placental membranes after birth. Sometimes there are accessory small hippomanes floating free in the fluid or attached to the chorio-allantoic membrane.

And the hippomanes' function? No one knows. Most sources agree that a hippomanes is composed of cellular debris, essentially a collections of spare parts. They're found in 90 percent of all equine placental membranes.

Folks through the ages have struggled to explain this strange, mysterious mass of cells.

Two thousand years ago the Roman naturalist Pliny the Elder wrote, "A love poison called the horse frenzy is found in the forehead of horses at birth, the size of a dried fig, black in color." Old ideas die hard: dried, powdered "horse love" was peddled as a potent aphrodisiac throughout Medieval Europe.

Well into the 20th century, British horsemen believed that an unborn foal carried his hippomanes in his mouth and that its function was to shape his tongue. Old-time Suffolk horsemen who attended foaling mares were required to show the hippomanes to the farmer the next morning because the milt would prove that the attendant was actually on the spot and not asleep when the foal was born. "Because to get the milt you had to put your two fingers into the colt's mouth and prise it out just as it were a-coming out of the bag. Do you leave it a moment later, the colt would swallow it."

Nineteenth-century horse tamers, members of the Brotherhood of the Horse, carried salt-cured hippomanes in pouches, carefully enveloped in red cloth and anointed with secret "taming oils" to make them irresistible to horses.

CHAPTER 17

THOSE MULES

My favorite animal is the mule. He has more horse sense than a horse. He knows when to stop eating — and he knows when to stop working.

— President Harry S. Truman

Veteran pack burro racer and breeder Curtis Imre describes mules as a "hybrid cross of a jack and a mare: soul of the donkey and size and power of the mare." We think that says it all. History's first "mules" were probably onager hybrids. By tethering in-heat jennies where wild onager jacks could find and mate with them overnight, ancient humans gained the advantage of greater size and hybrid vigor. The wealthier classes probably kept captive onager jacks for just this purpose. Later, when horses came on the scene, humans switched to breeding domestic donkey jacks to mares (to breed mules) and to a lesser degree, stallions to domestic jennies (to create hinnies).

Origins

We know from surviving art and written records (including the Hebrew Old Testament) that mules were bred (and treasured) as early as 2800 BC, when the Sumerians began intentionally breeding onager/ass hybrids. The king of Urim, author of the Sumerian text *A praise poem of Culgi*, writes, "I am a mule, most suitable for the road. I am a horse, whose tail waves on the highway. I am a donkey of Cakkan, who loves running."

Between 2000 and 600 BC, Assyrian merchants in the ancient city-state of Aššur (located in what is now Iraq) inscribed more than 20,000 cuneiform tablets describing trade between Aššur and merchants in central Anatolia. From these records we know that a mule cost between two and four minas of silver — four

An onager hybrid, the offspring of an onager and a domestic ass.

times the price of a sturdy ass — and that the Assyrians coined several words for describing mules: anše.*kunga* (a catchall word for mule), anše.*la.gu* (riding mule), and anše.*gír.nun.na* (a fancy mule used for drawing the royal chariot).

The ancient Egyptians had mules, although the earliest were likely onager hybrids. Bas-reliefs in the tomb of Khaemhet, royal scribe and "Overseer of the Granaries of Upper and Lower Egypt" during the Eighteenth Dynasty (1567 to 1320 BC) reign of Amenhotep III, depict a pair of ropy-tailed mules drawing a chariot. Colored carvings at a tomb at Thebes dating to 1400 BC depict a pair of similar white mules hitched to a chariot.

The Hittites also treasured the mule. According to *The Code of the Nesilim*, circa 1650–1500 BC, "A plow-ox costs fifteen half-shekels of silver, a bull costs ten half-shekels of silver, a great cow costs seven half-shekels of silver, a sheep one half-shekel of silver, a draft horse twenty half-shekels of silver, a mule one pound of silver, a horse fourteen half-shekels of silver." Another entry tells us that the price of a mule was one mina of silver (one Hittite mina equaled 40 shekels), so mules were the Hittites' most valuable livestock.

The Greek poet Homer immortalized mules in his epic poem *The Iliad*, penned sometime during the seventh or eighth century BC, when he described mules awarded as prizes: "For swift charioteers first he set forth goodly prizes, a woman to lead away, one skilled in goodly handiwork, and an eared tripod of two and twenty measures for him that should be first; and for the second he appointed a mare of six years, unbroken, with a mule foal in her womb," and "He brought out a strong mule, and made it fast in the middle of the crowd — a she-mule never yet broken, but six years old — when it is hardest of all to break them: this was for the victor, and for the vanquished he offered a double cup." Mule races became part of the Olympics about 500 BC, and Aristotle also wrote about mules.

The Roman legions knew a good thing in the mule. It carried so much weight so well that *mulus Marianus* (Marius's mule) was slang for a heavily laden

BRAY SAY: *Mane, forelock, and tail are triple gifts bestowed by the gods upon the horse for the sake of pride and ornament, and here is the proof: a brood mare, so long as her mane is long and flowing, will not readily suffer herself to be covered by an ass; hence breeders of mules take care to clip the mane of the mare.*

— Xenophon, 431–355 BC

legionary soldier. Every legion (4,200 men during the republican period of Rome and around 5,500 in the imperial period) had a baggage train of 500 to 550 mules or about one mule for every ten legionaries. The later Roman legions fitted their pack mules with *soleae ferreae,* or temporary shoes called "hipposandals," by wrapping their hooves and then placing them in shoelike iron contraptions secured by bands running through hooks and eyes at the front and back ends of the shoes (horse- and mule shoes as we know them weren't used at that time).

In AD 60, one L. Junius Moderatus Columella authored a multivolume work called *On Agriculture* in which he writes, "Horses are divided into three groups: the noble breed that provides animals for holiday races in the circus; the stock used for breeding mules, which fetches a price that puts it on a level with the noble variety; and the common breed, which produces ordinary mares and stallions."

Mules came to Europe with the Roman legions, where they soon became the favored mounts of high-ranking clerics. In AD 1116, the abbey of Burton-on-Trent kept three Spanish jacks and 36 broodmares specifically to breed mules for this purpose. The popes rode mules, as did Spanish kings, and the Doges of Venice. Everyone who was anyone rode a mule.

The 20-inch by 230-foot Bayeux Tapestry commemorating the 1066 Norman invasion of England depicts Count Guy de Ponthieu astride a handsome mule while out hunting with a falcon, and in another panel, a mule is shown pulling a plow.

As clerics rode mules, so did fine ladies throughout Great Britain and Europe, including the ladies of King Arthur's fictional court as described in *The High History of the Holy Graal*: "Thereupon, behold you, a damsel of surpassing great beauty that cometh, sitting on a right costly mule, full well caparisoned. She had a gilded bridle and gilded saddle, and was clad in a right rich cloth of silk. A squire followed after her that drove her mule from behind. She came before King Arthur as straight as she might, and saluted him right worshipfully, and he made answer the best he might."

Fine mules were raised throughout the Eastern world as well. Marco Polo praised the mules he encountered in central Asia; little wonder, their dams being fine Turkoman (Akhal-Teke) mares! Mules plied the Silk Road as well as lesser trading routes such as the Ancient Road of Mules and Horses, built during the thirty-third year of the reign of the first emperor of the Qin Dynasty (214 BC), and an important line of communications that linked the central plain area to southern China. During the Tang Dynasty (AD 618 to 907) pack mules carried tea, silk, salt, and opium along the Tea-Horse Road through mountain passes and over gorges via rope bridges to and from India by way of Burma to Tibet and then to central China.

Mules and mule breeding came to North America via two distinct routes: up through the Southwest and Southeast with the Spaniards (Coronado's party included mules when it entered Arizona in 1540) and from Europe to the United States. George Washington's fine imported jacks, Royal Gift and Knight of Malta, were the fathers of the American mule *(see page 49).*

Aristotle on Mules

Greek philosopher Aristotle penned his multivolume classic, *History of Animals,* around 343 BC. One of the many hundreds of birds, fish, and animals he describes is the mule:

The mule lives for a number of years. There are on record cases of mules living to the age of eighty, as did one in Athens at the time of the building of the temple; this mule on account of its age was let go free, but continued to assist in dragging burdens, and would go side by side with the other draught-beasts and stimulate them to their work; and in consequence a public decree was passed forbidding any baker driving the creature away from his bread-tray.

Washington was so committed to his mule-breeding program that in 1788 he swapped his prize Arabian stallion, Magnolia, to Light Horse Harry Lee for 5,000 acres of land in the Kentucky territory. In a letter to Light Horse Harry Lee, Washington wrote: "I am willing to confirm the bargain because it is my intention to breed mules only." Soon after the Revolutionary War, he began replacing workhorses at his Virginia estate with mules; in 15 years, the Mount Vernon stables housed 58 head of mules.

Military Mules

Mules have a long association with the American military. They were widely used in battle as recently as World War I when more than 265,000 horses and mules died on the Western Front. At the close of the war, more than one million horses and mules served with American and Commonwealth armies.

Mules didn't play a significant role in the American Revolutionary War because they were scarce at the time. Things were different at the beginning of the Civil

PLINY THE ELDER ON MULES

Mules have indeed been around for a long time! These words from Pliny the Elder were written in AD 77 in his *Natural History*, chapter 69: "The Nature of Mules, and of Other Beasts of Burden" (1855 translation by John Bostock):

From the union of the male ass and the mare a mule is produced in the thirteenth month, an animal remarkable for its strength in laborious work. We are told that, for this purpose, the mare ought not to be less than four years old, nor more than ten. It is said also that these two species will repulse each other, unless the male has been brought up, in its infancy, upon the milk of the other species; for which reason they take the foals away from the mare, in the dark, and substitute for them the male colts of the ass. A mule may also be produced from a horse and a female ass; but it can never be properly broken in, and is incorrigibly sluggish, being in all respects as slow as an old animal.

If a mare has conceived by a horse, and is afterwards covered by an ass, the first conception is abortive; but this is not the case when the horse comes after the ass. It has been observed, that the female is in the best state for receiving the male in the seventh day after parturition, and that the males are best adapted for the purpose when they are fatigued. A female ass, which has not conceived before shedding what are called the

milk-teeth, is considered to be barren; which is also looked upon as the case when a she-ass does not become pregnant after the first covering.

The male which is produced from a horse and a female ass, was called by the ancients 'hinnulus,' and that from an ass and a mare 'mulus.' It has been observed that the animal which is thus produced by the union of the two species is of a third species, and does not resemble either of the parents; and that all animals produced in this way, of whatever kind they may be, are incapable of reproduction; she-mules are therefore barren. . . .

It is said in the works of many of the Greek writers, that from the union of a mule with a mare, the dwarf mule is produced, which they call 'ginnus.' From the union of the mare and the wild ass, when it has been domesticated, a mule is produced which is remarkably swift in running, and has extremely hard feet, and a thin body, while it has a spirit that is quite indomitable. The very best stallion of all, however, for this purpose, is one produced from a union of the wild ass and the female domesticated ass.

This postcard shows an artillery mule being unloaded during World War I. It's labeled in English and French (in the lower left corner), "The War — Discharging of a Gun."

War. Both sides required draft horses and mules to pull wagons and move equipment, but the mules were especially valuable. According to records of the day, between 50 and 90 percent of the draft animals purchased by both sides were mules — and for good reason. At the height of the Civil War, M. C. Meigs, the quartermaster general of the U.S. Army, reported that "mules bear the exposure and hardship of the campaign much better than horses, and they are used to a great extent in the trains." They were also more abuse resistant than horses, a boon at military installations of the day.

Mules were mainly used for pulling wagons and heavy artillery and for packing small mountain howitzers. However, they were replaced by horses before hostilities commenced because mules didn't do well in the heat of battle itself. Describing his battery mules' reaction to fire at the Battle of Port Arthur, Confed-erate Brig. Gen. John D. Imboden said, "The mules became frantic. They kicked, plunged and squealed. It was impossible to quiet them, and it took three or four men to hold one mule from breaking away. Each mule had about three hundred pounds weight on him, so

Did You 'Ear?

Baghlet el Qebour. According to Moroccan folklore, Baghlet el Qebour ("mule of the graves") is an evil spirit who, in the shape of a mule, roams the Houz of Marrakech seeking lazy men who are away from their homes in the evening. When she finds one, she buries him alive.

BRAY SAY: *While more than 68,000 of them serving with our armies in World War I passed to the great beyond in silent agony, and while many of them now sleep on the gentle slopes made beautiful by the poppy's bloom, no white crosses, row on row, mark their last resting places.*

— Anna M. Waller writing about army mules, 1958

securely fastened that the load could not be dislodged by any of his capers. Several of them lay down and tried to wallow their loads off. The men held these down and that suggested the idea of throwing them all to the ground and holding them there. The ravine sheltered us so we were in no danger from the shot or shell which passed over us."

Mules also served in the cavalry on occasion. Colonel Able Streight of the 51st Indiana Infantry led a raiding force of roughly 1,600 men, most of whom were mounted on mules, and it's said that many of Confederate General James Longstreet's troops rode mules, too.

At the beginning of the American Civil War there were approximately 100,000 mules in the North, 800,000 in the secessionist states, and 200,000 in Kentucky and Missouri. While no statistics pertaining specifically to mules exist, an estimated 1.5 million military horses and mules were killed, either on the battlefield or from starvation or disease, between 1861 and 1865. More than 3,000 perished at Gettysburg alone.

Fourteen thousand United States Army mules saw action in World War II, particularly in Tunisia, the rugged mountains of Italy, and in Burma. At the close of World War II, the 4th Field Artillery Battalion and the 35th QM Pack Co., equipped with mules and a small number of horses, were retained at Fort Carson, Colorado, for training purposes.

Legend has it that Hambone (Hamilton T. Bone) was the last army mule to be deactivated, an event that

The animal hitched to the buggy in this photo is a horse, but the others are mules.

occurred at Fort Carson, Colorado, on December 15, 1956, when Battery A of the 4th Field Artillery Battalion (Pack) was redesignated, the 35th Quartermaster Company (Pack) deactivated, and a total of 322 army mules went into retirement, replaced as it were by helicopters.

Hambone, a silvery-white mule, served a 13-year stint at Fort Carson, where he once carried the first sergeants of the 4th along the foothills of the Rockies to Cheyenne, Wyoming, for the Frontier Days Rodeo. The charismatic mule was so popular that in 1949, *Life* magazine ran a feature story about him.

After deactivation, Hambone spent his retirement as a star attraction in the Pikes Peak or Bust Rodeo and the Pikes Peak Range Ride. In 1970 he returned to Fort Collins, where he died on March 29, 1971. Hambone

SHELBY'S MULE

Mules are dependable and steady, while horses are all prance, fart, and no sense.

— General Joseph "Jo" Shelby,
Confederate States of America (who rode a mule)

In 1831, Joseph Orville Shelby was born to a wealthy, aristocratic Southern family in Lexington, Kentucky. As a young man he moved to Missouri, where he set himself up as a hemp-farming and rope-making entrepreneur. He became one of the wealthiest landowners in Missouri.

When the Civil War broke out, Selby joined the cavalry and quickly became known for his courage, his flamboyance — and his mule. In 1862, Shelby and his 800-man "Iron Brigade" rode 1,500 miles through the heart of Union-occupied territory, from the Arkansas line to the Missouri River and back again in 34 days. When the war ended, Selby refused to surrender and took his men to Mexico for two years before returning to Missouri to take up farming again.

During the war, Joseph Leddy, a popular minstrel-show star, wrote and performed a tune called "Shelby's Mule", which was, at least in Missouri, more popular than "Bonnie Blue Flag" or even "Dixie." Folk groups still perform it today.

The Union folks away up North are getting much afraid,
'Bout something coming from the South, they think it
    is a raid.
Now I will tell you what it is, if you will just keep cool —
It has long ears, and a long slick tail, it looks like Shelby's
    Mule.

Chorus (after each verse):
Shout boys, make a noise, the Yankees are afraid
That something's up and hell's to pay when Shelby's on
    a raid.

Once this mule went on a spree, up close to Lexington,
And every time he gave a snort he made the Blue Coats
    run.
Coming back through "Old Saline" he got into a trap,
He seared Old Brown, kicked up his heels, and came
    back safe to Pap.

Once I went to see Old Abe and found him in a rage,
Because this mule had started north, and just then
    crossed 'sage.
Indeed, his anger knew no bounds, says I, "Sir, pray
    keep cool."
"I can't," said he, "I've lost so much to Shelby's long-
    tailed mule."

was buried with military honors in front of Division Artillery Headquarters under a headstone quarried on the military reservation.

Although many believed that Hambone was the last army mule deactivated, another mule lays claim to that distinction. According to papers submitted to the U.S. Army Quartermaster Corps by former soldier Kimberly A. Cook, Wind River, a molly mule who retired alongside Hambone in 1956 joined a Shriners group and made appearances at various local, national, military, and private functions until her second retirement in 1978. Wind River died a few months later at age 46.

But was Wind River truly the last living army mule? No, indeed. That honor falls to Trotter, a four-gaited pack mule (he walked, trotted, cantered, and paced) two or three years her senior.

Trotter, along with the rest of the 35th Quartermaster Pack Troop, landed in Calcutta in early 1940, assigned to assist the 3rd Battalion, 475th Infantry Regiment on the Pacific Front. During his time in active duty, Trotter carried a 75mm pack howitzer on the unit's historic 500-mile trek from Calcutta to Nampakka in Japanese-occupied Burma.

After leaving active service, Trotter became West Point's Army Mule (sharing the limelight with several other mules) from 1957 through 1972, until he became "testy and difficult to handle." Then he retired for a second time, to the New York City Health Department farm near Otisville, New York, then functioning as a retirement home for New York City police horses. Trotter died at the ripe old age of 49.

Francis the Talking Mule

Francis the talking mule starred in seven immensely popular comedy films done by Universal Studios and released between 1950 and 1956. They were loosely based on a popular 1946 novel of the day, *Francis the Talking Mule,* by David Stern.

In the first six films, Francis, a sage and sarcastic army mule (voiced by veteran character actor Chill Wills), and Peter Stirling (played by actor Donald O'Connor), the naive young soldier who befriends him, romped through a series of adventures both in and out of the army. Throughout each film, Francis spoke only to Peter, thus creating mayhem (and a good bit of mental scrutiny) for his hapless human friend. At the end of each film, Francis invariably spoke up and saved the day.

After O'Connor abandoned the series (peeved, it is said, because Francis garnered more fan mail than he did), the seventh and final film, *Francis in the Haunted House*, paired a similar character, David Prescott (portrayed by Mickey Rooney), with the sardonic, wisecracking mule (voiced by Paul Frees).

The mule that portrayed Francis was in fact a female named Molly, purchased by Universal Studios for $350 and chosen because she was easy to handle. Horse trainer Les Hilton helped Molly to "talk" by feeding a thread through her mouth that, when tugged, caused Molly to wiggle her lips. (He later went on to train Bamboo Harvester, the palomino horse who played Mr. Ed.) For her acting efforts, Molly (as Francis) won prestigious American Humane Association Patsy Awards in 1950 (the first one awarded) through 1957, and others, including the award of excellence in 1953.

Today's Mules (and Hinnies)

On January 1, 1925, the American mule population peaked at 5,918,000; it dipped to a record low of about 10,000 head in the 1960s. Now mules are staging a comeback — a *spectacular* comeback — as America is rediscovering the mule. Today's designer mules do anything that horses do, and often better. They excel at everything from jumping to carriage competitions to endurance racing and more.

For a real look at what mules can do, join more than 30,000 spectators at next year's Bishop Mule Days in Bishop, California, and watch 700 fine mules compete in events as diverse as chariot racing, reining, dressage, cattle sorting, cutting, packing, team roping, trail, pleasure (English, Western, sidesaddle, driving, and gaited), and halter classes — and those are just the tip of the iceberg.

Or would you prefer pari-mutuel mule racing sanctioned by the American Mule Racing Association or draft-mule pulling contests, in which competing teams of draft mules pull heavy metal sleds for at least 15 feet? Perhaps you'd like coon mule jumping competitions (a timed event where mules, saddled or otherwise, emulate raccoon-hunting saddle mules by jumping to astounding heights from a standstill) at shows and county fairs? And there are Miniature mules, too, capable of doing anything that Miniature horses and donkeys can do. The list goes on and on!

So if you're looking for the "soul of the donkey and size and power of the mare," get a mule. We bet that you'll be glad you did.

BRAY SAY: *You got to have smelt a lot of mule manure before you can sing like a hillbilly.*

— Hank Williams

CHAPTER 18

MAKING MONEY WITH DONKEYS (MAYBE)

*'Tis better to be an ass's head
than a horse's bottom.*

— Irish proverb

It's certainly possible to make money with a donkey enterprise, but perhaps not precisely the way you think. People equate moneymaking livestock ventures with selling young stock; if done right, that's a good ploy. But there are additional, less common donkey-based businesses ideal for imaginative donkey entrepreneurs. One might be just the ticket for you. This chapter contains a few that you might consider.

Breeding to Sell

We touched on this lightly in chapter 16, but it's so important that we'll say it again: Please don't breed donkeys unless the end result is top-quality, saleable foals. To achieve that worthy goal, there are many details to consider.

Start with an in-demand, registered breed (Miniatures, Mammoths, Poitou Asses, or another threatened ass breed from abroad) with built-in sales appeal. Set up your farm as a credible breeding operation. Your facilities needn't be fancy, but buyers will be visiting to view your stock; things should be kept up, safe, and

This donkey postcard was sent from Colorado in 1906.

tidy. If you plan to stand a jack at stud, provide safe, separate quarters for visiting mares and jennies.

Carefully study the breed standard and then buy the best foundation stock you can find and afford. You'll show more profit selling foals from two stellar jennies than you will from six nondescript jennies or a whole slew of jennies with major faults. Start small if you must, and expand your operation by retaining your best jenny foals for your own herd.

Don't breed solely for short-term market fads like popular bloodlines or color, but do take them into consideration. Quality first, *then* the frosting on the cake: That should be every breeder's goal.

Work with your farrier and vet to establish hoof care, deworming, and vaccination schedules tailored for your herd and your locale. Maintain your animals in tip-top condition — no excuses. Keep careful health, breeding, and training records on each animal and be ready to provide copies to buyers. Likewise, register foals as soon as they're born, so that their papers are in order when buyers come looking.

Don't breed jennies until they're physically and mentally mature: no younger than three years old for Miniatures and Standards and three or four years old for Mammoth jennies. Geld most of your jack foals, keeping only the crème de la crème of each year's jack foals intact. *Never* sell a less-than-breeding-quality jack intact, particularly if he carries your farm prefix. If unsure about an individual, geld him anyway or keep him a while until you know for certain.

Be totally, painfully ethical at all times. Represent your donkeys in a positive light but point out major failings, particularly when selling sight unseen. Stand behind the stock you sell. Use a sales contract/written guarantee for every sale or service. Offer after-sales support. Satisfied customers are your best advertising.

Actively promote your farm and your donkeys; without this step you simply can't succeed (*see chapter 19*). Take it to heart. Buyers may want what you have to sell, but if they don't know that you have it or they can't find you, you're both out of luck.

When you sell a donkey, make it a memorable event. Provide the new owner with a new, quality halter and lead rope for the donkey, along with a portfolio containing written information about what the donkey is eating, complete health, farrier, and breeding (if applicable) records, a photocopy of the donkey's registration certificate and a copy of his extended pedigree, and for foals that you've raised, at least one picture of each parent.

Mail the transfer to the registry yourself (this is why you've provided a copy of the registration certificate

BRAY SAY: *And so we Gipsies always burn an ash-fire every Great Day. For the Saviour was born in the open field like a Gipsy, and rode on an ass like one, and went round the land a begging his bread like a Rom. And he was always a poor wretched man like us, till he was destroyed by the Gentiles. And He rode on an ass. Yes. Once he asked the mule if he might ride her, but she told him no. So because the mule would not carry him, she was cursed never to be a mother or have children. So she never had any, nor any cross either.*

Then he asked the ass to carry him, and she said "Yes"; so he put a cross upon her back. And to this day the ass has a cross and bears young, but the mule has none. So the asses belong to the Gipsies.
— from *English Gipsies and Their Language,* by Charles G. Leland, 1874

in lieu of the original); buyers often forget to do it, and over time a lot of papers have been misplaced and valuable pedigrees lost. Have the transferred papers sent directly to the new owner; make certain that the buyer understands this policy.

HOW-TO
Writing a Contract

Use a well-worded sales contract, and use it every time. While it's tempting to use an across-the-board contract borrowed from a book or found online, don't do it! Before you sell your first donkey, hire a lawyer to draw up a contract specifically for your farm and its needs. It should address:

1. **The precise date of sale.**
2. **The identity of both parties in detail.** That's you (the seller) and the buyer, including full names, addresses, phone numbers, and Social Security or federal tax identification numbers.
3. **The identity of the donkey in detail.** Include his name, age, color, markings, and registration number, and the names and registration numbers of his sire and dam.
4. **Price and terms of sale.** State the sale price. If the buyer pays a deposit, when is the balance due, and if the buyer backs out, will the deposit be returned or forfeited? If there are special contingencies to the sale (it's a trade or partial trade, or you expect one or more foals back, or you're retaining stud-service privileges for a specified amount of time), spell them out in detail. When selling for time payments, indicate the interest rate (if any), payment schedule, and who will retain possession of the donkey and/or his registration papers until the transaction is completed. Also spell out what will happen if the buyer misses payments or defaults.
5. **Additional costs.** Specify how long you will board the donkey at no additional cost and, if that

date passes, what the consequences will be. Will the buyer pick up the donkey or do you deliver? If you deliver, what is the cost? If the donkey needs special vet work prior to shipment, who pays?

6. **Attendant risk.** Indicate when the buyer assumes responsibility for the donkey's illness, injury, or death, or for injuries or damage caused by the donkey. This usually occurs when the buyer signs the contract or when he takes possession, but get the specifics in writing.
7. **Warranties.** Whatever sort of guarantees you offer, spell out the specifics in minute detail. Leave *nothing* to chance; sooner or later you're sure to be glad that you included this clause.
8. **Insurance.** If you sell on time payments, insist on insurance at least to the extent of the unpaid balance, at the buyer's expense, with yourself named as the loss payee.
9. **Signatures.** Everyone named in the contract should sign it.

Marketing Donkey Milk

Yes, donkey milk! Entrepreneurs seeking an unusual, satisfying cottage industry should consider milking jennies for profit.

Donkey milk has been used as food, as medicine, and in cosmetics for thousands of years. The ancient Libyans were the first (that we know of) to keep stables of jennies specifically for their milk. Wealthy ladies bathed in it, applied facial masks of bread soaked in donkey milk, and splashed it onto their bodies to tone their skin. Some other notable historical uses include the following:

✗ Cleopatra bathed in donkey milk and kept a retinue of 300 she-asses to provide for her needs. Wherever the Queen of Egypt traveled, so did a convoy of jennies.

✗ Poppaea, wife of the infamous Roman emperor Nero (AD 37–68), maintained a collection of 500 milking jennies; they traveled with their tony mistress, too.

✘ Pliny the Elder (AD 23–79) named three uses for donkey milk: beverage (the upper classes imbibed it sweetened with honey), cosmetic, and medicine.

✘ Centuries later, when French king Francis I (AD 1494–1547) returned ill and broken from war abroad, none of his physicians could restore his health. Then a Jewish physician from Constantinople suggested that he drink donkey milk; the king quickly recovered, and asses' milk was soon the rage in Paris.

By the beginning of the early nineteenth century, asineries (donkey dairies) were established in most French cities, where milkmen herded jennies and their foals door to door and milked the jennies to order. Then, for a time, interest in donkey milk waned, until Dr. Parrot, a French physician who supervised the nursery for ailing babies at the Hôpital de Enfants Assistés in Paris, discovered that donkey milk was the elixir of life for sickly newborns.

Rubber baby-bottle nipples weren't invented until 1845 and weren't commonly used until the early 1900s, making bottle-feeding difficult indeed. In addition, considering the hygiene of the day, bottle-feeding was not a good thing. However, when Dr. Parrot's diseased babies suckled wet nurses, the nurses became infected as well. He thought that the answer was live milk from animal donors, and that perhaps infants who suckled directly from the teats of a goat or an ass could bypass the impurities of the bottle. And there was another bonus: They'd be ingesting milk at perfect body temperature.

Despite the protests of his peers, the Ministry for Public Assistance allowed Dr. Parrot's experiment. Over the course of one year he supervised the feeding of 86 sickly or diseased babies. Six infants were given cow milk from a bottle (one survived), 42 ingested goat milk straight from the goat (eight survived), and of the 38 who suckled milk straight from she-asses' teats, an astounding 28 survived.

Thus vindicated, Dr. Parrot set up his donkey stable adjacent to a feeding room. In his own words (taken from the *Bulletin de l'Académie de Medecine*, 1882), "The stables where the donkeys are kept are clean, healthy and well-aired; they open onto the nursing infants' dormitory. Treated gently, the donkey easily lets itself be suckled by the baby presented to it. Its teat is well adapted to the baby's mouth for latching on and sucking. The nurse sets on a stool to the right of the animal near its hindquarters. She supports the child's head with her left hand, with his body resting on her lap. With her right hand she presses the udder from time to time to help the milk to flow, especially if the baby is weak. The babies are nursed five times during the day and twice during the night. One donkey can feed three infants for five months."

Did You 'Ear?

In January of 1825, Brigadier Richard Charlton was appointed the first active consul to the Sandwich (Hawaiian) Islands. His ship departed Falmouth in England carrying the first three donkeys to set hoof on Hawaiian soil; others followed. It wasn't until 1879 that donkeys from California arrived on the Islands.

In 1858, the *Pacific Commercial Advertiser* reported that the sailing ship *L. P. Foster* departed Hawaii for Portland, Victoria, Australia, carrying "175 bbls. [barrels] Hawaiian beef; 15 kegs of sugar; 5 bbls. tallow; 30 jackasses."

Although domestic donkeys descend from the African Wild Ass, there were no donkeys in South Africa until the Dutch imported some in 1652.

MAMMALS' MILK

This chart shows milk composition of various dairy species, including human breast milk. The numbers represent approximate percentage per weight. The number for "energy" represents the calories available per 100 g.

SPECIES	WATER	PROTEIN	FAT	LACTOSE	TOTAL SOLIDS	ENERGY
Human	87.1	1.1	4.5	6.8	12.6	72
Ass (donkey)	88.3	1.7	1.2	6.9	10.2	44
Camel	86.5	3.7	4.9	5.1	14.4	70
Cow/Holstein	87.3	3.1	3.5	4.9	12.2	66
Cow/Jersey	---	3.9	5.5	4.9	15.0	---
Goat/Saanen	86.7	3.1	3.5	4.6	12.0	70
Horse	88.8	2.7	1.6	6.1	11.0	52
Reindeer	66.7	10.3	22.5	2.5	36.7	214
Sheep	82.0	5.5	5.3	4.6	16.3	102
Water Buffalo	82.8	5.9	10.4	4.3	21.5	101

The babies in Dr. Parrot's care thrived (as have countless donkey-milk-fed human infants since the dawn of time) because the composition of donkey milk is remarkably like that of human breast milk *(see Mammals' Milk above)*. Plus, it's a significant source of protein, vitamins (A, B1, B2, B6, C, D, and E), and minerals (calcium, magnesium, phosphorus, iron, and zinc) as well as fatty acids, immunoglobulins, and alkylglycerols. It's whiter than cow's milk and considerably lower in fat (0.6 grams of fat compared to 3.7 grams in cow's milk), and since the bacteria count in donkey milk is practically nil, there is usually no need to denude it of nutrients via pasteurization.

But donkey milk isn't just for infants. *The Guinness Book of World Records* recognized Maria Ester de Capovilla, who died in her native Ecuador in August of 2006 at 116 years of age, as the world's oldest woman. The secret of her longevity? Drinking donkey's milk.

Cleopatra was right on the money when she bathed in donkey milk. Fatty acids and vitamins A and E nourish skin treated with donkey milk. It's also a natural tensor. Today, companies in Chile, France, and Cyprus manufacture cosmetics containing donkey milk. Up to one-third of the milk produced by the world's donkey dairies is used in soaps, shampoos, creams, and lotions formulated to make skin feel smooth and silky, to moisturize dry skin and banish wrinkles, and to soothe and heal eczema, psoriasis, and acne.

But drinking donkey milk may fight disease as well; immunoglobulins found in donkey milk are known to boost the human immune system. And, according to a medical revue conducted in 1988 (as reported in "Asses' Milk — A Remedy Even in Modern Times" in the September/October 2001 issue of *The Brayer*), donkey milk, taken internally, is useful in the treatment of

Donkey Milk Soaps and Cosmetics

It can be done! Thanks to America's love affair with milk-based soaps, plenty of literature exists to pave your way. Best bets include *Soapmaker's Companion: A Comprehensive Guide with Recipes, Techniques & Know-How,* by Susan Miller Cavitch (Storey, 1997) and *Milk-Based Soaps: Making Natural, Skin-Nourishing Soap,* by Casey Makela (Storey, 1997). There's a niche market for specialty products such as these. You and your donkeys could fill it.

gastrointestinal problems, bronchitis, chest infections, and nervous exhaustion.

Putting It All to Use

If you're seeking an unusual and lucrative way to make money with donkeys, consider niche-marketing donkey milk as a beverage or as an ingredient in specialty soaps and skin conditioners.

Before marketing donkey milk as a beverage, investigate your state's dairy regulations in detail. Ask your County Extension agent which agency oversees dairy operations in your state; regulations vary widely from state to state.

Keep in mind that jennies don't give a lot of milk per milking, so they must be milked more frequently than cows or goats. Most European donkey dairies milk three times a day. Also, some jennies refuse to let down their milk unless their foals are present, so you may have to share with her baby. To do this, pen the foal separately but within sight of his dam overnight, and milk Mama first thing in the morning; however, be sure that the foal is readily consuming his own feed of hay and grain before you begin milking his dam. The average Standard-size jenny gives one pint to two quarts of milk per day.

> ### Ancient Wisdom
>
> As Pliny the Elder indicated in the first century AD, donkey milk has been used in medicine for thousands of years, so its benefits really shouldn't come as a surprise. Here are some of his observations.
>
> *It is generally believed that asses' milk effaces wrinkles in the face, renders the skin more delicate, and preserves its whiteness.*
>
> • • •
>
> *The sweetest, next to woman's milk, is camels' milk; but the most efficacious, medicinally speaking, is asses' milk.*
>
> • • •
>
> *Instances are cited, also, of persons who have been cured of gout in the hands and feet, by drinking asses' milk.*
>
> • • •
>
> *For persons suffering from asthma, the most efficient remedy of all is the blood of wild horses taken in drink; and next to that, asses' milk boiled with bulbs.*
>
> • • •
>
> *Poisons are neutralized by taking asses' milk. We shall have to mention many other uses to which asses' milk is applied; but it should be remembered that in all cases it must be used fresh, or, if not, as new as possible, and warmed, for there is nothing that more speedily loses its virtue.*

HOW-TO

Hand Milk a Jenny

Milking a jenny is like milking a goat except that her teats are smaller and harder to reach. Some jennies initially (and vigorously) object to being milked, but if you persist (clicker training or treats for good behavior win most jennies' cooperation), most adapt to the process surprisingly quickly.

This worker kneels to hand milk a jenny at a French asinerie.

Hand milking is a team effort between a milker and the creature that he milks. When the milker preps his animal by washing her udder, the hypothalamus in her brain signals her posterior pituitary gland to release oxytocin into her bloodstream, causing tiny muscles around those milk-holding alveoli to contract. In other words, she "lets down her milk." Milk letdown usually lasts for five to eight minutes, and milking should be completed during that time.

However, if the animal becomes excited or frightened, or experiences pain, her adrenal gland secretes adrenaline, which constricts blood vessels and capillaries in her udder and blocks the flow of oxytocin needed for effective milk letdown. Therefore, good hand milkers are efficient and patient. They approach milking in a low-key manner, and they practice good milking technique. Whether milking a jenny, goat, sheep, or cow, the same basic protocol applies. You will need:

- Squeaky-clean hands with short fingernails
- A recently sterilized, seamless, stainless-steel milking pail
- Udder wash and paper towels (or unscented baby wipes)
- Teat dip and a teat dip cup (or an aerosol post-milking spray like Fight Bac)
- A strip cup with a dark, perforated insert or a screen
- A clean milking area with a token feeding of grain waiting in the feed box

Here's what you do:

1. Secure the jenny in the milking area.
2. Wash her udder using your favorite prepping product. Make sure that her udder is dry, then massage it for 30 seconds to facilitate milk letdown.
3. Squirt the first few streams of milk from each teat into your strip cup and examine it for strings, lumps, or a watery consistency that might indicate mastitis. This is rare in donkeys, but you simply can't be too careful.
4. Sit or squat beside the jenny holding your milking container in one hand while milking the jenny, one teat at a time, with the other hand.
5. Trap milk in the teat by wrapping your thumb and forefinger around its base. Squeeze with your middle finger, then your ring finger (or just your middle finger if the jenny's teats are really small), in one smooth, successive motion to force milk trapped in the teat cistern out into your pail (never pull on a donkey's teats!). Relax your grip to allow the cistern to refill and do it again.
6. As each teat deflates and becomes increasingly more flaccid, gently bump or massage the udder to encourage additional milk letdown. Don't finish by stripping the teats between your thumb and first two fingers; this hurts and annoys the jenny, possibly causing her to kick.
7. Pour enough teat dip into the teat cup to dip each teat in fresh solution, and allow the teats to air dry. Alternately, spritz the end of each teat with Fight Bac.

Processing Donkey Milk

The trick to producing great-tasting milk is to keep it clean during the milking stage and to process it as soon as you're finished. Always milk into an easily sanitized stainless-steel container; milk in a tidy, relatively dust-free area separate from your donkeys' living quarters; and keep your hands spotlessly clean. These things really make a difference!

Process the milk by straining it through a standard paper milk filter (a coffeemaker filter will do in a pinch) inserted in a stainless-steel strainer, then pour the milk into sterilized bottles and submerge them in ice water until they're cool. Refrigerate and use within 5 days or freeze for up to 8 months.

You'll probably milk your jennies by hand, although there are options. A group of Italian scientists who authored *Composition and Characteristics of Ass's Milk* describe milking jennies using a wheeled trolley-type commercial milking unit with a sheep cluster set at vacuum level 42 kPA, pulse ratio 50 percent, and pulse rate of 120 cycles per minute.

`HOW-TO`

Manufacture Donkey Treats

You may have heard of Mrs. Pasture's Cookies for Horses. Why not compete for your share of the market with _____'s (fill in your name or your donkey's name) Donkey Delights?

There are thousands of horse-treat recipes for grabs online (search for *horse treat recipe* using your favorite search engine), or create your own. Manufacture and sell them through your Web site, listserv sales days, and online classifieds, or stock the saddlery shops, feed stores, and veterinary offices in your area. Stress in your advertising that horses love your yummy treats, too.

Ishtar's Donkey Treats

These recipes were gleaned from various sources, tested, tweaked, and approved by a panel of asses, including my girl Ishtar, the taste-test queen.

Easy Donkey Cookies

5 cups oatmeal
2 cups flour
1 cup grated carrots
½ cup corn oil

Combine the ingredients in a bowl. Form into small balls, flatten lightly with a spoon, then place on a microwavable dish and microwave on high for six minutes per batch.

Oven-Baked Donkey Delights

1 cup oatmeal
1 cup whole-wheat flour
2 tablespoons molasses
½ cup water
¼ cup grated carrot
¼ cup diced apple

Preheat the oven to 350°F (175°C). Combine ingredients in the order listed. Form the batter into small balls and place them on a greased cookie sheet. Bake for 8 to 10 minutes.

Peppermint-Chip Cookies

3 cups pelleted feed, soaked in 3 cups of hot water for 1 hour
1 cup corn meal
1 cup molasses
1 cup brown sugar
1 cup corn oil
10 peppermints, coarsely crushed
1 tablespoon salt
1 teaspoon baking soda
3 cups oatmeal

Preheat the oven to 325°F (160°C). Mix all ingredients except oatmeal, then add oatmeal 1 cup at a time until mixture is very doughy. Form into small balls, flatten lightly with a spoon, press peppermint bits well into their surfaces, and bake on a greased cookie sheet for 10 minutes or until well browned.

Crunchy Donkey Biscuits

1 cup molasses
½ cup brown sugar
4 large carrots, shredded
1 cup applesauce
2 cups Grapenuts cereal
1 cup sweet feed
1 cup oatmeal

Preheat the oven to 300°F (150°C). Combine molasses, brown sugar, carrots, and applesauce in one bowl. In another, mix the dry ingredients. Slowly combine the molasses mixture with the dry ingredients, adding just enough molasses mixture to form thick dough (add more Grapenuts if it's too mushy). Drop tablespoon-size globs of batter onto a greased cookie sheet and flatten slightly. Bake for about 1 hour; turn and bake for an additional 45 minutes, until they are dried out (keep checking to make sure that they don't burn).

Manufacture Donkey Equipment

I would love to buy a barefoot boot like Old Macs or Boa Boots that truly fit a donkey's hoof. Other buyers search the Internet for snap-crown bridles for Standard and Miniature Donkeys and saddles engineered to fit a donkey's back.

Dig out the latest tack catalogs, then put on your thinking cap and figure out how to tailor those products specifically for donkeys. Open an online or catalog business. Let me know — I want a copy of your catalog!

> **BRAY SAY:** *I must tell you what a coster is. Costers are people who go to the great London market, called Covent Garden, and buy cheap vegetables and fruits and flowers, and sell them in the poorer parts of the city. The coster men dress in velveteen suits trimmed with rows and rows of pearl buttons, which they call "pearlies." They are very proud of these costumes. The women wear bright, gaudily coloured dresses, and very big hats, covered with feathers. They hawk their wares about in barrows or little carts, drawn by such a tiny donkey (a "moke" as the costers call it), that you wonder how he is able to pull a whole family of costers as well as a big load of vegetables, as they often do.*
>
> *— Blanche McManus, in Our Little English Cousin, 1905*

Donkey Trekking

If you do an Internet search for *donkey trekking*, the first of many Web sites that you'll find is that of Fédération Nationale Anes et Randonnées (FNAR; the National Association for Donkey-Trekking), a French organization that "specializes in introducing people who like country walking to donkeys who like people." Entrepreneurs who have at least two years' experience hire out donkeys for accompanied or unaccompanied walking tours, manning more than 80 FNAR-affiliated member centers throughout France (*see* Resources).

The Fédération Nationale Anes et Randonnées Web site says it so well: "We work hard for the promotion and preservation of donkeys. Through the increasing numbers of people who enjoy holidays with donkeys, they get new status in a motorized world . . . we encourage people to prefer exploring the by-ways, rather than the highways. No record-breaker for speed, a donkey is a docile and intelligent companion, who gives a special rhythm to walking; you'll have the time to appreciate the beauty of landscapes and meet inhabitants . . . most FNAR centers are located in lonely places

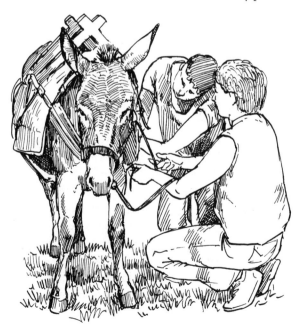

Accredited trekking centers provide every necessity, including a donkey.

where economic activities are rare. They are family businesses, well integrated into the local community." Their goal is to give their customers "a way to avoid a nervous breakdown: to discover fabulous places, to be moved by charming donkeys, to enjoy the conviviality of donkey-owners."

Why don't we have businesses like this in North America?

Donkey trekking is popular not only in France but also in Crete, Spain, Egypt, Tanzania, and Australia. The concept is simple: Entrepreneurs furnish clients with a pleasant, well-trained donkey and his gear for a specified length of time, along with instruction on how to care for the animal during the journey. Donkeys are allowed to carry no more than 30 to 40 kilos (66 to 88 pounds) of gear, although a small child rider can be included in the tally. FNAR describes several types of family treks.

One-day treks. Nearly all affiliated centers offer one-day treks along marked paths. Some provide packed meals to eat along the way.

Guided treks. The center provides all equipment and a knowledgeable guide (or two or three) who describes the history of the area and identifies its flora and fauna. Lodging along the route is furnished and meals provided. These are multiple-day group treks involving groups of people and several donkeys.

All-in treks. The center provides everything except a guide; these treks can be up to 10 days long.

Free treks. Experienced hikers who can read a map and use a compass plan their entire holiday including how long they want to walk; where they eat or whether they'll tent-camp, sleep in mountain shelters, or find lodging in villages along the route is totally their call.

Other options. These include short treks for mentally or physically challenged hikers, discovery classes for schools, and children's live-in donkey-trekking holidays.

Donkey trekking would be an ideal enterprise for donkey owners living in scenic, out-of-the-way America or for anyone living near state or provincial parks that offer hiking and riding trails. Type *llama packing* or *goat packing* in the search box at your favorite search engine to discover how outfitters handle similar pack-animal trekking ventures (mostly in the mountainous western states) in North America.

Donkey Seat

Have you ever wondered why donkey riders in some parts of the world ride perched over their donkeys' loins instead of on their backs behind their withers? What you're seeing is "donkey seat," adopted in ancient times so that full-size riders could ride relatively tiny beasts. Seated thus, there is articulated limb support for the rider's weight; the animal's hind legs can carry more weight than his unsupported back. Ridden long distances, bareback, a low-headed, downhill donkey's gaits felt smoother to the rider as well. Donkeys so ridden were (and are) often severely overweighted, but they could perform, and that was the name of the game. Old habits die hard, so donkeys are still sometimes ridden in donkey seat today.

Be a Traveling Photographer

In years gone by, in an era when the common person didn't own even the simplest box camera, traveling photographers went town to town, door to door, with their camera equipment in tow, photographing children (and occasionally adults) posed for the occasion on a well-behaved donkey or pony or holding the reins while seated in a vehicle drawn by a long-horned goat. Many times these were the only images a family had of their children growing up. (*Several examples are shown in this book, such as on pages 251 and 294.*)

If you're a creative donkey owner who loves children and you're seeking a unique way to earn money with a long-eared friend, consider replicating those photos today, either traveling door to door the old-fashioned way or through a booth at fairs, festivals, reunions, and similar soirees. It's a snap to re-create the look of bygone days by combining antique saddlery and an array of easily adjustable period costumes with today's digital cameras and photo-imaging software. Parents and grandparents (who may remember the traveling children's photographers of yore, some of whom plied their craft well into the 1950s) love these nostalgic images, and with a laptop and a quality printer you can print them on the spot to order.

In addition to props and photo equipment, you will need a well-behaved, friendly donkey, a backdrop to use when you can't find a bank of trees to serve as a background, and an attractive portfolio or booth to display your work.

Portray Sinterklaas

Another part-time, paying job for folks who love donkeys and children: portray Saint Nicholas or Father Christmas over the winter holidays. Stores that hire a mundane Santa may leap at the chance to offer something unique, especially if a colleague shoots instant-print digital photos of youngsters on Saint Nick's sweet donkey's back. Another option: Present educational programs (a friend's retired father earns several thousand dollars each year doing Father Christmas programs — sans donkey — in Minnesota's public schools). Here's what you need to know.

This 1910 postcard of a European Santa and his donkey was said to be printed in Belgium. It is titled "Joyeux Noel" and addressed to "Mon petite Maurice."

Did You 'Ear?

According to Wang Shih Chen (who wrote about events occurring in China between the second and fifth centuries AD), when nobleman Wang Chuang Suan (who dearly loved to hear his donkeys bray) died, the great Emperor Wen attended his funeral. As a parting eulogy to his good friend, the emperor ordered all attending to sound a series of farewell brays.

Many of China's early emperors where enamored with asses. During the Han Dynasty (206 BC to AD 9), four white donkeys were used to pull the emperor's carriage. Then horses replaced the donkeys, but Ming Tu (late 1st century AD) ordered the donkeys reinstated. Emperor T'sin (AD 424–451) decreed that all Chinese officials replace their riding horses with asses. Perhaps the greatest donkey lover of them all, Emperor Ling (AD 508–554), so adored pure white asses that he kept hundreds of them in the Imperial stables.

COLLECTING VINTAGE PICTURE POSTCARDS

Most of the old-time pictures in this book are actually vintage postcards; most were less than $20, and all came from eBay. Collecting vintage donkey ephemera is a fun, rewarding, and relatively inexpensive hobby. Here's what you need to know to get started.

TO SPECIALIZE — OR NOT. Decide what type of card you'd like to collect. I prefer any sort of real picture postcards (RPPC) depicting donkeys in positive situations. You might choose comic cards, traveling photographers' cards of children and donkeys, cards from a favorite country or era, or cards depicting beach donkeys, Western burros, or any other topic that strikes your fancy.

DATING PICTURE POSTCARDS. Postcard collectors refer to the period between 1900 and 1918 as the Golden Age of postcards. A majority of vintage donkey cards featured on eBay were printed during this time period. In 1903 Kodak introduced the No. 3A Folding Pocket Kodak, the first postcard camera. With it, travelling photographers produced their own postcards, including most of the one-of-a-kinds that you see in this book.

Cards produced between 1901 and 1906 had undivided backs — only the address was allowed on the back, so messages were often scrawled on or around the postcard's picture image. Postcards became so popular during this time that the number published doubled about every six months.

On March 1, 1907, the divided-back postcard came into circulation. They quickly became more popular because they allowed senders to write the address and a short message on the back of the card. Publishers printed millions of cards during this era; most were printed in Germany, the world leader in lithographic processes until the beginning of World War I.

When German imports ground to a halt at the beginning of the First World War, American entrepreneurs began printing postcards to pick up the slack.

Pieces missing, creasing, staining, and brown spots (a mold known as foxing) usually render a vintage card of little monetary value, and lesser deterioration significantly reduces the value of a card. However, some collectors (myself included) think that a certain amount of aging lends character, and we're in it for the love of donkeys, not investment.

CARING FOR YOUR COLLECTION

To make your collection last (and to preserve its monetary value):

- Each item should be encased in its own archival-quality plastic holder made of uPVC. This material prevents the formation of acidic oils that can destroy old cards. Holders are available at stores that sell collectors' supplies or from catalog companies like Light Impressions, Get Smart Products, and ArchivalUSA (*see* Resources).

- Next, place them in archival-quality cardboard postcard boxes or albums. Select nonacidic paper album pages, paper dividers, inserts, and cardboard boxes as acids can ruin the colors and textures of old postcards. Choose papers that have a pH (acid level) of 7.0 or slightly higher.

- Store the boxes or albums in a cool, dry, smoke-free environment. Excessive heating and moisture causes cards to stick to surfaces and make removal difficult, if not impossible, without damage.

Should you frame a favorite card or photo? Die-hard collectors say not, but if you do so, use archival materials and hang it out of direct sunlight.

The real St. Nicholas was a fourth-century bishop of Myra (now part of Turkey). The Catholic church eventually designated him the patron saint of children (as well as virgins, pawnbrokers, pirates, thieves, brewers, pilgrims, fishermen, barrel makers, dyers, butchers, meatpackers, and haberdashers; he is also the guardian saint of Russia, Austria, Germany, Belgium, France, Greece, and Norway). Soon Europeans began presenting gifts to their children on the eve of St. Nicholas's feast day (December 6) in his honor.

St. Nick was initially depicted as a tall, slender man who dressed in traditional garb befitting a bishop. In the early days there were no eight tiny reindeer. Instead there was a solid and faithful little donkey whom (depending on the countries where his legends were told) he rode or led.

Children in France, Belgium, Luxemburg, Holland, and Switzerland believed that he visited their homes accompanied by a donkey laden with baskets of treats. Youngsters put straw, carrots, or other toothsome donkey treats in their shoes or clogs and placed them by the stove or fireplace before going to bed; in the morning, the donkey grub was gone, replaced by presents. St. Nicholas (a.k.a. Sinterklaas) came to America with Dutch immigrants in the late 1600s, but it wasn't until much later that he gained a lot of weight and swapped his donkey for deer.

Men of the Royal Scots Fusiliers pose with their regimental mascot in Cologne, Germany, in 1920.

DONKEY TAILS

Endearing Mascots

During the Civil War, the 3rd Louisiana Company, Confederate States of America, had a donkey mascot whose name wasn't recorded for posterity but whose claim to fame was his habit of repeatedly shouldering his way into the company commander's tent in the heart of the night, after apparently mistaking the tent for that of his owner.

Lady Moe, a Tunisian donkey, was the mascot of the British 96th Bombardment Group during World War II. A B-17 crew taking part in the Regensburg–North Africa shuttle mission found Lady Moe starving in an Algerian slum; she accompanied the crew back to England on August 24, 1943, on the return leg of the mission. En route the crew bombed Bordeaux, France, making Lady Moe the only known donkey to fly a combat mission. She adapted well to her new home at Snetterton Heath, where she participated in many patriotic events. She also served as the army mule mascot at the Army-Navy football game at London's White City Stadium on November 12, 1944.

A debonair donkey named El Paso Henry, who was the former burro mascot of the University of Texas at El Paso Miners, was also a member of the Kappa Sigma fraternity. His predecessor was a burro called Clyde, whom then-President Joseph Ray drummed out of service in 1966, declaring him "a sorry-looking pot-bellied creature not fit to represent the Miners." An animated cartoon miner named Paydirt Pete eventually replaced the Miners' flesh-and-blood burro mascots.

Present-day long-eared mascots include donkeys Nugget (Western Museum of Mining and Industry), Fanny (Big Ass Fans), Irene (the Alabama Democratic Party), the famous West Point army mules, and a mule named Abbedale (University of Central Missouri's athletic teams).

This postcard, taken by a traveling photographer, is inscribed, "Taken on good old Rutger St. or Hickory St. about 1904 or '05."

In 1822, Dr. Clement Clark Moore, a professor of divinity, penned a poem called *A Visit From St. Nicholas* (it later became *The Night Before Christmas*) that transformed the gaunt, ancient bishop into "a jolly old elf" whose belly "shook when he laughed like a bowl full of jelly," and his conveyance into a "miniature sleigh" drawn by "eight tiny reindeer." *A Visit From St. Nicholas* was published in the Troy, New York, *Sentinel* on December 23, 1823, and the rest, as they say, is history.

Images of Sinterklaas and his donkey prevailed in Europe until fairly recent times. Vintage Yuletide cards depicting St. Nick and his toy-laden donkey remain favorite collectibles on eBay.

Throw a Donkey Party

Hosting parties is another way to combine a love of little folk and donkeys. There are two approaches: you can furnish complete donkey party packages including live donkeys for donkey rides, donkey games (pin-the-tail-on-the-donkey and a donkey piñata), donkey favors (Shrek's Donkey and Eeyore to the rescue), and a donkey-themed cake — or you can simply provide donkey rides.

BRAY SAY: *Sint Nikolaas Vooravond or La Vielle de Saint Nicolas (Saint Nicholas' Eve), on December 5: Saint Nicholas rides a donkey and is attended by his assistant. The saint sees everything. He knows everything, but no child has ever seen or known him. The children leave carrots and pieces of bread in the chimney corner for Saint Nicholas' donkey who surely will be hungry from journeying across village housetops.*

— Dorothy Gladys Spicer,
Festivals of Western Europe, 1958

CHAPTER 19

DONKEY MARKETING MOJO

The ass is jealous of the horse until he learns better.

— Turkish proverb

The trick to achieving success in any business venture is in letting folks know what you have to sell. Here are some great ways to do just that, and you needn't break the bank in the process.

Web Sites

If you produce show and breeding stock, offer a service (donkey birthday parties or donkey trekking), or sell products like donkey-milk cosmetics or donkey gear of any kind, you definitely need a Web site. In today's hectic world the Internet is a customer's quickest way to locate the goods that he needs. Make it easy for him to find you and he'll buy.

What can a Web site do for your donkey enterprise?

- ✗ You can showcase your stock or service and present your message to an unlimited number of people from all parts of the world without leaving your home or paying for a single high-priced ad.
- ✗ You can give potential buyers more information than you could in an ad or typical brochure,

and you can change it whenever you want, from home, 24 hours a day.
- ✗ You'll automatically target an audience interested in donkeys, because if they weren't interested in them, they wouldn't visit your site.
- ✗ On the Internet, your backyard business can appear every bit as impressive as your biggest competitors' — even better, perhaps, if you're an able photographer and a skilled Web designer.
- ✗ You can display all of your merchandise rather than highlighting just a few choice items.

Maintaining an Internet presence is a remarkably cost-effective way to market and promote your donkeys and donkey-related products, especially if you do the work yourself. A comprehensive, do-it-yourself Web site costs about an hour or two a week and two or three hundred dollars per year to build and maintain, although you'll have to master certain skills to do it well. If you're a technophobe, you can go the alternate route and hire a developer to create and maintain your site. Whichever avenue you choose, make sure that your site is a good one.

Work with a Pro

Before hiring a Webmaster to build your Web site, surf the Internet and look at lots of sites. Pick out those you like, and contact whoever built them. This information is generally found at the bottom of the Web site's home page, but if it isn't, e-mail its owners and ask.

Collect a list of potential designers, keeping in mind that Web designers charge all over the board — from pizzas and soda to fuel your teenage nephew while he builds your Web site, to $80 an hour and up for some professionals (and your nephew might build a better Web site).

Ask prospective designers how much bang you'll get for your bucks. Their base fee gets you how many pages — and how many pictures per page? What else is included in the base price, and what, if any, features can be had at additional cost? How many updates are included? How much are additional updates? Will the developer work with you or insist on doing things his way (some techies are tied into bells and whistles, but they're a liability, not an asset, on sites designed to sell). How long will it take to get your Web site up on line? Will the developer list your site with major search engines, or is that your job? Does he offer a satisfaction guarantee? Ask for URLs (Web site addresses) to other sites that he's built. E-mail their owners. Are they happy with his work, and, if they aren't, why not?

Do It Yourself?

Unless you're computer illiterate and tremendously strapped for time, it's best to build your own Web site. You know precisely what you want, and you're much more likely to get it when you do it yourself. Besides, building a better Web site is easier than you probably think.

Check with schools and libraries in your locale. Many offer free or inexpensive weekend or evening computer classes, including classes on site building and Internet commerce. County Extension services hold Internet-farm-marketing seminars, and the Small Business Administration (*see* Resources) will also help newbies to get started. Or choose one of the scores of up-to-date, easy-to-understand books designed to guide beginners through the intricacies of building a Web site (you'll find some good ones listed under Resources).

It costs less to bypass professional designers and build your own site; updates are easy and free; and the satisfaction of designing it yourself is priceless. Are you game? If so, read on for a few things that you will need for the job.

BRAY SAY: *It is good for a man to eat thistles and to remember that he is an ass.*

— E. S. Dallas, nineteenth-century writer and journalist

Start with a Plan

Web sites sell something or distribute information; the best of them do both. Decide what you want yours to do. Plan each page on paper, based on the principles that follow. It's easier to build from a blueprint than to backtrack and fix things later.

Web Page Authoring Software

HTML is the programming language of the World Wide Web. Stop reading, go to your computer, and navigate to virtually any Web site. Choose View in your browser's tool bar and drop down to Source. Release your mouse to view the page's HTML. Though it looks complicated, you won't have to learn to write it if you use Web page building software; it's a must-have for novice site builders, and you needn't break the bank to buy the kind you need.

High-end programs like Dreamweaver and Front-Page build wonderful high-tech Web sites, but for selling purposes they're usually much more than you need. Free or inexpensive programs like WebPlus, Web-Master, WebStudio, and Ewisoft Website Builder work just fine. I build and maintain our half-dozen Web sites using Claris Home Page 3.0 software that was written in 1997 and purchased outdated but brand spanking new on eBay for 12 whole bucks. Similar plain-Jane but fully functional building programs can be downloaded direct from the Web.

A Great Domain Name

Your domain name is the "who" segment of your Web site's Internet address. Register it right now. Catchy names are purchased and taken out of circulation every day, so don't lose yours because you waited too long. Use your business name if you possibly can. If your business name is already taken, try customizing it. Let's use our former farm name, Bluestone Brayers, as an example. If that name is taken, try:

- ✘ Adding dashes: Bluestone-Brayers
- ✘ Adding underscores: Bluestone_Brayers
- ✘ Adding numbers: BluestoneBrayers1
- ✘ Adding or subtracting descriptive words: BluestoneMammoths, BluestoneBrayersFarm, or BluestoneBrayersRanch

If that still doesn't work, brainstorm another easy-to-remember donkey-related name. As we go to press these

This advertising trading card, described as "Victorian," reads, "Ladies' ready made Suits, Dry Goods, and Carpets."

catchy names are still up for grabs: Donkeys4You.com, WorldsBestJackstock.com, and TeenyTinyDonkeys. com — and I'll bet that you can think up something even better! If your first choice domain name is taken (let's say it's DonkeysRUs.com), you might still be able to register it as DonkeysRUs.net, DonkeysRUs.biz, or DonkeysRUs.us.

To see if your favorite choices are already taken, type *domain name check* in the search box of your favorite search engine and use any of thousands of free, online domain name searches to check it out. A tip: When typing domain names into a browser bar, you needn't stop to capitalize its components; *www.bluestonebrayers.com* works the same as *www.BluestoneBrayers.com.* However, when adding your URL to e-mail signature lines, directories, or printed promotional items such as business cards and brochures, capitalizing helps folks to recognize your name.

Before registering your domain with an independent registration service, here's a tip: Many hosting servers include domain name registration in their monthly service charge. Smaller servers generally expect their subscribers to pay an independent service to register their domain names — it isn't included in their monthly or annual hosting fees. Check on this before you subscribe; you might save yourself some bucks.

A Hosting Service

The cardinal rule of Web-site construction: Don't use freebie hosting services to host your business Web site. People hate the advertising banners and pop-ups that are part and parcel of sites on freebie services, and they freeze some computers. Certain freebie services are worse than others. I won't follow links to sites on several of the major free-hosting servers because they're notoriously slow to load, they add pop-ups to every page, they're prone to freezing Macs and older PCs, and some put spy- and ad-ware on visitors' computers. Avoid them!

WHICH SERVICE?

Choose the best hosting service you can find and afford. Here some things to consider.

- How much bandwidth do you get for your money? Bandwidth is the number of bytes (usually expressed in GB or gigabytes) your customers' visits may consume each month. More is always better. When you run through your monthly allotment, your site goes down until the next billing period begins.

- How fast does it connect to the Internet? Connection speed varies; some services are infinitely slower than others.

- What does their support system consist of? Is there well-written online documentation that you can refer to when you have questions? Can you contact support staff by e-mail or phone? How fast are your e-mails answered? And are tech support personnel willing to patiently walk newbie designers through problems, using everyday language instead of technical jargon?

An easy-to-use hosting service is a jewel beyond price. A hint: For your first foray into site building choose a major player like Yahoo's GeoCities. GeoCities' online Help feature is written so that the rankest beginner can understand it; their Web-building tools are superb; and GeoCities' tech support is helpful as can be. We started with them for their ease of use, and they still host our sites 10 years later.

Also, most freebie services generate long, cumbersome URLs, which your customer won't remember and most book and magazine editors won't print. Some of the best farm and equine information Web sites I know of have URLs that don't squeak past editors. What a loss!

Also, major search engines give priority to sites with bona fide domain names; this is another good reason to avoid freebie services. And finally, freebie services sometimes abruptly switch to paying status. If it happens, you'll have to shell out the bucks or see your site go down the drain. Is it worth it? We don't think so.

Effective Web Design

Whether you build your own Web site or a developer does it for you, you'll want it to be eye-catching, effective, and user friendly. Most Web sites aren't.

According to the United States Department of Agriculture Marketing Service publication "How to Direct-Market Products on the Internet," these are the factors most likely to influence repeat visitors to a business's Web site: 75 percent expected high-quality content, while 66 percent praised ease of use. Fifty-eight percent won't revisit slow-loading Web pages, and 54 percent avoid dated sites; only 12 percent revisit to view cutting-edge technology.

As Ellie Winslow, rural marketing maven (*see* Resources) and author of *Marketing Farm Products, Growing Your Rural Business From the Inside Out* and other books, points out at her workshops, "Over the last decade Web sites in general have gotten more glitzy. If the Web site is for business, don't distract your visitor and don't wear him or her out with all the moving icons, streaming banners and other technically advanced stuff that doesn't actually promote your marketing goals."

For selling purposes, Web sites should present products and facts in a plain but attractive, easy-to-use manner. Keep in mind the following important considerations.

Loading Times

According to recent studies, the average Internet user spends nine minutes a day — that's a full 55 hours per year — waiting for Web pages to load. For each person who patiently waits six minutes for your picture-rich and gizmo-laden Web page to load, another will allow 30 seconds, then leave.

Many Web developers assume that every Internet user has access to ultrafast Internet service. They don't! In many rural areas dial-up is still the norm, often via antiquated phone lines.

And because many rural folk hang on to dated computer equipment, preferring to use their dollars to buy donkeys or feed instead of faster hardware, it's vitally important to design fast-loading pages that grab visitors' attention lest they leave to peruse another seller's site.

KISS (Keep It Simple and Sell!) Web designers recommend that a Web page's elements add up to single-page downloads no greater than 150 to 200 K. This precludes huge pictures and techie frou-frou, so plan your layouts accordingly.

Or, if you love bells and whistles, build a dual Web site linked through a common home page. Let visitors click links to enter a no-frills version or one resplendent with music, glitz, and animations. Add a hit counter to each; you'll be surprised at which one most visitors choose.

Images

Resize large photos and other graphics so that they're attractive, yet quickly downloadable. Software such as Adobe PhotoDeluxe and Adobe PhotoShop, or any of the free and inexpensive photo-editing software downloadable online, will help you to do it with pizzazz. Resolutions in the 72 pixels per inch range are ideal for Web-site applications.

Plan your pages carefully to eliminate photo glut. Start with a spare, uncluttered introductory page for

One group of livestock producers that knows more than a thing or two about marketing is alpaca breeders. If you're serious about marketing donkeys for healthy prices, consider attending an alpaca-marketing seminar. Some of what you learn won't apply to donkeys, but most of it will.

- To find a suitable seminar, contact the Alpaca Registry, Inc., or the Alpaca Owners and Breeders Association.
- Another avenue to pursue: Subscribe to *Alpacas* magazine, the lush, full-color official journal of the Alpaca Owners and Breeders Association. Each fat issue features one or more extremely helpful, well-organized articles about marketing that are well worth the price of the magazine.
- Finally, visit your favorite Internet search engine and enter a search for alpaca marketing. You'll be amazed at the useful material you'll turn up!

each product and link to additional pages of photos and information.

The cardinal rule for using pictures on your Web site: Choose good ones! Anyone can take good animal pictures if they're patient and persistent. If you feel stuck, hire a professional to shoot clear, well-composed pictures of your stock.

A word about pictures and graphics: While it's extremely bad manners to "borrow" an image or graphic from another site without previously clearing it with its Webmaster, many site owners are happy to share. But call or e-mail and ask; don't assume that everything online is yours for the taking.

BRAY SAY: *Who wants an ass without fault must walk.*

— Italian proverb

Six serious-looking ladies and a donkey pose for the camera at the 1914 Chillicothe, Ohio, Fall Festival.

The Rest of the Story

Your Web site is in essence your retail store, so it behooves you to put up a good one. We can't cover all there is to know about Web design in this short chapter, but we will cover some basics here. It's a good idea to buy a book, visit a Web site, or take a class on effective sales ploys before you put up your site.

Don't assume that your visitor is Internet savvy. Many people aren't. If they don't understand mouse-over images and pull-down menus, they'll miss a lot of content, so place links where people can easily find them. If your Web site crashes a visitor's computer, he probably won't be back — another good reason to give animations, music, and other computer gee-gaws wide berth.

Don't use frames! Sites utilizing frames are cluttered and confusing, and they rarely print out well. They load poorly on older computers and small computer monitors, flooding the screen and sometimes blocking access to your site's best features. If you must use frames, don't place contact information near the bottom because that's the part that generally doesn't fit. A best bet for frames aficionados: Design two sites linked to your home page, where visitors can choose the version they prefer.

'Ear's a Tip

Stay Current. Don't let your site become dated. Tweak your sales lists, upload pictures of those cute foals, or add a new educational item — anything to keep your content fresh and bright. And spend time each week checking links, especially those that lead to outside sources. One hour a week spent on Web-site maintenance pays big dividends in customer approval.

Choose your fonts with care. Make certain that they're large enough to be easily read on both Macs and PCs and in all of the standard browsers. Avoid nonstandard fonts; if in doubt, use Ariel or Times New Roman.

Avoid patterned backgrounds; they tire visitors' eyes and copy gets lost in the morass. Strive for clear contrast between font and background colors. Light-colored copy against a dark background looks great online, but it contributes to eyestrain and doesn't print well in older versions of certain browsers.

Your home page should make it clear what your site is about and what it has to offer your visitors. It should be welcoming and uncluttered and load in a snap. You're allowed only one first impression, so make it count. Include a site map and provide a link to it from each of your other pages. Visitors stay longer when they know what there is to see.

Compose headlines and page titles that help to make sense of what's on each page. Search engines frequently link to internal Web pages; when that happens, without some description your visitor feels lost.

Because not every visitor enters your site via the home page, place your name, contact information, and logo on every page and link the logo back to your site map or home page. Don't omit your physical address. Connecticut buyers might not be interested in buying donkeys from Oregon, but if you're a county or two away from Joe Buyer, he's much more likely to call and arrange a visit.

Visitors sometimes type in URLs to access internal Web pages, so keep them simple. Use all lower-case characters and avoid special characters like under-scores (_) and tildes (~) whenever you can. Write good copy. Triple-check all copy for typos and misspellings. Don't make nitpickers grit their teeth.

Add educational content — the more the better. When donkey owners visit educational sites like those maintained by Lucky Three Ranch and Cooke Livestock Miniature Donkeys (*see* Resources) they often stay to find out what's for sale. They'll be back the next time they need information, and they'll tell their friends. Prime advertising? You bet!

Include a Links page and exchange links with others in the donkey business. List your Web site in equine-related directories all over the Web; listing is usually free. Don't wait for Web crawlers to find your site, go directly to the major search engines and sign it up.

If someone e-mails via your Web site, respond as soon as you can. When marketing via e-mail, punctuality counts. According to a study reported in "How to Direct Market Farm Products on the Internet," 40 percent of the top e-commerce Web sites took longer than five days to respond to e-mail, never replied, or simply weren't accessible via e-mail. People want results right now. Give it to them and sell!

Finally, people won't visit unless they know about your site, so aggressively promote it in every way you can. Add it to your e-mail signature line so that it appears on every e-mail that you send. Use it on your business cards, brochures, and road signs. Letter it on your truck and have it emblazoned on the T-shirts that you wear to the store. Swap links with the owners of other Web sites. As a donkey entrepreneur, your Web site is your most important marketing tool. Make it a great one and consider your time (and money) well spent.

Business Cards

Dollar for dollar, the most effective advertising in the world is a top-notch business card. They're inexpensive, exquisitely portable, and, if designed right, people tend to keep them for a long, long time. And there are scores of ways to promote your business with good cards. Never leave home without them!

✗ Tack cards to every bulletin board you encounter, be it at the supermarket, laundromat, feed store, or wherever. Use sturdy push pins so that you can stack them; this encourages interested parties to take one along. Check back often to restock; be sure that you don't run out.

INDUSTRY ADVERTISING

Don't overlook *The Brayer,* journal of the American Donkey and Mule Society. Each bimonthly issue runs 112 information-packed pages, and advertising costs are chicken feed! American Donkey and Mule Society membership also includes placement on the organization's computerized breeder's listings.

And take advantage of free or low-cost classified ad possibilities offered by online sites such as Longears Mall, gotdonkeys, and The Mule Store, as well as all-equine ad sites like HorseWeb, Horsetopia, and Equine.com (*see* Resources). When placing these ads, remember:

- Put your name in the ad. Many people don't answer ads placed by unknown persons. And don't omit your farm name; the more people see it, the more they'll remember your business.

- Include a phone number as well as an e-mail address.

- Indicate which state or province you live in. Potential purchasers are more likely to call if they know that you're within their buying range.

- Provide enough information to pique potential buyers' interest.

- Don't be hasty about canceling an ad once you've sold an animal; edit your ad to read, "Sold, but we have more like her!" or "Sold, but we're taking orders for next year's foals. Don't be disappointed; reserve one now!"

A line of beach donkeys adorns this undated postcard, which reads, "Just a line from Cleethorpes" (a beach resort located in northeast Lincolnshire in England).

- Purchase inexpensive business card holders and ask businesses to display your cards near their cash registers. Target veterinarians, tack shops, and farm stores in your area but place them in other local businesses, too.
- Tuck a business card into every piece of mail you send out — personal correspondence (Great-Aunt Tilly may know someone who'd like a pair of Miniature Donkeys or a dozen bars of donkey milk soap), invoices, even bank payments and the electric bill. Some will be discarded; others won't.
- Craft your own gift tags — on the back of your business card.
- Place them in library books to use as bookmarks, but clear it with the librarian before you do.
- Hand them to people whom you meet. Ask them to take several cards and pass the extras along to their family and friends.
- Buy a conference-style nametag holder, insert your card, and wear it on your lapel; it's certain to stimulate conversation.
- Scan and add it to your e-mail as an attachment.
- Use your card as a camera-ready ad for publications; you'll also save money on setup fees!

Build a Better Business Card

The only thing worse than not using business cards is distributing bad ones. Soiled, poorly designed, el cheapo cards affixed to bulletin boards or stacked on the counter at the tack store create an impression of your business that you definitely want to avoid.

'Ear's a Tip

Buy in Quantity. Don't buy small quantities of business cards — the more you order (usually in multiples of 500) the less your cost per card. One thousand cards is a working minimum if you plan to aggressively promote using business cards.

While it's tempting to settle for freebie cards you can order online, don't. The advertising on the back of freebie cards distracts from your image. "Cheap" and "shoddy" aren't words you want associated with your enterprise, so go the extra mile and pay for top-flight cards.

If you spring for quality cardstock (not perforated punch-outs) and you understand design, you can probably print your own. Many word processing and most desktop publishing programs offer create-a-card capability, or use downloadable templates available from paper suppliers like Paper Direct (*see* Resources). If you can do a professional-looking job, do it; if not, order business cards printed by the pros.

Before designing your card or hiring a professional to do it for you, collect business cards you especially like. They needn't be livestock-related; what you're looking for are elements that engage your eye and that can be incorporated into your own card's design.

- ✘ Include all contact information on the face of your business card: your name, farm name, mailing address, e-mail address, phone and FAX numbers, and addresses to all pertinent Web pages.
- ✘ Keep it simple. Avoid hard-to-read fonts and use one (or at most two) font type families per card.
- ✘ Your name (or farm name) should be the largest text element on your card. Make certain that additional text is easily readable.
- ✘ Don't use nontext elements unless they're good ones. Consider having a custom logo designed. Barring that, good sources of useable graphics are online clip-art sites or any of the 31 compilations of specialty clip art in the Donkey Clip Art Collections available from the American Donkey and Mule Society's Hee Haw Book Service.
- ✘ A bad photo is worse than no photo. Resolution needs to be spot-on. If necessary, hire a pro to shoot a top-flight photo for your card.
- ✘ Find a way to make your card stand out. Textured cardstock and interesting cutouts (use a

'Ear's a Tip

Business Card Color for a Pittance. Create color interest at minimal cost by carefully rubber-stamping a small design element on basic black and white business cards using colored or metallic ink.

scrapbooking paper punch) help to render your card one of a kind.

- ✘ Choose cardstock and font colors that contrast with, yet complement, one another. Color adds interest, but a little color goes a long way.
- ✘ Stick to standard-size, 3.5-inch by 2-inch cards. Because other sizes don't fit ready-made cardholders, they're likely to be dumped into the trash.
- ✘ People will save your card if you print something useful on the back. Be creative. Try: "New to Longears? To learn more, visit these online resources" followed by the Web addresses of your favorite info sites.
- ✘ Most important: Run a spellchecker over your finished card before you print. Nothing says "amateur" louder than advertising replete with a slew of misspelled words.

Business Card Etiquette

Practice handing out your card before you try it, especially if you tend to be shy. Don't apologize about it — just smile, make eye contact, and memorize a simple, catchy line to say. Hand it over face up, with pride. Show recipients that you're proud of your donkeys or donkey-related goods or services.

If someone hands you his card in return, don't immediately stuff it into your wallet or pocket. Examine it. Hold it in your hand while you converse. When you treat the other person's card with respect, he'll likely extend the courtesy.

And don't distribute bent, soiled, or otherwise tatty-looking cards that reflect poorly on your business. Take care of your business cards. Keep them looking crisp and clean in an attractive business card holder or tucked away in a pocket in your wallet.

Keep your business card up to date. Don't scratch through or cover up text and write in corrections — get new cards.

Flyers and Brochures

A flyer (sometimes called a circular) is usually printed on one side of standard 8½" × 11" paper and designed to be tacked to bulletin boards, taped to the inside of windows, and distributed as handouts. They're the ideal venue to announce an upcoming production sale, spotlight your jack, or simply indicate that you have donkeys (or donkey products or services) to sell.

A brochure (also called a pamphlet) is somewhat more. It can be as simple as an 8½" × 11" or 8½" × 14" single-sheet trifold or as detailed as a multipage, spine-stapled booklet. An effective brochure piques the recipient's interest and makes him want to open it and read what's written inside. It must be visually appealing, be well written, and satisfy a need or send readers a specific message.

Both flyers and brochures have a place in your advertising arsenal. While you can have these items custom printed (and if you don't care to master home publishing, you really should), in this age of desktop publishing software, high-end printers, and quick-print shops, you can create effective flyers and brochures right at home. Showing you how to do it is beyond the scope of this book, but keep the following tips in mind:

✗ Determine what you want your flyer or brochure to achieve and focus on that goal. If you raise Mammoth Jackstock, Kiko goats, and Longhorn cattle, don't try to market them all through a single brochure.

✗ Choose headlines that draw attention to specific points. Avoid artsy fonts that are difficult to read. Basic sans-serif fonts (serif fonts have small strokes at the end of each letter; sans-serif fonts don't) work best; some good ones are Helvetica, Arial, and Geneva.

✗ Text should be written in a standard serif font such as Times, Times New Roman, or New Century Schoolbook. Don't use script fonts of any sort.

✗ DON'T USE ALL CAPITALS; they're distracting and difficult to read.

✗ Don't crowd too much onto a single page. Increased space between lines of text improves readability. Use one and one-half or double spacing for most text. Adequate margins on all sides of each page add elegance and increase readability, too.

✗ Keep sentences short and to the point. Strive for simplicity and clarity in the copy you write. If you're unsure of what you've written, ask someone unfamiliar with donkeys (or your donkey-related services or products) to proofread it before it goes to print.

✗ Always run a spellchecker over everything you've written, but after you do, proofread to catch small errors the program might have missed.

Show Off Your Stock

A booth at your county fair can generate a lot of local interest in your donkeys, and one at your state horse expo, business galore. The trick is planning in advance, so that you make a favorable impression.

✗ Ask the management about last year's attendance including a breakdown, day by day. That way you'll know how much handout material to bring and how many helpers you'll need to man your booth.

✘ Clarify the details up front. How much will you have to pay to set up your display and what's included in the price? How much stall space do you get, and how much room for your display? Are tables furnished? Bedding for your donkeys? Will you have enough electrical outlets to power video players, fans for your animals, and any other electrical appliances you require? If you sign up early, can you choose your location or are you stuck wherever you're put? Don't be afraid to ask questions. After all, they're asking you for money, so you have a right to know what's afoot.

✘ Try to arrange block stabling with other donkey exhibitors, especially if they have a different type of donkeys from your own. Miniatures stabled next to Mammoths attract more attention than either size stabled in separate stall blocks.

✘ Ask to be stabled away from concession areas. Otherwise standing diners will surely block your display, your booth could become inundated with paper cups and sandwich wrappers, and unless they're watched closely, attendees will treat your donkeys to lots of junk.

Plan an attractive display. A booth at a major horse expo will need to be fancier (or at least more unique) than one bound for the nearest county fair, but neatness and originality always count. Keep it simple and colorful and in tune with your farm or ranch image, and be sure to set it up at home before expo day to make certain that everything works.

Arrange for attractive signage. Have it professionally printed or whip it up on your home computer, but by all means avoid handwritten signs. Always hang name signs on stall doors; visitors love to know your donkeys' names. If an animal is for sale, let the stall sign say so.

If you plan to set up a TV to show video footage (and this is often a major draw), place the TV about five to six feet off the ground, at or slightly above the crowd's eye level. If you set it on a box atop a table, drape the box with an attractive fabric throw.

When displaying photographs, make sure that they're large enough to be seen. Hanging photos should be 11 × 14" or larger, and photo-album images a full-page 8 × 10" size.

A Sufi Lesson

Jalaluddin Rumi, the famous 13th-century Sufi mystic, recounted this teaching story about a much-loved donkey.

A Sufi had been traveling on donkey-back for a long while when he stopped for the night at a roadside inn. He told the stable man to make sure he mixed lots of barley with the straw that made up his donkey's supper and to please soak the uncooked barley in warm water because the donkey was old and had worn teeth. Also, would he please remove the saddle gently and put salve on the sore on his donkey's back and could he currycomb the donkey because his donkey loved to be groomed?

The instructions went on and on and the servant became annoyed. In a huff, he informed the Sufi that he'd cared for thousands of donkeys and he knew what he was doing, thank you very much!

So the Sufi went up to bed but as dawn approached he had frightful dreams about his donkey in which it was attacked by wild beasts and then fell into a deep ditch. The Sufi raced down to the stable to check on the donkey and while his dreams weren't exactly true, indeed the donkey had been totally neglected and had spent the long night without food or water; instead of caring for the donkey as instructed, the servant chose, instead, to go out carousing with his friends.

The moral of the story: When entrusted with the care of another creature, don't delegate its care to someone who doesn't share your concerns — always see to the tasks yourself.

Choose display animals wisely. Friendly, people-oriented donkeys are your best sales personnel. For liability reasons, even if your jack is a laid-back gent, it's probably best to leave him at home. No matter how vigilant you are, a few people will still manage to unexpectedly put themselves into close proximity to your stock, so a cheerful gelding or easygoing jenny with an older foal at her side makes a better ambassador. Please leave dams with newborn babies at home; despite how much crowds love them, exhibitions can be very stressful for new moms and their neonatal foals.

And line up plenty of help. Plan to have at least one attendant with your animals and two manning your booth at all times (more is always better). Line up enough personnel to allow for plenty of rest breaks, and pack lots of food and beverages to keep morale running high.

Find out in advance when you're allowed to unload your animals and display materials and arrive early to get the job done right.

While the expo or fair is in progress, *market*. Don't sit by and expect passersby to initiate contact; be on your feet ready to greet them. Make eye contact, smile, and ask open questions like, "What can I tell you about _____ (our donkeys, Miniature Donkeys, donkey milk soap, donkey trekking, etc.)?" Hand them your card and brochure; ask them to sign your guest book and invite them to meet with one of your long-eared ambassadors.

Remember, positive word of mouth is your best, least-expensive advertising. You have a chance to meet thousands of prospective customers at these venues. Make a good impression!

Penny-Wise Promotion

It needn't cost the world to promote your business or farm. There are lots of easy, effective ways to do it on a shoestring. Here are some ploys my friends and I have used. I bet they'd work for your business, too.

- ✘ Invest in quality truck and trailer lettering. Incorporate your phone number, e-mail, and Web site address into the design. Turn nonbusiness vehicles into rolling billboards for your farm by affixing custom-designed magnetic signs.
- ✘ Order T-shirts, jackets, and hats printed or embroidered with your logo and farm name and wear them everywhere you go. Buy extras for your customers and friends.
- ✘ Order custom-printed checks, invoices, and other business forms imprinted with your logo and business information.
- ✘ Erect a large, legible road sign by your road gate. If you live out in the boonies, erect directional signs to point the way to your farm, but research signage laws and ask permission from landowners before you tack them up.
- ✘ If you live on a well-traveled thoroughfare, display donkeys in a pasture adjacent to the road. If your jennies have babies, put them by the road; nothing stops traffic like a passel of adorable foals.
- ✘ Take your animals out in public. Train a harness or pack donkey, then drive or lead him in parades with farm signs attached to his cart or panniers. Take a well-behaved donkey on visitations to hospitals and nursing homes. Take donkeys and a display booth to community gatherings, county fairs, and farm expos. Again, utilize foals' "cute factor"; nothing attracts attention like a baby donkey!
- ✘ Give talks and demonstrations (haul a friendly donkey along whenever you can). Many civic groups need speakers for meetings and events; let it be known that you're available and interested. Hold an open house. Sponsor a 4-H clinic or training seminar at your farm.
- ✘ Build a donkey-related float for the Fourth of July parade or sew a donkey costume (with your farm name on the back, of course) and dance along the parade route flinging candies.

- Raise money for your favorite donkey charity. Organize a sponsored donkey trek in a nearby state park, hold a used-tack rummage sale, raffle a handmade quilt, or take pictures of happy children posed on your donkeys like the donkey photographers of yore. Whatever you do, call the newspaper and get your farm name in print. The donkey charity wins and you do, too!

- Offer to read donkey stories to Head Start or kindergarten children or volunteer to read to them at your library's story hour. Take along a donkey and be sure to tip off your local newspaper in advance.

- Throw a donkey bash. Invite all of your customers and potential customers to a barbeque or Christmas shindig. Hand out favors like T-shirts printed with your logo or calendars featuring contact info and pictures of your donkeys.

Donkeys have inspired artists and craftspeople all over the world throughout history. This sweet fellow was made of gray felted wool fleece and decorated with colorful embroidery in the Republic of Georgia.

- Participate in donkey-related listservs and e-mail lists; be the first to field questions. This establishes you as a knowledgeable donkey person and costs nothing but your time. People tend to buy from producers whom they know and respect, and you can advertise on some lists for free.

- Configure your e-mail program to add a signature to your outgoing e-mail. At a minimum include your farm name and location, a tag line describing your services, and your farm's Web address.

- Write a column for your local newspaper in exchange for printing your farm name and Web address in each article's byline.

- Remember your customers with handmade Christmas cards featuring your donkeys. Purchase blank greeting cards at an art supply store (Strathmore makes the high-quality, deckle-edged cards we use). Carefully apply rubber cement to a holiday-related 4 × 6" color photo on the outside of the card (donkeys wearing Santa hats and beards work well). Address the envelopes by hand and hand-write a personalized greeting inside each card. Before you seal the envelopes, add a pinch of colorful Christmas confetti and a holiday edition of your business card inside each card (print these short-run cards on your home computer or buy from an eBay entrepreneur). Will your cards stand out in a sea of commercial greeting cards? You bet they will!

- Hold a contest for kids. Donate a foal to the youngster who writes the best essay on "Why I would like to show a donkey in 4-H" or award plush toy donkeys to the kindergarteners who dream up the cutest names for the foal you take to their school (but be sure to distribute consolation prizes to the rest of the class). Again, invite the newspapers when prizes are doled out. It's an everybody-wins situation!

Donkey-Men

Tales of creatures that are a cross between man and donkey abound throughout written history, as seen in the following examples.

According to *Credulities, Past and Present* (William James, 1880), the Carvararadonques caste of India claimed to be descended from asses.

Lucius Apuelius's bawdy novel, *Metamorphosis, or The Golden Ass*, written in the 2nd century AD, is the only such work to survive in Latin. It's the story of Lucius, an aristocratic young man, and his adventures in Thessaly, culminating in his metamorphosis into an ass as punishment for betraying a priestess of Isis.

In Greek myth, the onocentaur (*Onokentauros*) was a centaurlike animal with the torso of a man and the body of an ass.

King Midas (who was later granted the golden touch) was fond of the satyr Pan, who played lovely music on his pipes. When Pan challenged Apollo — who played the lute most beautifully — to a music contest with King Midas as judge, the king awarded the prize to Pan. In anger, Apollo changed the king's ears to those of an ass. Humiliated, Midas took to wearing a cap yanked down over his ears; only Midas's barber knew the truth, and he was sworn to secrecy. However, the barber was bursting to tell *someone*, so he dug a hole in a solitary place, whispered the secret into the hole, and then covered it up. The following spring willowy reeds grew on the spot where the barber had planted his secret. Wind whispering through the reeds repeated again and again: "Midas has asses ears, Midas has asses' ears, Midas has asses' ears."

Perhaps the best-known donkey-man of all is Nick Bottom, a principal player in William Shakespeare's *A Midsummer Night's Dream*, who, thanks to the antics of the mischievous Puck, wakes up with an ass's head in place of his own.

And finally, there is Pinocchio. *The Story of a Puppet,* by Carlo Collodi, contains this passage: "Everyone, at one time or another, has found some surprise awaiting him. Of the kind that Pinocchio had on that eventful morning of his life, there are but few. What was it? I will tell you, my dear little readers. On awakening, Pinocchio put his hand up to his head and there he found . . . during the night, his ears had grown at least ten full inches . . . he went in search of a mirror, but not finding any, he just filled a basin with water and looked at himself. There he saw what he never could have wished to see. His manly figure was adorned and enriched by a beautiful pair of donkey's ears."

THE FRIENDLY BEASTS

(Twelfth-century English Christmas carol)

Jesus, our brother, kind and good,
Was humbly born in a stable rude,
And the friendly beasts around Him stood;
Jesus, our brother, kind and good.

"I," said the donkey, shaggy and brown,
"I carried His mother up hill and down;
I carried her safely to Bethlehem town;
"I," said the donkey, shaggy and brown.

"I," said the cow, all white and red,
"I gave Him my manger for His bed;
I gave Him my hay to pillow His head;
"I," said the cow, all white and red.

"I," said the sheep with curly horn,
"I gave Him my wool for His blanket warm.
He wore my coat on Christmas morn.
I," said the sheep with curly horn.

"I," said the dove from the rafters high,
"I cooed Him to sleep, that He should not cry,
We cooed Him to sleep, my mate and I.
"I," said the dove from the rafters high.

And every beast by some good spell,
In the stable dark was glad to tell
Of the gift he gave Immanuel,
The gift he gave Immanuel.

Cairo. Arabic woman on a donkey.

APPENDIXES

APPENDIX A

ADMS Code of Ethics

The American Donkey and Mule Society is the only national breed organization in the United States for donkeys and mules of all types and zebra hybrids. It administers the Miniature Donkey Registry, The American Donkey Registry, The American Mule Registry, The American Zebra Hybrid/Bloodstock Registry and The American Mule Racing Registry. The American Donkey and Mule Society requests that all breeders subscribe to the following Code of Ethics to promote and foster the highest standards among breeders, owners and fanciers, and to encourage sportsmanship and cooperation in the advancement and protection of all donkeys, mules, and hybrids.

Records. I am familiar with and follow the registry requirements for my breed. I will keep accurate records and retain those records for a minimum of 5 years. These records will include registry paperwork and health and breeding paperwork for all animals in my herd. I will notify the registry of the death of an animal and will work with the registry to keep records current. I will report any person who falsifies a registration or knowingly misrepresents a pedigree to the registry.

Breeding. I shall plan each breeding with the paramount intention of improving the breed. I will select the sire and dam with an eye to conformation, temperament, and good health with a careful study of the breed standard and the principles of genetics. I will refrain from close breeding without careful study of pedigrees and individual animals. I will not breed any male or female until they are both physically and mentally mature. I will not breed any female before the age of two years. Before entering into any breeding arrangement I will scrutinize the pedigree, conformation, and health of both the sire and dam, keeping in mind the ideal of the breed. I have an obligation to refuse the breeding if in my opinion it is not in the best interest of the breed. As a responsible owner of a sire, I understand that should I refuse a breeding I will fully explain my reasons to the owner of the female. I subscribe to the policy that only sires and dams free of defects such as cryptorchidism, under and over bite, dwarfism, and other genetic defects shall be used for breeding. Being a responsible breeder I will refrain from using a breeding animal, although free from the above defects, that consistently produce afflicted offspring.

Health. I shall maintain high standards of health and care for my animals and will keep their feet properly trimmed. I will guarantee the health of animals I sell to others.

Sales. I will be discriminating in the sales of my animals and concerned with the type of homes in which they are placed. My animals will not be sold to wholesalers or meat buyers and I will humanely euthanize animals that must be put down. I will transfer all applicable registration papers at the time that the purchase agreement is completed and is suitable to all parties. I will not advertise animals as "registered" and then present the new owner with a filled out application and expect him to register the animals himself. Upon the sale of an animal I will provide the new owner with a diet record, an inoculation and parasite control record, and a health guarantee. I will refrain from releasing any young animal until it is at least 4 to 6 months old and properly weaned. I will inform all buyers that donkeys need companions, preferably other equines, in order to be happy inasmuch as they are herd animals, not solitary creatures.

Advertising. I agree that all advertising of animals will be factual and honest both in substance and implication. I will avoid encouraging buyers regarding the breeding potential of an animal without explaining that the breeding of any animal involves certain responsibilities and is not to be taken lightly. I will make sure that all show winnings listed in advertising are accurate.

Exhibitor/Breeder Relations. I understand that exhibiting animals is a sport and that I am expected to express good sportsmanship in all activities. As an exhibitor I will refrain from unnecessary criticism of other people's animals, or of the judge. I will not represent any animal I sell as being a sure winner, knowing that judges differ in their opinions. I will not demonstrate behavior that could be defined as serious abuse or harm to an animal either in training the animal or in showing it.

Implementation. Inasmuch as a Code of Ethics is a guideline, not rules, regulations, or legal documents, it does not carry an enforceable punishment. It should be enforced by breeders and buyers upon each breeder in a civil and responsible manner. Buyers should read the Code of Ethics and determine if it is being followed before buying an animal.

Farm Requirements

Farm owners must meet all local, county, and state government requirements for farm or stable operations.

Food and Water. Clean water must be available at all times. Care will be taken that water supplies are unfrozen in winter, and clean at all times. Animals should be neither too obese nor too thin and proper rations should be available every day.

Shelter. All animals will have access to clean, dry shelter from rain, wind, and snow and shade from heat.

Confinement. Fences will be sturdy, well built, and safe for the type of animal enclosed within. Space will allow for adequate freedom of movement and exercise.

Safety and Protection. More than one jack over 6 months old will not be allowed to be pastured with a group of females. Caution will be taken at all times that underage females will never be available to be bred. Jacks will be confined so that they will not damage each other or jennets or kill foals.

Health Care. I will take care that vaccinations and Coggins tests will be administered in accordance with local and state requirements and health needs. An internal and external parasite control program will be maintained. Sick animals and newly introduced animals will be isolated from the general population. Health papers and Coggins tests must be required for any animals brought to the farm for breeding or training. Hoofs of all stock will be kept properly trimmed.

Waste Disposal. Waste must be removed in accordance with accepted local and state requirements. Also waste must be handled in accordance with veterinarian prescribed health practices. Barn and stall areas must be cleared of waste daily.

Record Keeping. Health and breeding records will be kept for each animal. All breedings will be recorded and a sire's breeding certificate made out at the time of breeding. A sales record showing to whom each animal (registered or non-registered) is sold will be kept. All animals should be easily identified by the owner, and if possible should be identified with methods that anyone can use such as neckstraps, freeze brands, microchips, etc. Eartags are not recommended by the registry for donkeys or mules.

Reprinted courtesy of
The American Donkey and Mule Society.

Donkeys as Livestock Guardians

On guard! Some of you probably picked up this book because you're thinking about getting a livestock guardian for your sheep, goats, alpacas, Miniature horses or donkeys, or some other type of small livestock. You're wondering if a guardian donkey would fill your needs — and the answer is a qualified "maybe."

If you keep small livestock, predation is an issue. According to the National Agricultural Statistics Service, in 1999 alone some 273,000 lambs and sheep were killed by predators to a monetary tune of $16,502,000. At least 165,800 of these unfortunates were dispatched by coyotes, a species now found in burgeoning numbers throughout the 48 continental United States and in every Canadian province. Another 41,300 were killed by dogs. Combined, coyotes and dogs account for more than 75 percent of the sheep predation in America. Donkeys are effective against both species, but not all donkeys will protect another species. Selecting the right donkey is important.

Who Uses Guard Donkeys?

In 1989, Murray T. Walton and C. Andy Field of the Texas Department of Agriculture presented a paper, "The Use of Donkeys to Guard Sheep and Goats in Texas," at the Fourth Eastern Wildlife Damage Control Conference; after conducting two surveys and talking to numerous farmers and ranchers who use donkeys to guard small livestock, these were their conclusions:

Based on the results of the first survey an estimated 2,400 Texas sheep and goat producers tried guard donkeys and 1,800 were currently using them. Most respondents used guard jennies, some used geldings, and a few used a jenny with her foal.

Twenty-two percent of 275 sheep and goat producers who responded to the second survey reported guard donkey use, and 16 percent indicated that donkeys were being used at the time of the survey (for a total of 133 donkeys). Forty percent reported that their donkeys did a good or better job (ratings were excellent, good, fair, poor, failure, and unknown) of protecting livestock from coyote and dog predation. One responder indicated that his donkey killed more goats than predators; another said that his donkey was observed successfully fending off three coyotes trying to attack a group of sheep bunched up behind the donkey at a fence corner.

According to the 1999 Colorado State University bulletin, "Livestock Guard Dogs, Llamas and Donkeys," 3 percent of producers in Colorado used donkeys to protect sheep. During the same year, according to National Agricultural Statistics Service statistics, about 9 percent of sheep producers throughout the United States used donkeys to protect their animals from predators, primarily coyotes.

Since 1995, the Swiss Wolf Project has encouraged farmers to use guard donkeys to protect their flocks from wolf predation in the western Alps. The experiment is ongoing, but early results appear promising.

On a similar note, stock producers Down Under are successfully using guard donkeys to stem predation by wild dogs, dingoes, and foxes, the major predators of livestock in Australia.

The 1995 *Ontario Predator Study, Study 6: Donkeys As Mobile Flock Protectors*, by Fytche Enterprises, reported that about 70 percent of the donkeys being used were rated excellent or good in terms of providing flock protection. However, the donkeys' effectiveness ranged from total elimination of predation, to having absolutely no impact on predation while simultaneously causing other problems within the flock. In other words, success varies according to each situation.

Training a Guard Donkey

This stellar information on training guard donkeys is gleaned from the Alberta (Canada) Ministry of Agriculture and Food publication, "Protecting Livestock with Guard Donkeys" (*see* Resources):

- ✘ Guard donkeys may need several weeks to adjust to livestock, so introduce them to stock well before predation is likely to occur.
- ✘ Keep young donkeys with goats, sheep, or cattle after the donkey is weaned. Do not allow guard donkeys to run with other donkeys or horses. In this way the young donkey will think it is part of the flock or herd. Ideally, the donkey should be born in the flock or herd, and its dam should be taken away at weaning to let the young animal grow up with the stock.
- ✘ Place a new donkey on the other side of a common fence line with livestock. This gives the donkey and the livestock an opportunity to get to know one another safely.
- ✘ A week to 10 days following this socialization period, lead the donkey around the cattle or sheep where they can smell and touch each other. Then tether the donkey inside the pen with the stock and feed and groom it there for about a week. By this time both will have accepted each other. Allow the donkey to run loose in the pen or pasture and soon the stock will seek the donkey out in times of danger.
- ✘ Feed the donkey with the stock so it feels like a member of the flock or herd. If stock is fed from troughs, feed the donkey first from a separate feeder or bowl so the livestock can eat unhindered. Always feed the guard donkey something every time you feed the stock.

Jenny and the Rotties

The first time I saw a donkey take on a pack of dogs, I could hardly believe my eyes. It was a small pack — two happy-go-lucky male Rottweilers out on a lark — but on this particular day they were in the wrong place.

Our Minnesota neighbor raised Spotted Draft Horses and Standard donkeys; the two species shared a huge pasture abutting our property along one side. A string of weekend properties lined her fencerow far across the field, and that's where the dogs came from. These dogs weren't bent on predation, but to the little jenny who spotted them frisking across the pasture, intent didn't matter one bit. She waited until they were well within reach of the herd — 30 feet at the most — then she tucked back her ears, brayed a mighty battle cry, and plunged toward the dogs fit to kill.

The Rotties froze for a stunned instant, then spun and fled for their lives. Donkeys are infinitely faster than Rottweilers; the dogs didn't have a chance. They ran as though dragons were hot on their tails, but the jenny didn't give up the chase. She galloped full tilt through that pasture, striking their butts with both front feet, while ripping off a steady stream of blood-curdling brays. Both dogs closed on the barbed wire fence and flew through without slowing down. The jenny screeched to a halt, paused for a heartbeat, then jerked her head once as if to say, "Take that!" Then she wheeled and jogged back across the big pasture with a smile on her face — and those Rottweilers *never* came back.

✗ Keep all dogs away from donkeys and do not test the donkey by teasing it with a dog. Do not allow farm dogs to become friendly with the donkey. Avoid or limit the use of stock dogs around donkeys.

✗ If a donkey is aggressive toward or fears stock (or vice versa), remove it immediately.

Donkey, Dog, or Llama?

However, before choosing a donkey guardian, make certain that you want one; an LGD (livestock guardian dog) or llama might better suit your needs.

✗ If you live where predation by mountain lions or bears is a problem, forget about donkeys and llamas; neither species can adequately protect itself, much less its charges, against big, aggressive predators like these.

✗ If noise is an issue, don't get an LGD. Livestock guardian dogs bark all night; this is how they warn predators away. Because they're big dogs with deep, resounding voices, their barking isn't particularly annoying, although nearby neighbors might disagree. If you choose an LGD, play a radio while your dog is on duty or buy a box fan with a loud, soothing hum; turn it on high speed when you go to bed; and snooze away.

✗ If you feed your goats or sheep feeds containing anticoccidial agents like monensin (Rumensin) or lasalocid (Bovatec), please think twice about buying a guard donkey. According to the Ontario Ministry of Agriculture Food and Rural Affairs, a lethal dose of monensin for horses is 1–2 mg/kg of body weight, and of lasalocid, 21.5 mg/kg of body weight. That's not a lot of monensin, and though no tests have included donkeys, they're often more sensitive to chemicals than are horses. A single slipup could be fatal. Either choose a different guardian or do as we do and feed nothing but unmedicated feed.

✗ Each species has its own pros and cons. Before choosing, ask people who keep each type of livestock guardian for their advice. Or write for the free booklet "Using Guard Animals to Protect Livestock" (*see* Resources).

Guard Donkey Tips

- Guard donkeys should be selected from medium-to-large-size stock. Do not use extremely small or Miniature Donkeys.
- Use jennies or geldings. Do not use jacks as guard animals because they are frequently aggressive to other livestock and may kill sheep or goats.
- Use only one donkey or jenny and her foal per pasture.
- Isolate guard donkeys from horses, mules, and other donkeys.
- To increase the probability of bonding, donkeys should be raised from birth or placed at weaning with sheep or goats.
- Raise guard donkeys away from dogs. Avoid the use of herding dogs around donkeys.

- Monitor the use of guard donkeys at lambing or kidding as some donkeys may be aggressive to or overly possessive of newborns. Remove donkeys temporarily if necessary.
- For best results, use donkeys in small (less than 600 acres) open pastures with not more than 200 head of sheep or goats. Large pastures, rough terrain, dense brush, too large a herd, and scattered sheep or goats all lessen the effectiveness of guard donkeys.

— Murray T. Walton and C. Andy Field, "The Use of Donkeys to Guard Sheep and Goats in Texas," The Texas Department of Agriculture, 1989

Weighing and Body Scoring Donkeys

When doling out medicines and dewormers, it's important to know how much the recipient weighs. And is he fat, way fat, or simply healthy donkey size? We can figure that out through an easy process called body scoring.

Which Weigh Works Best?

There are several ways to weigh a donkey; some are approximate, and others are right on the nose. Unfortunately, easy-to-use weight tapes designed for horses don't accurately weigh donkeys, but here is a measurement method that does. Simply multiply your donkey's

height by his heart girth by his torso length, divide by 300, and there you are! *Hint:* To measure a hinny or mule, substitute this formula: Multiply height by heart girth by torso length, divide by 300, add 50 = weight.

- **Height.** Measure from the highest point of your donkey's withers straight to the ground. Use a measuring standard or hold a yardstick flat across his withers and measure straight down from that.
- **Heart girth.** This is the circumference of his body about 2½ to 4 inches (depending on his size, from Miniature to Mammoth) behind his front legs. Measure his heart girth using a seamstress's cloth tape measure and pull it tight enough to depress the flesh a bit. For more accurate results, take several measurements and average them.
- **Torso length.** Measure your donkey in a straight line from the point of his shoulder to his buttocks, using the diagram below as a guide.

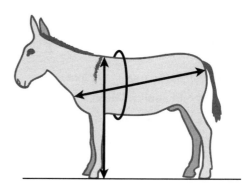

More Ways to Weigh Donkeys

Our friends at the Donkey Sanctuary in Sidmouth, in Devon, England, developed an accurate tape-measuring method for "weighing" donkeys in metrics (*see* Resources).

You can easily weigh a donkey foal on a household scale. Weigh yourself first, pick up the foal, step on the scale, then subtract your weight from the second number.

Most small-animal veterinary practices have large, step-on scales for weighing dogs; these can be used to weigh Miniature Donkeys. Some equine clinics have horse-size scales; if yours does, that's your best bet!

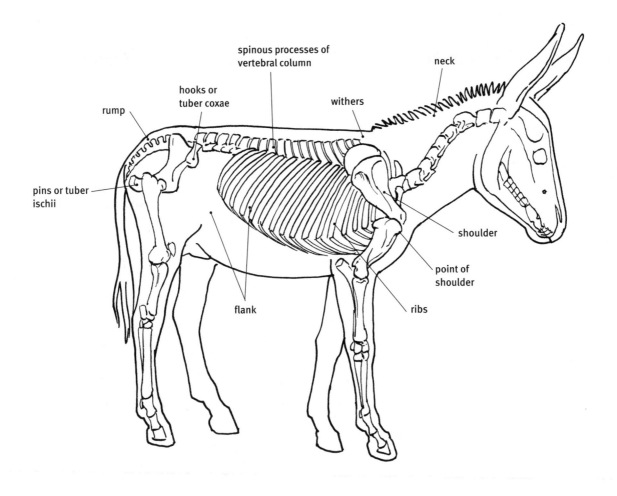

Labels (clockwise from top):

spinous processes of
vertebral column

neck

hooks or
tuber coxae

withers

rump

pins or tuber
ischii

shoulder

point of
shoulder

flank

ribs

Body Scoring — Bones to Blubber and Everything Between

While there are several different donkey body-scoring methods in use, including a five-point chart developed by the Donkey Sanctuary (*see* Resources), I prefer this one adapted from *A Guide to Live Weight Estimation and Body Condition Scoring of Donkeys,* by R. Anne Pearson and Mohammed Ouassat; it's more exacting, and this is a boon, especially for anyone involved in donkey rescue. Let's hope that you'll never see many ones or twos.

BODY SCORING A DONKEY

1. First decide to which of the three main categories he belongs. Walk around the donkey slowly, studying him from the side and back. Is he thin (1–3), fat (7–9), or someplace in between?

2. Next, examine the donkey more closely, looking at and feeling his neck, shoulders, back, ribs, flanks, and rump. Each area should be assessed and combined to reach a final score.

BODY SCORES

	SCORE	DESCRIPTION

  1. Very thin (emaciated)

This donkey is clearly emaciated. His bone structure is extremely obvious; there is little muscle present. His belly is severely tucked up, and there may be a cavity under his tail. The donkey is weak and lethargic.

  2. Thin

The donkey is emaciated. It's easy to see and feel individual spinous processes (the bony points rising from his backbone), ribs, hooks, pins, shoulder blades, and spine. His neck is thin, his withers are prominent, and his shoulders are sharply angular. He has very little muscle development.

  3. Less thin

The donkey's vertebral column, ribs, pins, and hooks are prominent, and individual spinous processes can be felt. There is little fat on this donkey, but superspinous muscling (the muscles along each side of his backbone) along the spinous processes is apparent. The loin and rump areas are concave, and there is little muscle or fat covering over his withers and shoulders.

  4. Less than moderate

The donkey's vertebral column is visible. His pins and ribs can be felt but can't be readily seen. His hooks are rounded but visible, and his rump is flat rather than concave. His neck, withers, and shoulders have some muscle and fat cover, and his shoulder blades are less clearly defined.

  5. Moderate (ideal)

This donkey's superspinous muscles are developed and readily apparent; the vertebral column can be felt. His hooks and rump are rounded. Some fat can be felt in the shoulder area and at the base of the donkey's neck. His ribs can be felt but not seen, and his pins are no longer visible.

  6. More than moderate (near ideal)

It's hard to feel this donkey's spinal processes. His back is becoming flatter and is well covered; his rump is convex and well muscled; and his hooks are barely visible.

  7. Less fat

This donkey's back is flat, and his spinous processes can't be felt; his hooks are barely visible. The fat on his neck and his shoulder area are beginning to expand; his flanks are filling and his neck thickening.

  8. Fat

This donkey is well covered. His neck is thick and the crest very hard. His back is quite broad, and his body is rounded with fat. His bones can't be seen.

  9. Very fat (obese)

This donkey is obese. His bones are buried in fat. He has large accumulations of fat on his neck (pones), and the crest may list over to one side. There are also large accumulations of fat over his shoulder area and ribs. His withers can't be felt, and his back is broad and flat and he may have a crease down the middle of his back. His flanks are fully filled with fat, and fat may overhang both sides of the tail head.

Emergency Euthanasia

Euthanasia is hard enough to face when a veterinarian does the deed by lethal injection, but if your donkey is seriously hurt and the vet is unavailable, what then? Should the unthinkable occur and you find you have to handle the sad task yourself, here are the things you need to know.

If your vet has entrusted you with prescription drugs such as Rompun or Banamine, give the donkey a hefty dose to sedate him (discuss this with your vet in advance, so that in an emergency you can administer the dose he recommends). If your donkey relaxes so much that he lies down, so much the better.

Use a .22-caliber long rifle, 9mm or .38-caliber firearm, or a shotgun (410 gauge or better) with a rifled slug. The muzzle of the gun should be held at least four to six inches away from the skull (never directly against the skull) when fired. The use of hollow-point or soft-nose bullets will increase brain tissue destruction and reduce the chance of ricochet. When performed skillfully, euthanasia by gunshot induces immediate unconsciousness, although muscle twitching will likely continue for a little while after the donkey is dead.

There is generally less bleeding than you'd probably expect. This method should be attempted only by individuals trained in the use of firearms and who understand the potential for ricochet. All humans and other animals should remain out of the line of fire.

Glossary

A

Abortion. Early (often spontaneous) termination of pregnancy.

Acre. A unit of measurement: 4,840 square yards or 43,560 square feet.

Action. The manner in which an animal moves his legs.

Afterbirth. The placenta and fetal membranes that are expelled from a jenny or mare after foaling.

ALBC. American Livestock Breeds Conservancy; a group dedicated to preserving and promoting rare and endangered breeds of American livestock and poultry, including asses.

Anestrus. The period of time when a jenny is not in estrus (heat).

Anthelmintic. A substance used to control or destroy internal parasites; a dewormer.

Antibodies. Circulating protein molecules that help neutralize disease organisms.

Antitoxin. An antibody capable of neutralizing a specific disease organism.

Artificial insemination (AI). A process by which semen is deposited within a jenny's or mare's uterus by artificial means.

Ascarids. Roundworms.

Aspirate. To pull back slightly on a syringe's plunger to draw fluid into the chamber.

Ass. Another (perfectly proper) word for *Equus asinus,* or donkey.

B

Back. To step an equine backward; backing is a two-beat gait in reverse.

Bag. Udder. (slang)

Bagging up. Enlargement of the udder prior to foaling.

Barefoot. Unshod.

Barn-sour. When an equine reacts badly to separation from a preferred area (such as the barn or feeding area).

Barren. Unable to conceive or bear foals.

Bars. The gap between an equine's front and back teeth where the bit on a bridled animal rests.

Bite. Occlusion; the manner in which the upper and lower teeth match up.

Bitting. Teaching an equine to properly carry a bit.

Blemish. A scar or deformity that diminishes an animal's beauty but doesn't affect his soundness.

Bloodlines. The ancestry of an animal; his family tree.

Body Condition Score. A value from one to five (or nine, depending on the system used) used to describe an animal's body condition.

Bolt. To devour feed quickly; also what an equine does when he gallops out of control with a rider on his back.

Bombproof. An extremely reliable, spook-proof equine.

Booster vaccination. A second or multiple vaccinations given to increase an animal's resistance to a specific disease.

Bosal (boh-ZAL). A specialized rawhide noseband used in lieu of a bit.

Bo-Se. An injectable prescription vitamin E–selenium supplement used in selenium-deficient areas to prevent and treat white muscle disease.

Bot flies. A beelike fly that lays eggs in an equine's hair.

Bot fly eggs. Minute yellow eggs deposited on the legs and chins of livestock and other hairy animals.

Bots. A type of internal parasite.

Box stall. A roomy, four-sided stall to house equines.

Bottom side. The dam's side of a pedigree.

Bray. The loud, forceful vocalization of a donkey or mule.

Brayer. Colloquialism referring to a donkey or mule.

Brayer, The. Journal of the American Donkey and Mule Society (ADMS).

Breast collar. A strap that passes around the front of an equine above its forelegs and attaches to the saddle; used to keep a saddle from sliding sideways or backward.

Breech birth. A birth in which the rump of the foal is presented first.

Breeching. *See* Britchen.

Breed. Animals of a color, body shape, and other characteristics similar to those of their ancestors that are capable of transmitting these characteristics to their offspring; the mating of animals.

Breeder. Generally speaking, anyone who breeds animals; more specifically, an animal's breeder is the person who owned the dam at the time of foaling.

Breeding class. A show class based on judging an animal's conformation and type.

Bridle. Riding headgear comprising the entire headpiece (headstall, reins, bit or bosal, and curb strap if one is used).

Britchen. Also spelled "britchin"; a series of straps attached to the saddle that drape over an equine's rump to prevent the saddle from sliding forward.

Broken crest. A crest that has permanently fallen over onto the side of an equine's neck. (*See also* Crest.)

Brood jenny (or broodmare). A female kept primarily for producing foals.

Browse. Morsels of woody plants including twigs, shoots, and leaves.

Bull pen. A sturdy, round enclosure used for training equines.

Burro. Western colloquialism referring to a donkey; donkey in Spanish is *el burro.*

Bute. Phenylbutazone, a prescription veterinary drug widely used for reducing pain.

By. Short for "sired by"; a donkey is by a jack and out of a jenny, a horse is by a stallion and out of a mare, a mule is by a jack and out of a mare, and so on.

C

Canter. A fast, three-beat gait that is slower than a gallop.

Castration. Removal of a male's testes.

Catheter-tip syringe. A syringe with a long, blunt tip used for the oral dosing of animals.

Cattle panel. A very sturdy, large-gauge, welded-wire fence panel; sold in various lengths and heights.

cc. Cubic centimeter; same as a milliliter (mL).

Cellulose. A component of plant cells that many animals, including donkeys, are unable to digest.

Cervix. The section of the uterus that protrudes into the vagina; it dilates during birth to allow foals to pass through.

Chestnut. The horny growth on the inside of an equine leg.

Chomping. The distinctive chomping mouth movements exhibited by a jenny in heat. (also called jawing or mouthing)

Chrome. Flashy white facial and leg markings. (slang)

Coarse. An animal lacking refinement.

Cob. In British terminology, a small, chunky horse. Halters and bridles marked "cob size" often fit Standard and Large Standard donkeys.

Coggins. A blood test used to detect carriers of equine infectious anemia; also the certificate indicating that an equine has been Coggins tested.

Colostrum. First milk a mammal gives after birth; high in antibodies, this milk protects newborns against disease.

Colt. An intact (uncastrated) male equine under three years of age.

Concentrate. High-energy, low-fiber, highly digestible feed such as grain.

Condition. The amount of fat and muscle tissue on an animal's body.

Conformation. The overall structure of an animal.

Congenital. Any condition acquired during development in the uterus and not through heredity.

Corpus luteum. The progesterone-producing mass of cells that form once an ovum (egg) is released from the ovary.

Cover. To breed (a jack covers a jenny or a mare).

Creep feeding. To provide supplementary feed to nursing foals in an enclosed area not accessible to larger animals.

Crest. The upper portion of an equine's neck stretching from the poll to the withers.

Cribber. An equine that grasps a solid object, expands his throat, and sucks in air; an animal that merely chews wood is *not* a cribber.

Crossbreed. An animal resulting from the mating of two entirely different breeds.

Crupper (KROO-per). A piece of tack consisting of a loop that encircles an equine's tail and a strap connecting the loop to the back of the saddle.

Cryptorchid. A jack or stallion with one or both testicles retained in the abdominal cavity; sometimes called a ridgeling or a rig. (*See also* Monorchid.)

Cull. To eliminate from a herd or breeding program.

D

Dam. The female parent.

Deccox. The brand name of decoquinate.

Decoquinate. A coccidiostat sometimes added to sheep or goat feed to control coccidiosis; not as toxic as Rumensin but it's still best to avoid allowing donkeys access to feed laced with this substance.

Deworm. The use of chemicals or herbs to rid an animal of internal parasites.

Dewormer. An anthelmintic; a substance used to rid an animal of internal parasites.

Dish-faced. Having a concave facial profile.

Disposition. The temperament of an animal.

Donk. Donkey. (slang)

Double. To turn an equine sharply so he turns on his haunches and heads back the way he came; a technique frequently used to stop runaways.

Drench. Giving liquid medicine by mouth; also a liquid medicine given by mouth.

Dust bath. A bare, sandy, or dusty spot where donkeys prefer to roll (dust bathe).

DVM. Doctor of Veterinary Medicine.

Dystocia. Difficulty in giving birth.

E

Ear shy. An animal that reacts unfavorably to having its ears touched is ear shy.

Easy-entry cart. A popular two-wheeled driving vehicle in which it's unnecessary to step over a support in order to be seated.

Easy keeper. An animal that easily maintains weight.

Emaciation. Loss of flesh resulting in extreme leanness.

Embryo. An animal in the early stage of development before birth; a fertilized egg.

Embryo transplant. Implantation of embryos into a surrogate mother.

Energy. A nutrient category of feeds usually expressed as TDN (total digestible nutrients).

Entire. An uncastrated male animal; a jack or stallion.

Equine. All members of the genus *Equus*, including horses, donkeys, zebras, and their hybrids.

Estrogen. Female sex hormone produced by the ovaries; estrogen is the hormone responsible for the estrus cycle.

Estrus. The period when a jenny or mare is receptive (will mate with a jack or stallion; e.g. she is "in heat") and can become pregnant.

Estrus cycle. A jenny or mare's reproductive cycle.

Euthanize. To humanely end an animal's life.

Extra-label. The use of a drug for a purpose for which it isn't approved (also called off-label).

F

Farrier. A skilled craftsperson who shoes equines and trims hooves.

Far side. An equine's right side (also called the off side).

Favor. To limp slightly.

Fecal egg count (FEC). The number of worm eggs in a gram of feces; sometimes written as EPG (eggs per gram).

Feed bag. A bag held on an equine's nose by a strap behind his ears that allows him to eat grain any time, anywhere, without a feed pan or other container; also called a nose bag.

Feral. A "wild" donkey or horse whose ancestors were domestic equines.

Field shelter. A basic shelter with a roof and three sides.

Filly. A female equine under three years of age.

Fitting. Preparing an animal for show.

Flake. One segment of a bale of hay.

Flehmen. Curling of the upper lip in order to increase the ability to discern scent.

Float. To file an equine's teeth to remove sharp edges and points.

Foal. An equine less than one year of age; also the act of an equine giving birth.

Foal heat. The first estrus that occurs after foaling.

Foaling colic. Abdominal pain experienced by a jenny or mare that has just given birth, caused by strong postfoaling contractions of the uterus.

Follicle. A fluid-filled sack (on an ovary) that contains an ovum (egg).

Follicle stimulating hormone (FSH). A hormone produced in the pituitary gland; used to stimulate the release of follicles by the ovaries.

Forage. Grass and the edible parts of browse plants that can be used to feed livestock.

Forb. A broad-leafed herbaceous plant (e.g., curly dock, plantain, and dandelion).

Forelock. The hair growing between a horse's ears that falls forward onto the forehead; donkeys don't have truc forelocks, but some mules and hinnies do.

Free choice. Available 24 hours a day, 7 days a week. Hay and mineral mixes are generally fed free choice.

FSW (color term). Frosted spotted white.

G

Gait. A pattern of foot movements such as the walk, trot, canter, and gallop.

Gaited horse (or easy-gaited horse). A member of a breed that performs specific, unusually smooth gaits (also called easy gaits) in addition to or in place of the standard walk-trot-canter, such as the Missouri Foxtrotter, Tennessee Walking Horse, Single-footing Morgan, or Peruvian Paso; gaited mares are used to produce gaited mules.

Gaited mule. A mule that performs at least one easy gait.

Gallop. A fast, three-beat gait; the fastest of the equine gaits.

Gelding. A castrated male equine.

Genotype. The genetic makeup of an animal or plant.

Gestation. The length of pregnancy.

Get. The progeny of a male animal.

Grain. Seeds of cereal crops such as oats, corn, barley, milo, and wheat.

Groundwork (or ground training). Training done on foot rather than in the saddle.

H

Halter. A piece of tack used to harness an animal's head to facilitate catching, leading, and tying.

Halter class. An event in which an animal's conformation and way of going are judged but the handler and his actions aren't taken into consideration.

Hand. Equines are measured in hands; one hand equals four inches. Equines are measured from the highest point of their withers to the ground. Fractions are shown as hands-point-inches, so that a 15.3-hand mule would be 15 hands and 3 inches tall.

Hand breeding. Breeding a jenny or mare to a jack or stallion under controlled conditions, usually with both animals haltered and controlled by skilled handlers.

Hard keeper. An animal that requires more than the usual amount of feed to maintain weight.

Haunches. Hindquarters.

Hay. Grass or legume greenery mowed and cured for use as off-season forage.

Head collar. British term synonymous with the American word *halter*.

Head shy. An animal that reacts unfavorably to having its head touched.

Heart girth. The circumference of an animal's chest immediately behind the front legs.

Heat. *See* Estrus.

Herd-bound. When an equine reacts badly when separated from other equines, he's considered herd-bound. (*See also* Barn-sour.)

Heritability. The degree to which a trait is inherited.

Heterosis. The increased performance of hybrids over purebreds; hybrid vigor.

Hinny. The sterile hybrid offspring of a stallion and a jenny.

Hybrid. The offspring of two separate species.

Hybrid vigor. *See* Heterosis.

Hypothermia. A condition characterized by low body temperature.

I

Immunity. Resistance to a specific disease.

Inbreeding. Mating closely related individuals such as father and daughter, mother and son, and full or half siblings.

In foal. Pregnant.

In milk. Lactating.

In season. In heat. (*See also* Estrus.)

Intramuscular (IM). Within the muscle.

Intravenously (IV). Within a vein.

J

Jack. An uncastrated male donkey.

Jackass. An uncastrated male donkey.

Jennet. The correct term for a female donkey.

Jennet jack. A jack used to breed jennies to produce more donkeys.

Jenny. A colloquial word for jennet (and the one generally used in this book).

Jog. Western riding term referring to a slow, relaxed trot. (*See also* Trot.)

John. A gelded male mule or hinny.

K

Keepers. Fixed loops used to keep elements of a saddle, bridle, or halter in place.

L

Lactation. The period when a jenny is giving milk; the secretion of milk in the udder.

Lameness. A condition in which an equine does not carry weight equally on all four legs due to disease or injury.

Large Standard donkey. In American Donkey and Mule Society terminology, a jenny 48.01 inches to 54 inches tall, measured at the withers, or a male (gelding or jack) 48.01 inches to 55 inches tall, measured at the withers.

Larvae. Immature stages of adult parasites; the term applies to insects, ticks, and worms.

Lead rope. A sturdy 7- to 10-foot rope with a snap on one end; used for leading equines.

Lead shank. A rope or strap with a short length of chain and a snap at one end; used for leading equines.

Legume. Plants such as alfalfa, clover, and lespedeza.

Libido. Sex drive; the desire to copulate.

Light points (color term). White on the nose, around the eyes, and on the belly of a donkey or donkey hybrid.

Line-breeding. The mating of individuals sharing a common ancestor.

Longear. Colloquialism for a donkey or mule.

Longe line. A length of cotton or nylon strapping, generally 30 feet in length, used to secure an equine as it moves in a circle around its handler.

Lope. Western riding term synonymous with a very slow canter. (*See also* Canter.)

Lutenizing hormone (LH). The hormone that triggers ovulation and stimulates the corpus luteum to secrete progesterone; it also stimulates testosterone production.

Lymph. A clear, watery, sometimes faintly yellowish fluid derived from body tissues; it contains white blood cells and circulates throughout the lymphatic system.

M

Maiden. A mare or jenny that has never foaled.

Mammoth Ass (or Jackstock). In American Donkey and Mule Society terminology, a jenny 54 inches tall or taller, measured at the withers or a male 56 inches tall or taller, measured at the withers. In American Mammoth Jackstock Registry terminology, a jenny 14 hands (56 inches or taller) or a male 14.2 hands (58 inches) or taller, measured at the withers; jennies must also measure at least 61 inches around the heart girth and 7½ inches around the cannon bone; jacks or geldings must measure 61 inches around the heart girth and 8 inches around the cannon bone. These asses are never properly referred to as donkeys or burros; they are Mammoth Jackstock, Jackstock, Jacks or Jennets, or Mammoths.

Mare. A female horse three years of age or over; in Britain, female donkeys three years of age or over are also called mares, rather than jennies or jennets.

Mare mule. A female mule.

Mastitis. Inflammation of the udder, usually caused by an infection.

Mecate (may-CAH-tay). A braided hair rope used to create the reins used with a bosal.

Meconium. The sticky, blackish fecal matter that a foal passes within a few hours after birth.

Mediterranean donkey. A Miniature Donkey.

Milk letdown. Release of milk by the mammary glands.

Miniature Donkey. The American Donkey and Mule Society/Miniature Donkey Registry registers donkeys standing up to 36 inches tall, measured at the withers; the International Miniature Donkey Registry registers Miniature Donkeys up to 38 inches tall.

Miniature Donkey (oversized or Class B). The American Donkey and Mule Society/Miniature Donkey Registry registers donkeys 36.01 to 38 inches tall in their Class B studbook.

Miniature Horse. In American Miniature Horse Association and American Miniature Horse Registry terminology, a horse standing 34 inches (AMHA) or 38 inches (AMHR), measured at the last hairs of the mane.

Miniature Mule. Miniature mules are the offspring of a Miniature Donkey jack bred to a Miniature Horse mare. The American Miniature Mule Society registers them in two sizes: Class A, under 38 inches; and Class B, 38 to 48 inches.

mL. Milliliter; the same as a cubic centimeter (cc).

Molly. A female mule or hinny.

Molly mule. A female mule.

Monensin. A coccidiostat sometimes added to sheep and goat feed to control coccidiosis; marketed under the brand name Rumensin, monensin is highly toxic to equines, including donkeys.

Monkey mouth. *See* Underbite.

Monorchid. A male with one descended testicle and one testicle retained in the abdominal cavity. (*See also* Cryptorchid.)

Motility. The ability of sperm to move.

Mouthing. *See* Chomping.

Mule. The sterile hybrid offspring of a male donkey (jack) and a female horse (mare).

Mule jack. A jack kept specifically for siring mules.

N

Near side. An equine's left side.

Neck roll. Wide, hard deposits of fat on an obese donkey's neck; also called neck pones.

Necropsy. A postmortem (after death) examination.

Nematode. A type of internal parasite; a worm.

Nymph. Immature insects and ticks that lack developed sex organs.

O

Off feed. Not eating as much as usual or at all.

Off-label. *See* Extra-label.

Off side. An equine's right side (also called its far side).

Open jenny (or open mare). An equine that isn't pregnant.

Out of. An equine is "out of" its dam; a donkey is by a jack and out of a jenny, a horse is by a stallion and out of a mare, a mule is by a jack and out of a mare, etc.

Ovary. One of a pair of egg- and hormone-producing glands in a female animal.

Overshot (parrot) mouth. When the lower jaw is shorter than the upper jaw so that the animal's upper teeth protrude ahead of the lower teeth.

Over the counter (OTC) drugs. Nonprescription drugs.

Ovulation. The release of an egg from the ovary.

Ovum. An egg (also called an ova or oocyte).

Oxytocin. A naturally occurring hormone important in milk letdown and muscle contraction during the birthing process.

P

Pace. A two-beat lateral gait in which an equine moves both front and rear legs on the same side at the same time.

Paddock. A small, enclosed area used for grazing.

Palpation. Examining something with one's hands.

Papers. A registration certificate.

Parrot mouth. *See* Overshot (parrot) mouth.

Parturition. The act of giving birth.

Pasture breeding. When a jack or stallion runs loose with a group of jennies or mares and breeding occurs without human intervention.

Pathogen. An agent that causes disease, especially a living microorganism such as a bacterium or virus.

Pecking order. The social hierarchy within a group of animals.

Pedigree. A certificate documenting an animal's line of descent.

Perennial. A usually herbaceous plant that doesn't die at the end of its first growing season but regrows from year to year.

pH. pH is a measure of the activity of hydrogen ions in a solution and therefore its acidity or alkalinity.

Pharmaceutical. A substance used in the treatment of disease.

Phenotype. An individual's observable physical characteristics.

Picket line. A rope stretched between two solid objects onto which equines are tied.

Placenta. *See* Afterbirth.

Pneumonia. Infection in the lungs.

Pointing. Standing with one foreleg extended farther than normal; a sign of lameness.

Poitou Ass (pwah-TOO ass). A race of large, longhaired asses originating in the Poitou region of France.

Pones. Lumpy fat deposits on an obese donkey.

Pony mule. A small mule produced by crossing a jack with a pony mare.

Postpartum. After giving birth.

Predator. An animal that survives by killing and eating other animals.

Prepartum. Before giving birth.

Prey animal. An animal belonging to a species preyed upon by predators.

Probiotic. Living organisms used to influence gut health.

Produce. The progeny of a dam.

Progeny. Offspring.

Progesterone. A hormone secreted by the corpus luteum and produced by the placenta during pregnancy.

Proliferate. To vastly multiply in numbers, usually over a short span of time.

Prolific. Producing more than the usual number of offspring.

Protein. A nutrient category of feed used for growth, milk production, and repair of body tissue.

Puberty. When a young animal becomes sexually mature.

Purebred. An animal of a recognized breed that has been kept pure for a specified number of generations.

Put down. A euphemism for euthanizing, or humanely destroying, an animal.

Q

Quarantine. To isolate or separate an individual from others of its kind.

Quidding. Drooling globs of partially chewed feed, particularly hay.

R

Radiograph. X-ray.

Ration. Total amount of feed given an animal during a 24-hour period.

Reabsorb. Spontaneous absorption of an early embryo back into a mother's system.

Registered. An animal that has a registration certificate and number issued by a breed association or registry.

Rehydrate. To replace body fluids lost through dehydration.

Roman nosed. Having a convex facial profile.

Rotational grazing (or browsing). Moving grazing or browsing animals from one paddock to another before plant growth in the first is fully depleted.

Roughage. Plant fiber.

Round pen. *See* Bull pen.

Roundworm. A parasitic worm with an elongated, round body.

Rug. A horse blanket. (*See also* Turnout blanket or rug.)

Rumensin. The brand name for monensin; Rumensin is highly toxic to all equines (never, under any circumstances, allow donkeys access to products containing Rumensin).

S

Sardinian donkey. A Miniature Donkey.

Scours. Persistent diarrhea in young animals.

Scrotum. The external pouch in which testicles are suspended.

Selection. Choosing superior animals as parents for future generations.

Serology test. Blood test.

Settle. Impregnate.

Shank. *See* Lead shank.

Sheath. The outer skin covering protecting a male equine's penis.

Sheet. An unlined horse or donkey blanket primarily used to keep an animal clean or dry.

Show. A jenny or mare is "showing" when she indicates she's receptive to being bred.

Sicilian donkey. A Miniature Donkey.

Silent heat. When a jenny or mare is in heat (estrus) but showing no outward signs.

Sire. The male parent.

Slippers (or sultan's slippers). The long, turned-up toes of a neglected, foundered donkey, or hooves that haven't been trimmed in a very long while.

Smegma. Accumulated dead skin cells, fatty secretions, and grime found inside a male equine's sheath and, to a lesser degree, between a female equine's teats.

Smooth mouth. Aged; approximately ten years of age and over.

Sound. Having no defects that affect serviceability.

Sow mouth. *See* Underbite.

Spooky. Easily startled.

Square. Having a boxy appearance with "a leg in each corner," a very desirable trait in Miniature Donkeys.

Stallion. An uncastrated male horse three years of age or over. In Britain and a few other parts of the world, uncastrated male donkeys three years of age or over are also called stallions (instead of jacks).

Standard Donkey. In American Donkey and Mule Society terminology, a donkey 40.01 inches to 48 inches tall, measured at the withers.

Standing heat. The period during estrus (heat) when a jenny or a mare allows a jack to breed her.

Stud. A term used incorrectly to refer to a stallion or jack; is actually the term for an establishment where equines are kept for breeding purposes.

Studbook. A compilation of information maintained by a registry about individual breeding animals.

Stud fee. The charge for breeding to a jack (or stallion).

Subcutaneous (SQ). Under the skin.

Substance. The strong, stocky-built characteristic of an animal with good bone.

Suckling. A foal that is still nursing its dam.

Surcingle. A belt that encircles an equine's heart girth; used to provide rings and terrets for training purposes, or to fasten a saddle or pack to the animal's back.

Sweet feed. A commercial mixture of grains with added molasses.

Swish. The longhaired lower section of a donkey's tail (also called a switch).

Switch. *See* Swish.

Systemic. Affecting the entire body.

T

Tack. Equipment used for riding, driving, and caring for equines.

Tail wrap. Material (nowadays usually four-inch-wide, self-stick, disposable bandage) used to wrap and protect an equine's tail during trailering, breeding, and foaling.

Tapeworm. A segmented, ribbonlike, intestinal parasite.

Teaser. A stallion or jack used to discern a mare or jenny's sexual receptiveness.

Testosterone. A hormone that promotes the development and maintenance of male sexual characteristics.

Topline. The area between an animal's withers and the beginning of the tail.

Top side. The sire's side of a pedigree.

Total Digestible Nutrients (TDN). A standard system for expressing the energy value of feed.

Trace minerals. Minerals needed only in minute amounts.

Trachea. The windpipe; a tube leading from the throat to the lungs.

Tree. The basic structure upon which a saddle is constructed.

Trimester. One-third of a pregnancy.

Trot. A two-beat, diagonal gait in which the right foreleg moves forward at the same time the left hind leg moves forward, and vice versa.

Turnout. When an equine is let out of its stall into a field, pasture, or exercise area.

Turnout blanket or rug. A warm, one-piece, unfitted winter covering designed to be worn by horses or donkeys kept outdoors during inclement weather.

Twitch. A means of restraint whereby a loop of rope or chain is tightened around an equine's upper lip. Twitching releases natural endorphins that help keep an animal calm.

U

Udder. The mammary system with a teat or a nipple.

Ultrasound. A procedure in which sound waves are bounced off tissues and organs; widely used to confirm pregnancy in female mammals.

Underbite. When the lower jaw is longer than the upper jaw so that the lower teeth extend out beyond the teeth of the upper jaw; also known as monkey mouth or sow mouth.

Underline. The undercarriage of an equine stretching from the elbow to the sheath or udder.

Unsound. Lame or otherwise unable to function in the capacity for which an animal is kept.

USDA. United States Department of Agriculture.

Uterus. The female organ in which fetuses develop; the womb.

V

Vagina. The passageway between the uterus and the external genital opening; the birth canal.

Vascular. Pertaining to, or provided with, vessels; usually refers to veins and arteries.

VMD Veterinary Medical Doctor.

Vulva. The external parts of the female genital organs.

W

Weanling. A young horse less than one year of age that has been weaned from his dam.

Winking. The opening and closing of the vulva by a jenny or mare in heat.

Withers. The slight rise in a horse's back just before the mane starts. This is where height is measured on equines.

Wolf teeth. Vestigial premolars.

Y

Yard. In British terminology, a dry lot where animals are kept.

Yearling. A young equine of either sex that is one to two years of age.

Z

Zebrass. A zebra and donkey hybrid; zebrass is the term accepted by the American Donkey and Mule Society, other sources refer to this cross as a zedonk or zonkey.

Zebroid. A zebra hybrid of any sort.

Zedonk. *See* Zebrass.

Zonkey. *See* Zebrass.

Zony. A zebra and pony hybrid.

Zorse. A zebra and horse hybrid.

Resources

Books, Magazines, Articles, and Videos

Donkey-Specific How-To Books

Gross, Bonnie R. *Caring for Your Miniature Donkey*, 2nd ed. Westminster, MD: *Miniature Donkey Talk* Magazine. 1998.

Hodges, Meredith. *Donkey Training*. Fort Collins, CO: Lucky Three Ranch, 1999.

——. *Training Mules and Donkeys: A Logical Approach to Longears*. Crawford, CO: Alpine Publications, 1993.

Hutchins, Betsy and Paul, revised and edited by Leah Patton. *The Definitive Donkey: A Textbook of the Modern Ass*. Lewisville, TX: American Donkey and Mule Society, 1981, revised 1999.

Jones, Peta. *Donkeys for Development*. ATNESA and the Agricultural Research Council of South Africa, 1997.

Kokas, Christine, Berry and Jo-Anne. *Donkey Business III: A Guide for Raising, Training, Managing and Showing Donkeys*. Self-published, 1998.

Morris, Dorothy. *Looking After a Donkey*. Wiltshire, UK: Whittet Books, 1997.

Svendsen, Elisabeth D. *The Professional Handbook of the Donkey*, 3rd ed. Wiltshire, UK: Whittet Books, 1998.

Donkey History

Brookshier, Frank. *The Burro*. Norman, OK: University of Oklahoma Press, 1974.

Clutton-Brock, Juliet. *Horse Power: A History of the Horse and the Donkey in Human Societies*. Cambridge, MA: Harvard University Press, 1992.

Dean, Faisal A. "Lost Forever: The Onager of Arabia," *Arabian Wildlife* 1, no. 2.

Dent, Anthony Austen. *Donkey: The Story of the Ass from East to West*. Edinburgh: Harrap, 1972.

Additional Recommended Reading

Browning, David G., and Peter M. Scheifele. "Vocalization of *Equus asinus*: The Hees and Haws of Donkey Brays." *Acoustical Society of America Journal* 115, no. 6: 2485.

Clay, Jackie. *Build the Right Fencing for Horses*. Storey Country Wisdom Bulletin A-193. North Adams, MA: Storey Publishing, 1999.

Damerow, Gail. *Fences for Pasture & Garden*. North Adams, MA: Storey Publishing, 1992.

Ekarius, Carol. *How to Build Animal Housing*. North Adams, MA: Storey Publishing, 2004.

Foley, Sharon. *Getting to Yes: Clicker Training for Improved Horsemanship*. Mail Neptune, NJ: TFH Publications, 2007.

Hayes, Karen. *Emergency! The Active Horseman's Book of Emergency Care*. Boonsboro, MD: Half Halt Press, 1995.
This comprehensive, easily referenced first-aid manual belongs on every equine owners bookshelf.

Hayes, Karen E. N. *The Complete Book of Foaling: An Illustrated Guide for the Foaling Attendant*. Indianapolis, IN: Howell Book House, 1993.
The only foaling guide you'll ever need; keep it in your foaling kit

Henry, M., S. M. McDonnell, L. D. Lodi, and E. L. Gastal. "Pasture Mating Behaviour of Donkeys (*Equus minus*) at Natural and Induced Oestrus." *Journals of Reproduction & Fertility* (1991) 44: 77–86.

Hill, Cherry. *Becoming an Effective Rider: Developing Your Mind and Body for Balance and Unity*. North Adams, MA: Storey Publishing, 1991.
My favorite book on riding technique, even after all these years

——. *How to Think Like a Horse*. North Adams, MA: Storey Publishing, 2006.

Karrasch, Shawna, Vinton Karrasch, and Arlene J. Newman. *You Can Train Your Horse to Do Anything! On Target Training: Clicker Training and Beyond.* North Pomfret, VT: Trafalgar Square Books, 2000.

Kurland, Alexandra. *The Click That Teaches: A Step-By-Step Guide in Pictures.* Delmar, NY: The Clicker Center, 2003.

———. *Clicker Training for Your Horse.* Waltham, MA: Sunshine Books, 2007.

McDonnell, Sue M., and Amy Poulin. "Equid Play Ethogram." *Applied Animal Behaviour Science*, no. 78 (2002): 263–90.

Miller, Robert M. *Imprint Training of the Newborn Foal.* Colorado Springs, CO: *Western Horseman*, 2003.

Missouri Department of Conservation. "Using Guard Animals to Protect Livestock. "Download for free at *http://mdc.mo.gov/249).*

Pearson, R. Anne, and Mohammed Ouassat. *A Guide to Live Weight Estimation and Body Condition Scoring of Donkeys.* Edinburgh: University of Edinburgh, Centre for Tropical Veterinary Medicine, 2000.

Poe, Rhonda Hart. *Trail Riding.* North Adams, MA: Storey Publishing, 2005.
    The best book for trail riders whether riding horses, donkeys, or mules

"Protecting Livestock with Guard Donkeys"
    This stellar information on training guard donkeys is gleaned from the Alberta Ministry of Agriculture and Rural Development.
    Available through their Web site:
    www.agric.gov.ab.ca

Rees, Lucy. *The Horse's Mind.* New York: Arco: 1985.
    Salimei, Elisabetta, Francesco Fantuz, Raffaele Coppola, Biagina Chiofalo, Paolo Polidori, and Giogio Varisco.

"Composition and Characteristics of Ass's Milk"
    Presented at the 4th Congress of the Società italiana di ippologia in Campobasso, Italy, July 2002.

Sponenberg, D. Phillip. *Equine Color Genetics*, 2nd ed. Indianapolis, IN: Wiley-Blackwell Publishing, 2003.
    Includes a section on donkey color genetics

———. *Horse Color: A Complete Guide to Horse Coat Colors.* Emmaus, PA: Breakthrough Publications, 1992.
    An excellent color guide for mule owners

Tellington-Jones, Linda. *The Ultimate Horse Behavior and Training Book: Enlightened and Revolutionary Solutions for the 21st Century.* North Pomfret, VT: Trafalgar Square Books, 2006.
    This comprehensive TTEAM (Tellington-Jones Equine Awareness Method) and TTouch guide addresses working with horses, but the logic and exercises work exceedingly well for donkeys and mules, too.

Tillman, Peggy. *Clicking with Your Dog.* Waltham, MA: Sunshine Books, 2006.

Tobias, Michael, and Jane Morrison. *Donkey: The Mystique of Equus Asinus.* Tulsa, OK: Oak Council Books, 2006.
    History and personal essays combined with oodles of wonderful color illustrations make this a must-have for donkey lovers everywhere.

Donkey-Related Magazines

Asset
National Miniature Donkey Association
315-336-0154
www.nmdaasset.com
    Miniature Donkey publication

The Brayer
American Donkey and Mule Society
972-219-0781
www.lovelongears.com
    Every donkey owner should subscribe!

Miniature Donkey Talk
719-689-2904
www.web-donkeys.com
    Also tack, books, and gifts

Mules and More
573-646-3934
www.mulesandmore.com
    Also books and gifts

Rural Heritage
931-268-0655
www.ruralheritage.com

Western Mule Magazine
417-859-6853
www.westernmulemagazine.com

Donkey-Specific Videos

Basic Donkey Care
Basic Donkey Health Care
An Introduction to Driving Your Donkey
    These 30- to 40-minute DVDs are available only from the Donkey Sanctuary in Great Britain.
+44-0-1395-578222
www.thedonkeysanctuary.org.uk

Donkey Training with Crystal Ward

This excellent production in both VHS and DVD formats covers donkey care, leading, standing tied, handling the feet, ground driving, introducing the saddle, and the first ride. It's a good one!
Available through Crystal Ward at Ass-Pen Ranch
530-295-0292

www.asspenranch.com

Training Donkeys and Mules: A Logical Approach to Longears

A fantastic series by Meredith Hodges. Contains 10 donkey- and mule-training videos in both VHS and DVD formats. Each video comes with a workbook and laminated field cards to take with you as you train your donkey or mule.

#1 Foal Training
#2 Preparing for Performance: Groundwork
#3 Preparing for Performance: Driving
#4 Basic Foundation for Saddle
#5 Intermediate Saddle Training
#6 Advanced Saddle Training
#7 Jumping
#8 Management, Fitting, and Grooming
#9 Donkey Training: Introduction and Basic Training
#10 Donkey Training: Saddle Training and Jumping
Available through Meredith's company, Lucky Three Ranch
800-816-7566

www.luckythreeranch.com

Organizations

For information and resources on international donkey rescue, type *donkey sanctuaries* into your favorite Internet search engine

Adoptions

BLM
866-4MUSTANGS

Cañon City Facility
Cañon City, Colorado
719-269-8539

Delta Wild Horse and Burro Facility
Delta, Utah
435-743-3100

Eastern States Wild Horse and Burro Facility
Ewing, Illinois
800-370-3936

Elm Creek Wild Horse and Burro Center
Elm Creek, Nebraska
308-856-4498

Kingman Regional Wild Horse and Burro Facility
Kingman, Arizona
928-718-3700

Litchfield Wild Horse and Burro Corrals
Susanville, California
800-545-4556

National Wild Horse and Burro Center at Palomino Valley
Palomino Valley, Nevada
775-475-2222

Pauls Valley Adoption Center
Oklahoma City, Oklahoma
405-238-7138

Ridgecrest Regional Wild Horse and Burro Corrals
Ridgecrest, California
760-384-5765

Salt Lake Regional Wild Horse and Burro Center
Herriman, Utah
877-224-3956

Wild Horse and Burro Adoption Program
Bureau of Land Management
866-468-7826
www.blm.gov

Associations

UNITED STATES/DONKEYS AND MULES

American Council of Spotted Asses
Wentzville, Missouri
636-828-5955
www.spottedass.com

American Donkey and Mule Society
Lewisville, Texas
972-219-0781
www.lovelongears.com
Registers donkeys and donkey hybrids

American Mammoth Jackstock Registry
Johnson City, Texas
360-868-2357
www.amjr.us

American Mule Association
Yerington, Nevada
775-463-1922
www.americanmuleassociation. com
   Registers donkeys and mules

American Livestock Breeds Conservancy
Pittsboro, North Carolina
919-542-5704
www.albc-usa.org

International Miniature Donkey Registry
Westminster, Maryland
410-875-0118
www.miniaturedonkeyinfo.com

Missouri Miniature Donkey Breeders Association
Festus, Missouri
636-937-4095
www.mmdba.com
   Offers a free downloadable show manual to teach you the ins and outs of showing Miniature Donkeys

National Miniature Donkey Association
Rome, New York
315-336-0154
www.nmdaasset.com

North American Saddle Mule Association
Boyd, Texas
940-389-5608
http://nasma.us
   Registers donkeys and mules

Western Pack Burro Ass-ociation
Denver, Colorado
303-688-5104
www.packburroracing.com
   Governing body of pack burro racing

UNITED STATES/MULES

American Gaited Mule Association
Shelbyville, Tennessee
931-684-7649
www.americangaitedmule.com

American Miniature Mule Society
Canton, Illinois
309-647-7162
www.miniaturemulesociety.com

American Mule Racing Association
Sacramento, California
916-263-1529
www.muleracing.org

INTERNATIONAL

Asociación Nacional de Criadores de la Raza Asnal Andaluza
Estepona, Spain
+34-952790511
www.ancraa.org

Association of Breeders of Andalusian Donkeys
Niort, France
+33-05-49-35-22-68
www.baudet-du-poitou.fr/Sabaud. htm

Canadian Donkey and Mule Society
Langley, British Columbia
604-857-4990
www.donkeyandmule.com

Donkey Breed Society
Kent, United Kingdom
+44-1732-864414
www.donkeybreedsociety.co.uk

Local Domestic Breeds of Catalonia
www.rac.uab.es

Miniature Mediterranean Donkey Association
Devon, United Kingdom
+44-01647-281642
www.miniature-donkey-assoc.com

Donkey Rescues and Sanctuaries

UNITED STATES

Burro Rescue-Rehab-Relocation Onus (BRRRO)
Cheney, Washington
509-235-2255
http://brrro.org

Forever Home Donkey Rescue & Sanctuary
Benson, Arizona
520-212-5300
www.foreverhomedonkey.com

Hacienda de los Milagros
Chino Valley, Arizona
928-636-5348
www.haciendadelosmilagros.org

Longhopes Donkey Shelter
Bennett, Colorado
303-644-5930
www.longhopes.org

Peaceful Valley Donkey Rescue
Tehachapi, California
866-366-5731
www.donkeyrescue.org

Turning Pointe Donkey Rescue
Dansville, Michigan
517-623-0000
www.turningpointedonkeyrescue. com

Wild Burro Rescue
Olancha, California
760-384-8523
www.wildburrorescue.org

CANADA

Donkey Sanctuary of Canada
Guelph, Ontario
519-836-1697
www.thedonkeysanctuary.ca

PrimRose Donkey Sanctuary
Roseneath, Ontario
905-352-2772
www.donkeyinfo.ca

UNITED KINGDOM

Donkey Sanctuary
Devon, United Kingdom
+44-0-1395-578222
www.thedonkeysanctuary.org.uk

Donkey Sanctuary in Ireland
Mallow, Ireland
+353-022-48398
www.thedonkeysanctuary.ie

Fluffsfield Donkey Sanctuary
Fyvie, Scotland
+44-0-1771-644770
www.fluffsfield.co.uk

Freshfields Donkey Village
Peak Forest, United Kingdom
+44-0-1298-79775
www.donkey-village.org.uk

Island Farm Donkey Sanctuary
Wallingford, United Kingdom
+44-0-1491-833938
www.donkeyrescue.co.uk

Scottish Borders Donkey Sanctuary
Melrose, Scotland
+44-0-1835-823468
www.donkeyheaven.org

Tamar Valley Donkey Park
Gunnislake, United Kingdom
+44-0-1822-834072
www.donkeypark.com

Donkey (and Mule) Riding and Packing Gear and Information

Bates Australia
Perth, Australia
www.batesaustralia.com.au
Wintec saddles

Buford Saddle and Tack
Pryor, Oklahoma
918-824-2668
www.bufordsaddle.com

Crest Ridge Saddlery
Norfolk, Arkansas
888-297-1261
www.crestridgesaddlery.com

Fitting a McClellan Saddle to Your Horse
Ninth Virginia Cavalry, Inc.
http://9thvirginia.com/fitting.html
Perhaps of more use to mule aficionados than donkey riders, this page is a treasure trove of saddle-fitting information.

Hands-On Horse Care
Diana Thompson
Fulton, California
707-542-4646
www.dianathompson.com
Sidepull bridles

The McClellan Military Saddle-Society of the Military Horse
http://militaryhorse.org/studies/mcclellan
Everything you wanted to know about the McClellan saddle — a favorite with donkey riders everywhere

Equipment

Missouri Mule Company
Springfield, Missouri
417-833-9399
www.missourimuleco.com

Northwest Pack Goats
Weippe, Idaho
888-722-5462
www.northwestpackgoats.com
High-quality wooden goatpacking sawbuck kits that fit or can be easily modified to fit most donkeys; also has a pocket-style pad that provides additional padding

Outfitters Supply
Columbia Falls, Montana
888-467-2256
www.outfitterssupply.com

Queen Valley Mule Ranch
Queen Valley, Arizona
602-999-6853
www.muleranch.com

Reed Tack
Washington, Iowa
319-653-6901
www.reedtack.com

Rowan's Bookshop
Victoria, Australia
http://stores.lulu.com/rowan
The only up-to-date, all-donkey publications in print and e-books, including "Make a Donkey Packsaddle," and *Rowan's Guide to Packing with Donkeys*

Thorowgood Saddles
Bloxwich Walsall, United Kingdom
+44-0-1922-711676
www.thorowgood.com

Tuff Enuff
Greenwood, Arkansas
866-477-9731
www.tuffenuff.org
    Carries burro packing gear

Valley Mule Company
Corvallis, Oregon
541-754-3266
www.valleymulecompany.com

Other Donkey-Related Items

Best Friend Equine Supply
New Holland, Pennsylvania
800-681-2495
www.bestfriendequine.com
    Grazing muzzles

Cashel Company
Chehalis, Washington
800-333-2202
www.cashelcompany.com
    Fly masks for donkeys and mules

Dinky Donkey Shop
Mount Barker, Australia
+61-8-9851-1562
www.dinkydonkeyshop.com.au
    Apparel, mugs, jewelry, and note
    cards

The Donkey Sanctuary
Devon, United Kingdom
+44-0-1395-578222
www.thedonkeysanctuary.org.uk
    Books, videos, gifts, donkey-care
    items as well as an accurate tape-
    measuring method for "weighing"
    donkeys

HeeHaw Book Service
American Donkey and Mule
Society
Lewisville, Texas
972-219-0781
www.lovelongears.com
    Books, videos, and reprints

Kicking Donkey Products
Oro Station, Ontario
888-366-5397
www.kickingdonkeyproducts.com
    Gifts and home accessories

Livingston Productions
Casi Cielo Farm
Greenville, Florida
850-342-1193
www.casicielofarm.com
    Blankets, books, apparel, and gifts

Lucky Three Ranch
Fort Collins, Colorado
800-816-7566
www.luckythreeranch.com
    Books, videos, apparel, art, and gifts

The Mule Store
Montague, Michigan
877-654-6853
www.themulestore.com
    Books, videos, apparel, art, and gifts

Newman Enterprises
Omro, Wisconsin
888-685-2244
www.bitingflies.com
    HorsePal horsefly trap

Ozark Mountain
Gassville, Arkansas
888-775-6446
www.minitack.com
    A full line of Standard and smaller
    donkey supplies

WXICOF
Wentzville, Missouri
888-499-4263
www.wxicof.com
    Halters, books, and gifts

Veterinary and Health Care

American Association of Equine
    Practitioners
Lexington, Kentucky
859-233-0147
www.aaep.org

American Farrier's Association
Lexington, Kentucky
859-233-7411
www.americanfarriers.org

American Veterinary Medical
    Association
Schaumburg, Illinois
847-925-8070
www.avma.org

American Holistic Veterinary
    Medical Association
Bel Air, Maryland
410-569-0795
www.ahvma.org

Chamisa Ridge
Santa Fe, New Mexico
800-825-9120
www.chamisaridge.com
    Full line of holistic products

Donkey Sanctuary
Devon, United Kingdom
+44-0-1395-578222
www.thedonkeysanctuary.org.uk
    If your donkey is sick or injured
    and your veterinarian doesn't know
    how to proceed, ask him to contact
    the Donkey Sanctuary for further
    information.

Equilite, Inc
Pottstown, Pennsylvania
800-942-5483
www.equilite.com
Full line of holistic products

Fias Co Farm
Mooresburg, Tennessee
http://fiascofarm.com
Learn to run your own fecal egg counts

Flower Essence Services
Nevada City, California
800-548-0075
www.floweressence.com

Herb'n Horse
Ames, Iowa
800-267-6141
www.herbnhorse.com

Herbs for Horses
Guelph, Ontario
888-423-7777
www.horseherbs.com

Herbsmith
Oconomowoc, Wisconsin
800-624-6429
www.herbsmithinc.com
Chinese herbs

The Holistic Horse
Perkasie, Pennsylvania
215-249-1965
www.holistichorse.com
Free online article archives

Holistic Horsekeeping
Bear Creek Veterinary Clinic
Austin, Texas
303-575-1170
www.holistichorsekeeping.com
Many archived articles and free e-zine

Lake Immunogenics
Ontario, New York
800-648-9990
www.lakeimmunogenics.com
Immunoglobulin (IgG)

Laminitis Trust
Mead House Farm
Wilts, England
+44-0-9051-051051
www.laminitis.org

Mg Biologics
Ames, Iowa
515-769-2340
www.mgbiologics.com
Immunoglobulin (IgG)

National Sweet Itch Centre
Horses Etc.
Flintshire, England
+44-0-1352-771718
www.sweet-itch.co.uk

Natural Horse
Soquel, California
831-479-1289
www.thenaturalhorse.net
Herbs and aromatherapy

***Natural Horse* Magazine**
Leesport, Pennyslvania
800-660-8923
www.naturalhorse.com

Nelsons
Wilmington, Massachusetts
800-319-9151
www.nelsonbach.com
Flower essences

Premier1 Supplies
Washington, Iowa
800-282-6631
www.premier1supplies.com
SuperLube

Schreiner's Herbal Solution Restoration Products Company
The Dalles, Oregon
800-223-4325
www.schreiners.com
Holistic liquid wound dressing

Sera, Inc.
Central Biomedia, Inc.
Shawnee Mission, Kansas
800-552-3984
www.seramune.com
Immunoglobulin (IgG)

Wendals Herbs
Massillon, Ohio
800-321-0235
www.wendalsusa.com
Excellent online remedy finder and A–Z herb guide

Whole Horse
Oakhurst, California
559-683-4434
www.wholehorse.com
Chinese herbs, acupressure charts, and books

Other Web Sites of Interest

About the Poitou

Asinerie due la Baie
www.baudet-du-poitou.fr

Baudet du Poitou
www.baudetdupoitou.fr

Le Baudet du Poitou
http://baudetdupoitou.online.fr

BLM Burro Resources

BLM BURRO INFORMATION

BLM National Wild Horse and Burro Program
www.wildhorseandburro.blm.gov

KBR Horse Net
www.kbrhorse.net
     Lots of BLM burro information

Western States Wild Horse & Burro Expo
www.wildhorseandburroexpo.com

BURRO RACING

Burro Days
www.burrodays.com
Pack burro racing

Western Pack Burro Ass-ociation
www.packburroracing.com
     Pack Burro Racing, schedule, rules, and training tips

Clicker Training

The Clicker Center
www.theclickercenter.com

Karen Pryor Clickertraining
www.clickertraining.com

On-Target Training
Shawna Karrasch
www.on-target-training.com

Zen Clicker Horsemanship
www.zenhorsemanship.com

Donkey Classifieds and Directories

NOTE: Most association Web sites also offer breeder directories.

Equine.com
Source Interlink Media
www.equine.com

Gotdonkeys Breeders List
Gails Mini-Donkey Ranch
http://gotdonkeys.com

HorseTopia
www.horsetopia.com

HorseWeb
www.horseweb.com

International Miniature Donkey Directory
www.donkeydirectory.com

LongearsMall.com
http://longearsmall.com

The Mule Store
www.themulestore.com

Mule Trader
www.muletrader.ca

Donkey Milk

Anès Beauté en Gascogne
www.anesbeaute-cosmetiques.com
     French makers of donkey-milk cosmetics

Asinerie de Feillet
www.asinus.fr
     French donkey dairy

Asinerie d'Embazac
www.embazac.com
     French donkey dairy

Asinerie Lisane
www.lisane.be
     Belgian makers of donkey-milk cosmetics

Bohemia Style — Andean Wares
www.bohemia-style.com
     Donkey-milk cosmetics and nutritional supplements

L'Asinerie du Pays des Collines
www.asineriedupaysdescollines.be
     Belgian donkey dairy

Donkey Trekking

Fédération Nationale Anes et Randonnées
National Association for Donkey-Trekking
www.ane-et-rando.com
     A complete guide to donkey trekking in France

Hike with a Donkey
www.hikingwithdonkey.com

Horse and Donkey Driving

The American Driving Society
www.americandrivingsociety.org
     The primary driving organization serving the United States

Animal Traction Information Gateway
Animal Traction Network for Eastern and Southern Africa (ATNESA)
www.animaltraction.net
     Click on "Donkeys, People and Development"

CarriageDriving.net
www.carriagedriving.net

Asinerie du Bocage
www.asineriedubocage.com
> A French Web site with lots of fascinating information in English

BlindHorses.org
Rolling Dog Ranch Animal Sanctuary
www.blindhorses.org

Blue Mountain Farm
www.oregonvos.net/~jrachau/index.htm
> Care, breeding, and training — outstanding!

Canyon Colorado Equid Sanctuary
www.canyoncolorado.com
> Wild conservation group

Chaffhaye
www.chaffhaye.com
> U.S. manufacturer of bagged haylage

Cooke Livestock Miniature Donkeys
www.miniaturedonkeys.com
> Excellent FAQs

Farm Service Agency
U.S. Department of Agriculture
www.fsa.usda.gov
> Manages HayNet, a list of hay suppliers

Foaling Resources
FoalStory
www.foalstory.com/resources.html
> An outstanding collection of equine foaling resources

Longears Mall
http://pgv.longearsmall.com/pgv

Lucky Three Ranch
www.luckythreeranch.com

Robinson Ranch
www.donkeys.com

Save Your Back and Pack Your Ass
www.glaciertoyellowstone.com/donkeys

Donkey Listservs

Click Ryder
http://pets.groups.yahoo.com/group/ClickRyder

Donkey Click
http://groups.yahoo.com/group/DonkeyClick

Mammoth Donkeys
http://pets.groups.yahoo.com/group/mammothdonkeys

Recreational Equine Driving
http://sports.groups.yahoo.com/group/RecreationalEquineDriving
> The largest, most active driving list at YahooGroups; actively encourages donkey drivers to join their friendly ranks

Miscellanea

ArchivalUSA
www.archivalusa.com
> Collector's supplies

Beyond the sidewalk
http://beyondthesidewalk.com
> Small-farm marketing books and seminars

Clovelly Donkeys
www.clovellydonkeys.co.uk

Get Smart Products
www.pfile.com
> Collector's supplies

The Horse
www.thehorse.com
> The Horse is an online companion to the outstanding equine health management magazine of the same name. Applying for an online account gives visitors access to thousands of articles covering every possible aspect of equine ownership; it's a must for anyone owning equines of any kind.

John Henry
http://john.henry.org
> The famous mule's very own Web site

Light Impressions
www.lightimpressionsdirect.com
> Collector's supplies

Names by Chinaroad
http://lowchensaustralia.com/names.htm
> Naming Web site

Paper Direct
www.paperdirect.com
> Downloadable templates, including sales contracts

The Quagga Project South Africa
www.quaggaproject.org
> This site chronicles the ongoing reconstruction of the extinct Quagga

Small Business Administration
www.sba.org
Helps beginning entrepreneurs

Index

Bold text indicates breed profiles; *italic* text indicates an illustration or photograph.

Other Storey Titles You Will Enjoy

Barnyard in Your Backyard, **edited by Gail Damerow.**
Expert advice on raising healthy, happy, productive farm animals.
416 pages. Paper. ISBN 978-1-58017-456-5.

Fences for Pasture & Garden, **by Gail Damerow.**
Sound, up-to-date advice and instruction to make building fences a task
anyone can tackle with confidence.
160 pages. Paper. ISBN 978-0-88266-753-9.

How to Build Animal Housing, **by Carol Ekarius.**
An all-inclusive guide to building shelters that meet animals' individual
needs: barns, windbreaks, and shade structures, plus watering systems,
feeders, chutes, stanchions, and more.
272 pages. Paper. ISBN 978-1-58017-527-2.

Keeping Livestock Healthy, **by N. Bruce Haynes, DVM.**
A complete guide to disease prevention through good nutrition, proper
housing, and appropriate care.
352 pages. Paper. ISBN 978-1-58017-435-0.

Livestock Guardians, **by Janet Vorwald Dohner.**
Essential information on using dogs, donkeys, and llamas as a highly
effective, low-cost, and nonlethal method to protect livestock and their
owners.
240 pages. Paper. ISBN 978-1-58017-695-8.
Hardcover. ISBN 978-1-58017-696-5.

Oxen: A Teamster's Guide, **by Drew Conroy.**
The definitive guide to selecting, training, and caring for the mighty ox.
304 pages. Paper. ISBN 978-1-58017-692-7.
Hardcover. ISBN 978-1-58017-693-4.

Small-Scale Livestock Farming, **by Carol Ekarius.**
A natural, organic approach to livestock management to produce healthier
animals, reduce feed and health care costs, and maximize profit.
224 pages. Paper. ISBN 978-1-58017-162-5.

Trail Riding, **by Rhonda Hart Poe.**
Fundamental instruction and detailed advice on every aspect of preparing
for and executing a pleasurable trail ride.
336 pages. Paper. ISBN 978-1-58017-560-9.

These and other books from Storey Publishing are available
wherever quality books are sold or by calling 1-800-441-5700.
Visit us at *www.storey.com*.